IET CONTROL, ROBOTICS AND SENSORS SERIES 129

Signal Processing to Drive Human–Computer Interaction

Other volumes in this series:

Volume 8	**A History of Control Engineering, 1800–1930** S. Bennett
Volume 9	**Embedded Mechatronics System Design for Uncertain Environments: Linux-based, MATLAB® xPC target, PIC, Arduino and Raspberry Pi approaches** C.S. Chin
Volume 18	**Applied Control Theory, 2nd Edition** J.R. Leigh
Volume 20	**Design of Modern Control Systems** D.J. Bell, P.A. Cook and N. Munro (Editors)
Volume 28	**Robots and Automated Manufacture** J. Billingsley (Editor)
Volume 33	**Temperature Measurement and Control** J.R. Leigh
Volume 34	**Singular Perturbation Methodology in Control Systems** D.S. Naidu
Volume 35	**Implementation of Self-Tuning Controllers** K. Warwick (Editor)
Volume 37	**Industrial Digital Control Systems, 2nd Edition** K. Warwick and D. Rees (Editors)
Volume 39	**Continuous Time Controller Design** R. Balasubramanian
Volume 40	**Deterministic Control of Uncertain Systems** A.S.I. Zinober (Editor)
Volume 41	**Computer Control of Real-Time Processes** S. Bennett and G.S. Virk (Editors)
Volume 42	**Digital Signal Processing: Principles, devices and applications** N.B. Jones and J.D.McK. Watson (Editors)
Volume 44	**Knowledge-Based Systems for Industrial Control** J. McGhee, M.J. Grimble and A. Mowforth (Editors)
Volume 47	**A History of Control Engineering, 1930–1956** S. Bennett
Volume 49	**Polynomial Methods in Optimal Control and Filtering** K.J. Hunt (Editor)
Volume 50	**Programming Industrial Control Systems Using IEC 1131-3** R.W. Lewis
Volume 51	**Advanced Robotics and Intelligent Machines** J.O. Gray and D.G. Caldwell (Editors)
Volume 52	**Adaptive Prediction and Predictive Control** P.P. Kanjilal
Volume 53	**Neural Network Applications in Control** G.W. Irwin, K. Warwick and K.J. Hunt (Editors)
Volume 54	**Control Engineering Solutions: A practical approach** P. Albertos, R. Strietzel and N. Mort (Editors)
Volume 55	**Genetic Algorithms in Engineering Systems** A.M.S. Zalzala and P.J. Fleming (Editors)
Volume 56	**Symbolic Methods in Control System Analysis and Design** N. Munro (Editor)
Volume 57	**Flight Control Systems** R.W. Pratt (Editor)
Volume 58	**Power-Plant Control and Instrumentation: The control of boilers and HRSG systems** D. Lindsley
Volume 59	**Modelling Control Systems Using IEC 61499** R. Lewis
Volume 60	**People in Control: Human factors in control room design** J. Noyes and M. Bransby (Editors)
Volume 61	**Nonlinear Predictive Control: Theory and practice** B. Kouvaritakis and M. Cannon (Editors)
Volume 62	**Active Sound and Vibration Control** M.O. Tokhi and S.M. Veres
Volume 63	**Stepping Motors, 4th Edition** P.P. Acarnley
Volume 64	**Control Theory, 2nd Edition** J.R. Leigh
Volume 65	**Modelling and Parameter Estimation of Dynamic Systems** J.R. Raol, G. Girija and J. Singh
Volume 66	**Variable Structure Systems: From principles to implementation** A. Sabanovic, L. Fridman and S. Spurgeon (Editors)
Volume 67	**Motion Vision: Design of compact motion sensing solution for autonomous systems** J. Kolodko and L. Vlacic
Volume 68	**Flexible Robot Manipulators: Modelling, simulation and control** M.O. Tokhi and A.K.M. Azad (Editors)
Volume 69	**Advances in Unmanned Marine Vehicles** G. Roberts and R. Sutton (Editors)
Volume 70	**Intelligent Control Systems Using Computational Intelligence Techniques** A. Ruano (Editor)

Volume 71 **Advances in Cognitive Systems** S. Nefti and J. Gray (Editors)
Volume 72 **Control Theory: A guided tour, 3rd Edition** J.R. Leigh
Volume 73 **Adaptive Sampling with Mobile WSN** K. Sreenath, M.F. Mysorewala, D.O. Popa and F.L. Lewis
Volume 74 **Eigenstructure Control Algorithms: Applications to aircraft/rotorcraft handling qualities design** S. Srinathkumar
Volume 75 **Advanced Control for Constrained Processes and Systems** F. Garelli, R.J. Mantz and H. De Battista
Volume 76 **Developments in Control Theory towards Global Control** L. Qiu, J. Chen, T. Iwasaki and H. Fujioka (Editors)
Volume 77 **Further Advances in Unmanned Marine Vehicles** G.N. Roberts and R. Sutton (Editors)
Volume 78 **Frequency-Domain Control Design for High-Performance Systems** J. O'Brien
Volume 80 **Control-Oriented Modelling and Identification: Theory and practice** M. Lovera (Editor)
Volume 81 **Optimal Adaptive Control and Differential Games by Reinforcement Learning Principles** D. Vrabie, K. Vamvoudakis and F. Lewis
Volume 83 **Robust and Adaptive Model Predictive Control of Nonlinear Systems** M. Guay, V. Adetola and D. DeHaan
Volume 84 **Nonlinear and Adaptive Control Systems** Z. Ding
Volume 86 **Modeling and Control of Flexible Robot Manipulators, 2nd Edition** M.O. Tokhi and A.K.M. Azad
Volume 88 **Distributed Control and Filtering for Industrial Systems** M. Mahmoud
Volume 89 **Control-Based Operating System Design** A. Leva *et al.*
Volume 90 **Application of Dimensional Analysis in Systems Modelling and Control Design** P. Balaguer
Volume 91 **An Introduction to Fractional Control** D. Valério and J. Costa
Volume 92 **Handbook of Vehicle Suspension Control Systems** H. Liu, H. Gao and P. Li
Volume 93 **Design and Development of Multi-Lane Smart Electromechanical Actuators** F.Y. Annaz
Volume 94 **Analysis and Design of Reset Control Systems** Y. Guo, L. Xie and Y. Wang
Volume 95 **Modelling Control Systems Using IEC 61499, 2nd Edition** R. Lewis and A. Zoitl
Volume 96 **Cyber-Physical System Design with Sensor Networking Technologies** S. Zeadally and N. Jabeur (Editors)
Volume 99 **Practical Robotics and Mechatronics: Marine, space and medical applications** I. Yamamoto
Volume 100 **Organic Sensors: Materials and applications** E. Garcia-Breijo and P. Cosseddu (Editors)
Volume 102 **Recent Trends in Sliding Mode Control** L. Fridman, J.P. Barbot and F. Plestan (Editors)
Volume 104 **Control of Mechatronic Systems** L. Guvenc, B.A. Guvenc, B. Demirel and M.T. Emirler
Volume 105 **Mechatronic Hands: Prosthetic and robotic design** P.H. Chappell
Volume 107 **Solved Problems in Dynamical Systems and Control** D. Valério, J.T. Machado, A.M. Lopes and A.M. Galhano
Volume 108 **Wearable Exoskeleton Systems: Design, control and applications** S. Bai, G.S. Virk and T.G. Sugar
Volume 111 **The Inverted Pendulum in Control Theory and Robotics: From theory to new innovations** O. Boubaker and R. Iriarte (Editors)
Volume 112 **RFID Protocol Design, Optimization, and Security for the Internet of Things** A.X. Liu, M. Shahzad, X Liu and K. Li

Signal Processing to Drive Human–Computer Interaction

EEG and eye-controlled interfaces

Edited by
Spiros Nikolopoulos, Chandan Kumar and
Ioannis Kompatsiaris

The Institution of Engineering and Technology

Published by The Institution of Engineering and Technology, London, United Kingdom

The Institution of Engineering and Technology is registered as a Charity in England & Wales (no. 211014) and Scotland (no. SC038698).

First published 2020

The Institution of Engineering and Technology
Michael Faraday House
Six Hills Way, Stevenage
Herts, SG1 2AY, United Kingdom

www.theiet.org

British Library Cataloguing in Publication Data
A catalogue record for this product is available from the British Library

ISBN 978-1-78561-919-9 (hardback)
ISBN 978-1-78561-920-5 (PDF)

Typeset in India by MPS Limited

Contents

About the editors

Spiros Nikolopoulos is a senior researcher at the Centre for Research and Technology Hellas (CERTH) in the Information Technologies Institute (ITI) Greece. Brain–computer interfaces based on EEG analysis is classified among his main research interests, leading the brain-related activities of his lab. He is the co-author of 22 papers in refereed journals, 9 book chapters and more than 55 conference papers in international conferences.

Chandan Kumar is a post-doctoral researcher, leading the working group of Interactive Web and Human Computing at the Institute for Web Science and Technologies, Koblenz, Germany. His research interests combine the interdisciplinary fields of eye tracking, Web, geovisualization, and human–computer interaction. His research work has been recognized at several international venues with more than 40 publications at peer-reviewed conferences, journals and workshops.

Ioannis Kompatsiaris is a senior researcher at the Centre for Research and Technology Hellas (CERTH) in the Information Technologies Institute (ITI) Greece, where he leads the Multimedia, Knowledge and Social Media Analytics Lab. He is the co-author of 129 papers in refereed journals, 46 book chapters, 8 patents and more than 420 papers in international conferences. He is a senior member of IEEE and member of ACM.

Preface

The cornerstone of writing this book has been the genuine interest of its editors on two futuristic domains of research, namely, brain research and research on eye-tracking. It was somewhere within 2014 when a research team from Greece with the infrastructure and expertise to collect and analyse brain signals has decided to join forces with a research team from Germany experienced in capturing and analysing signals collected from eye movements. The ground of this collaboration has been the conception of an ambitious project that would answer the call of the European Commission for multimodal and natural computer interaction and more specifically, the development of multimodal, adaptive interfaces assisting people with disabilities. In this direction, we have envisioned the research and development of 'MAMEM – Multimodal Authoring using your Eyes and Mind' as the means to enable people who have lost their fine-motor skills to operate the computer, create multimedia content, share it through social networks and become more integrated in a constantly digitizing society. To make this vision a reality, the aforementioned teams were complemented by two small–medium enterprises specialized in brain research and eye-tracking, one team of social scientists and three organizations fostering the potential end users of MAMEM. In 2015, MAMEM became a reality by receiving funding from the European Commission and eight partners from around Europe and Israel started on a research journey that lasted for 3.5 years.

During this journey, we had to face a number of challenges and overcome important barriers but, in the end, produce significant achievements. The purpose of this book is to provide an overview of these achievements and introduce the reader in a futuristic world where human interaction will no longer require the use of mouse, keyboard or even hand-gestures but could rely on a combination of mental commands and eye-gaze. In introducing this world, we initially study the computer use requirements of people with (dis)abilities (Part I), we present the algorithms and interfaces that have been developed for interacting through eyes and mind (Part II) and conclude with the description of prototype multimodal interfaces integrating both modalities (Part III). The reader may use this book either as a general introduction of future human–computer interaction and how this is expected to evolve by including more modalities and becoming more natural or as a selected list of state-of-the-art algorithms and interfaces assisting people who have lost their fine-motor skills to interact with a computer.

In concluding this preface, we would like to have a special acknowledgment to Prof. Maria Petrou who has been the visioner behind this research and she is no longer with us to see how far her vision has brought us!

Chapter 1
Introduction

Spiros Nikolopoulos[1], *Chandan Kumar*[2], *and Ioannis Kompatsiaris*[1]

1.1 Background

Traditionally, human–computer interaction has been grounded on the principle of a healthy neuromuscular system allowing access to conventional interface channels like mouse and keyboard. Recently, in an effort to make human–computer interaction more natural other types of control mechanisms have been brought in the forefront of interest, such as gesture-based (e.g. touchscreens and gesture recognition using cameras or wearables) or speech-driven interfaces (e.g. Google Assistant, Apple's Siri). However, the potential of these interfaces is also limited by either the good condition of the neuromuscular system in the case of gestures or the use of an application that can be easily operated through spoken commands. It has been only very recently that the evolution of devices recording accurate information about eye movements, brain electrical signals and bio-measurements has given a new perspective on the control channels that can be used for interacting with a computer application. The necessity of using these alternative channels has been mainly motivated in the context of assisting people with disabilities. Loss of the voluntary muscular control but with preserved intellectual functions is a common symptom of neuromuscular conditions (i.e. muscular dystrophy, multiple sclerosis, Parkinson's disease and spinal cord injury) leading to functional deterioration and poor quality of life. Among a great variety of functional deficits, the most affected people may also lose their ability to operate computer applications that require the use of conventional interfaces like mouse, keyboard or touchscreens. As a result, they are marginalized and unable to keep up with the rest of the society in a digitized world.

During the last decade, a radically new perspective on natural computer interaction has gained momentum aiming to deliver the technology that will allow people to operate computer applications using their eyes and mind. By 'eyes', we refer to eye movements captured through eye-tracking devices following the user's gaze, whereas by 'mind' we refer to brain electrical signals captured through electroencephalography (EEG) devices reflecting the brain's state. The fundamental assumption underlying

[1]The Multimedia Knowledge and Social Media Analytics Laboratory, Information Technologies Institute, Centre for Research and Technology-Hellas (CERTH), Thessaloniki, Greece
[2]Institute for Web Science and Technologies, University of Koblenz-Landau, Koblenz, Germany

this radically new perspective is that the signals (i.e. eye movements and brain electrical signals) obtained from people with motor impairment that are in good mental health can be reliably used to drive the interface control of a computer application. Good mental health does not always guarantee that, due to psychological or other reasons, our users will not shake his/her head involuntarily, or that his/her brain activity will not be affected by his/her medical treatment (misleading the translation of brain activity into mental commands). Moreover, simply providing the technology is not enough, since people with disabilities are usually reluctant (as with most forms of therapy and intervention) to adhere to practice a new technology even though they may be totally aware of its benefits. Therefore, there is also a need for making this technology persuasive and provide the principles for designing interfaces that will effectively stimulate their potential users to use them and keep on using them. Thus, the modelling of users based on their (dis)abilities, interaction behaviour, emotions and intentions is also a critical aspect for the adoption of this technology.

Finally, it is important to mention that although the use of interfaces based on eye gaze and mental commands has been primarily motivated in the context of assisting people with motor impairment, their effective use can be also foreseen in cases where the user can be considered conditionally impaired, in the sense that his/her hands are occupied in performing another critical task. This could be the case, for example, of piloting an aircraft in a critical situation where both hands are engaged in the control panel and the pilot could make the use of an additional modality to issue even a simple on/off command.

1.2 Rationale

This book reflects on the knowledge and outcomes generated in the context of the 3-year research and innovation action 'MAMEM' – 'Multimedia Authoring and Management using your Eyes and Mind' (mamem.eu) that has been funded by the H2020 programme of the European Commission under the call of 'Multi-modal and natural computer interaction'. In the following, we present the rationale that has been adopted in the context of this project that also governs the way of organizing this book into different book chapters. More specifically, the adopted rationale can be considered to extend along the following axes: (a) review the existing literature on the benefits of using brain–computer interfaces (BCIs) for the communication and rehabilitation of people with motor impairment, study their requirements on computer use and model their (dis)abilities through a set of personas coupled with persuasion and compliance strategies; (b) design and implement a number of signal-processing algorithms for interaction control through eyes and mind and (c) develop multimodal, prototype interface applications that can be operated through eyes and mind.

In reviewing the existing literature for the benefits of BCI applications for communication and rehabilitation, we have focused on BCI research that deals with noninvasive, EEG-based BCIs, which are used with the intention to gain part of the lost autonomy through multiple communication and rehabilitation strategies. A total of 45 studies reporting noninvasive EEG-based BCI systems have been systematically reviewed by summarizing the underlying operational mechanisms and reporting the clinical evidence and outcomes. In studying the computer use requirements of people with motor impairment, a thorough review of the related literature has been initially performed, followed by a set of focus groups with the participation of medical experts targeting aspects like computer use, internet literacy as well as factors limiting the endorsement of applications that can be operated through eyes and mind. In addition, a set of questionnaires and interviews was used with the intention to engage the patients and their caregivers into the process of defining requirements, following an established methodology on how to formulate, collect and analyse the responses of the questionnaires. Finally, in defining a set of personas (modelling our end-users) coupled with the appropriate persuasion strategies, we have worked along two parallel tasks. The first task had to do with evaluating the patient group's attributes, needs and habits, as obtained from the aforementioned user group studies, with the intention to perform clustering among the patients and, in this way, identify a set of user profiles constituting the personas. The second task had to do with the definition of a persuasion strategy aiming to introduce a novel assistive device into the every-day life of people with motor impairment. Towards this direction, we present the theoretical background of persuasion strategies by going through the available persuasion theories (i.e. self-determination, self-regulation, social cognitive learning theory, theories of emotions and theory of planned behaviour), as well as the existing practices in translating these theories into design decisions (i.e. functional triad, persuasive systems design model and design with intent).

With respect to the design and implementation of signal-processing algorithms for interaction control, we have worked on eye-tracking and EEG-based interaction. More specifically, with respect to eye-tracking-based interaction, we have studied how to acquire the user's eye gaze information in real-time by an eye-tracking device that can be used to generate gaze events and to analyse the data for deducing more high-level events. In detecting these events, we have relied on accumulated dwell time selection, task-specific threshold, object size, position and several other interface-optimization algorithms for eye-tracking signals. In implementing these optimization algorithms, we had conducted studies to analyse different types of selection methods with eye-tracking, also taking target size into account. With respect to EEG-based interaction, our emphasis has been placed in three directions: (a) steady-state-visual-evoked potentials (SSVEPs), which are known from the literature to provide the most accurate setting for EEG-based BCI applications, (b) EEG-based BCIs that rely on motor-imagery, which are far less disturbing than SSVEPs but at the expense of considerable lower accuracy and (c) error-related potentials (ErrPs), which are potentials that have been reported to distinguish themselves in cases where an error has been observed by the user.

Finally, with respect to the development of multimodal interface applications, we present an error-aware gaze-based keyboard and a reinvented version of the popular TETRIS game that can be played through eyes and mind. The error-aware gaze-based keyboard uses the EEG signals to correct the erroneous actions performed through eye-tracking-based typewriting. While the keyboard interfaces based on eye-gaze typically include special buttons for undoing (in the 'selecting boxes paradigm') or backspacing (in the 'keyboard entry paradigm'), the use of these buttons is time-consuming and becomes tedious. On the other hand, EEG offer an easy and natural solution for undoing/backspacing based on ErrPs. ErrPs are brain signals that appear when the user detects and error, either by an erroneous action from the interfaces or by an error made by himself. In this sense, ErrPs have been used and evaluated as an automatic 'backspace' operation in a gaze-based keyboard. Finally, the multimodal reinvention of the popular TETRIS game has been properly modified to be controlled with the user's eye-movements, mental commands and bio-measurements. In the proposed version of the game, the use of eye-movements and mental commands works in a complementary fashion, by facilitating two different controls, the horizontal movement of the tiles (i.e. tetriminos) through the coordinates of the gaze and the tile rotation through the detection of senso-motoric brain signals, respectively. Additionally, bio-measurements provide the stress levels of the player, which in turn determine the speed of the tiles' drop. In this way, the three modalities smoothly collaborate to facilitate playing a game like TETRIS.

1.3 Book objectives

Among the objectives of this book, we may classify the advocacy of BCI applications as a novel means to facilitate the communication and rehabilitation of people with motor impairments. In Chapter 2, the reader will have the opportunity to get familiar with the existing efforts in this field in a structured and systematic way and realize that despite its great potential, the full capabilities of BCI technology remain rather unexplored. Understanding the difficulties that may be faced by people with motor impairment, when using computer applications, fails also within the objectives of this book. Computer-use habits and difficulties differing significantly from that of able-bodied as well as requirements for assistive interfaces with unexpected priorities from people with motor impairment synthesize the content of Chapter 3. This content is further extended to include objective measurements and qualitative feedback on the home usage of an assistive interface for a period of one month. Finally, the instruments to model the (dis)abilities of people with motor impairment, as well as the study of persuasion theories and intervention frameworks to stimulate a positive attitude towards the usage of eye- and brain-assistive interfaces, are the subject of Chapter 4 that concludes *Part I* of this book.

The objective of *Part II* is to allow the interested reader to become familiar with the details of some of the most common algorithmic challenges and methodological approaches when interacting with the computer using the eyes and mind. In particular, Chapter 5 provides a detailed walk-through of existing eye-tracking technologies, highlights the main problems in eye-controlled interaction and describes potential solutions in the context of adapted multimedia interfaces. Of particular interest is the description of a browser that can be operated solely with the eyes, called GazeTheWeb. GazeTheWeb is a custom-made browser that incorporates a number of layout and functional elements that have been specifically designed to facilitate its use through eye-tracking. It overrides the basic browser functionalities, such as scrolling, zooming, back button, history and favourites, by incorporating a parser for HTML5 that allows to make interventions on the way the web-page content is displayed to the user. Chapter 6, on the other hand, reviews and elaborates different evaluation methods used in gaze interaction research, so the readers can inform themselves of the procedure and metrics to assess their novel gaze interaction method or interface. In this context, the objective of this chapter is to address the following questions: how efficiently can pointing and selection be performed? Whether common tasks can be performed quickly and accurately with the novel interface? How different gaze interaction methods can be compared? What is the user experience while using eye-controlled interfaces?

Switching to the EEG modality, the objective of Chapter 7 is to provide an introductory overview of the machine-learning techniques that are typically used to recognize mental states from EEG signals in BCIs. Particular emphasis is placed in describing the available techniques for noise reduction and spatial filtering of EEG signals with the intention to extract robust features for mental state recognition. The presentation of the most effective classification approaches for performing this recognition is also within the objectives of Chapter 7, which concludes with some future directions on adaptive, transfer and deep learning. Chapters 8–10 offer a deep dive into the algorithmic details of BCIs for the cases of SSVEPs, motor imagery sensorimotor rhythms and near-infrared spectroscopy (NIRS) signals, respectively. In particular, Chapter 8 performs a comparative evaluation of the most promising SSVEP-based algorithms existing in the literature and also describes four novel approaches to improve accuracy under different operational contexts. Chapter 9 introduces the concept of endogenous BCIs that can be operated via movement imagination of one limb and are considered ideal for self-paced implementations. Next to the methodological details and the experimental findings on the use of sensomotoric rhythms for BCIs based on motor imagery, of particular interest in Chapter 9 is the study investigating whether the progressive loss of fine-motor skills in people suffering from neuromuscular dystrophy brings any significant difference (compared to healthy individuals) on their functional brain organization that can be considered to favour the Motor Imagery – BCI paradigm. Finally, the objective of Chapter 10 is to present how the emerging field of graph signal processing can be used to benefit the accuracy of BCIs. In proving this claim, a method based on graph Fourier transform is applied on NIRS signals so as to extract robust features capturing the spatial information characterizing this type of signals. The presented approach is evaluated in the context of a mental arithmetic task and achieves classification rates that compare favourably with the state-of-the-art methodologies.

In the last part of this book (Part III), our objective is to demonstrate how the modalities of eye-gaze and mental commands can be effectively combined to drive human–computer interaction. More specifically, in Chapter 11, we initially introduce the concept of ErrPs and describe how they can be effectively used to automatically recognize erroneous human–computer interactions and facilitate their correction. ErrPs are motivated as a passive correction mechanism in human–machine interaction leading towards more user-friendly environments. Their effectiveness as an error-aware mechanism is validated in two different settings: an error-aware SSVEP-based

BCI and an error-aware gaze-based keyboard. Another impressive demonstrator of the effective collaboration between eyes and mind is presented in Chapter 12 through the hands-free version of TETRIS. The natural sensory modalities, i.e. vision, brain commands and stress levels, are integrated into a single visceral experience that allows simultaneous control of various interface options. In Chapter 12, the reinvented version of TETRIS is presented in terms of gameplay design, associated algorithmic challenges and experimental validation.

Finally, this book concludes with Chapter 13 that wraps up the main knowledge and outcomes presented in the previous chapters, while also identifying the open questions and suggesting some future perspectives on making human–computer interaction more intuitive and natural.

Part I

Reviewing existing literature on the benefits of BCIs, studying the computer use requirements and modeling the (dis)abilities of people with motor impairment

Chapter 2

The added value of EEG-based BCIs for communication and rehabilitation of people with motor impairment

Ioulietta Lazarou[1], Spiros Nikolopoulos[1], and Ioannis Kompatsiaris[1]

People with severe motor impairment face many challenges in communication and control of the environment especially in late stages, while survivors from neurological disorders have increased demand for advanced, adaptive and personalized rehabilitation. In the last decades, many studies have underlined the importance of brain–computer interfaces (BCIs) with great contribution in medical fields, ranging from communication restoration to entertainment (e.g. brain games) and motor rehabilitation. Through this chapter we shed light in BCI research that focuses on non-invasive, electroencephalography (EEG)-based BCIs, which is used with the intention to support people with motor impairment in gaining part of their lost autonomy through multiple communication and rehabilitation strategies. This type of approach is primarily intended to help severely paralyzed and locked-in state people by using slow cortical potentials, sensorimotor rhythm and P300 as operational mechanisms. Moreover, they can assist people with spinal cord injury and chronic stroke by applying novel methods for restoration and reorganization of their movement. In this review, 45 studies reporting noninvasive EEG-based BCI systems, which have been tested and operated by individuals with neuromuscular disorders, were identified in the literature. Our review systematically organizes these studies, summarizing the underlying operational mechanisms of each study while reviewing the clinical evidence and outcomes reported between 2000 and 2016. The general conclusion from this review is that BCI systems could successfully serve as a means of communication and control for people with motor impairment, as well as help those with incomplete motor function to regain their motion. However, BCIs need first to be validated in long-term studies of systematic everyday usage by people with severe motor disabilities, as well as to take into consideration the differences between able-bodied and motor-impaired subjects during evaluation.

[1]Centre for Research and Technology Hellas (CERTH), Information Technologies Institute (ITI), Multimedia Knowledge and Social Media Analytics Laboratory (MKLab), Thessaloniki, Greece

2.1 Introduction

Nowadays, several advancements in the fields of clinical neurophysiology and computational neuroscience have been contributed toward the development of promising approaches based on brain–computer interfaces (BCIs) that pave the way for effective communication and efficient rehabilitation of people with motor disabilities.

BCI systems can provide end users with communication means operated from the human brain. In this way, the social exclusion can be avoided and BCI users can feel more active and capable of interacting with their friends and relatives. Among the earlier BCIs for communication, the Wadsworth Center's "Right Justified Box" [1], the participants learned how to choose specific target by controlling their mu rhythm through motor imagery (MI). In the later ones, we may include the electroencephalography (EEG)-based thought translation device (TTD) [2] that is based on regulating slow cortical potentials (SCPs) to move a cursor up or down, which could be interpreted as "yes" or "no." More recent solutions use speller implementations with BCIs, which measure the P300 evoked response when letters of the alphabet arranged in a matrix are flashed in random order [3–5]. A number of tools have been developed for accessing the Internet, with popular examples the BCI-controlled web browsers, such as "Descartes" [6] and "Brain Browser" [7]. In particular, the Brain Browser [8] was based on modulating mu rhythms over motor cortex to select browsing commands such as "next" and "previous." The most advanced BCIs that are currently available allow users to select web links by increasing their mu amplitude with MI or to move to the next link. A more recent version of the Brain Browser that is currently in experimental phase uses P300-based BCIs for direct selection [9].

Perhaps one of the most significant and innovative applications for BCIs that is currently under investigation involves the creation of therapies to regain the lost motor control for people suffering from neurological disorders, such as stroke. Mobility rehabilitation is a form of physical rehabilitation applied to people with mobility problems to restore their motor functions and regain previous levels of mobility or to be adapted to their disabilities. Its implementation lies in the fact that the central nervous system (CNS) has the ability to restore lost functions through the CNS plasticity [10]. Based on the aforementioned hypothesis, various recent studies have shed light in the BCI-based rehabilitation training and motor retraining through real, virtual [11] and augmented approaches and have endeavored to identify the characteristics of an MI-based BCI for rehabilitation. In this common vein, we can group in two categories the approaches for gaining movement in paralyzed participants by using BCIs: (i) approaches that train the patient to produce better motor brain signals and (ii) approaches that train the patient in activating a device that assists movement and as a result to eventually improve their motor function. In the latter case, the goal is to produce sensory input which leads to restoration of normal motor control for those who have their limbs but the disorder or trauma has severely impaired motion capabilities [12]. Preliminary work has shown that MI is an area of high research interest within the field of poststroke rehabilitation. Moreover, "MI engages the autonomic nervous system as if an actual movement was under way. In turn, that force implies metabolic demands which allow humans to adapt to the ongoing conditions" [13]. Considering the beneficial effects of MI on motor rehabilitation, this control task has been the one used for BCIs with therapeutic purposes of recovery. As a result, MI training has rapidly become a key novelty to reactive sensorimotor networks which have been affected by an acute stroke and as a result restore the motor function [14]. So far, preliminary studies have shown that individuals who have had a stroke or other motor difficulties could gain control of specific EEG features [15,16]. A recent systematic review has shown the benefits and the clinical efficacy of the BCI in stroke rehabilitation through the combination of an MI-based BCI, Functional Electrical Stimulation (FES), physiotherapy sessions and robotic assistive orthotic devices for motor recovery poststroke [10].

In most cases of neuromuscular disorders (NMDs), there is a gradual loss of muscle activity that affects speaking, walking and execution of fine motor tasks, and results in the deterioration of quality of life. Although the symptoms of the diseases are easier to cope with during their early stages, they become much more severe in the late stages. Thus, specific priorities must be set forth in order to develop systems that will be deployable for each stage of the disease, even for late stages. The majority of people with NMDs indicate a preference for portable solutions, such as a tablet or a laptop [17]. Nevertheless, one common issue that may arise from these devices is their inability to adapt to patient's needs, especially at late stages of the disease where many abilities have deteriorated. Even though many technological adaptive devices are currently used, such as head tracking technology (e.g. "SmartNav*") and eyetracking technology (e.g. Tobii Dynavox†), they have several limitations [17]. In detail, for people in extreme pathological conditions (e.g., those who are in complete locked-in state without any remaining muscle control), the use of such systems may not be possible.

On the contrary, recent studies have underlined the importance of BCI systems in rehabilitation and restoration of motor function [18–20]. In a nutshell, BCI systems promise to offer a unique and multimodal solution for both communication and rehabilitative therapy that will overcome the shortfalls of the aforementioned approaches [21]. The idea of operating an external device with one's thoughts is a highly promising option for people whose functions such as speech or motion are impaired. Many clinical studies on BCI research field have highlighted not only the potential utility and integration of these innovative technological approaches into the life of people with motor impairment, but also the positive impact occurring by translating scientific knowledge and experimental design into clinical benefits by enabling a novel real-time communication between the user and the external world [22].

2.2 BCI systems

Nowadays, there are several techniques and operational systems that are used on a daily basis in order to record brain activity and obtain useful brain signals. In detail,

*https://www.naturalpoint.com/smartnav/

†https://www.tobiidynavox.com/en-us/about/about-us/how-eye-tracking-works/

BCI systems can be based on invasive recordings, i.e., implantable electrodes or grid recordings from the cortex (ECoG) [23], multi-unit activity (MUA) or single-unit activity (SUA) [24], local field potentials (LFPs) [25], as well as noninvasive recordings such as EEG, magnetoencephalography (MEG) [26], functional magnetic resonance imaging (fMRI) [27] and near-infrared spectroscopy (NIRS) [28]. Furthermore, other approaches concerning combinations of the aforementioned recording methods with additional electrical stimulation have also been suggested, such as electrical/magnetic stimulation [29–31]. Figure 2.1 depicts all brain recording and stimulation technologies used for BCI systems.

Among the aforementioned types for BCI system realization, EEG recordings have been so far more extensively tested and deployed [32–34] due to their effective implementations, noninvasiveness, and portable and adaptable manifestation [10]. In this systematic review, we focus on the EEG-based BCI systems. In this direction, there are multiple types of EEG-based BCI systems for communication, environmental control and rehabilitation depending on different modalities of the EEG, namely SCPs, sensorimotor rhythms (SMRs), P300 event-related potentials and

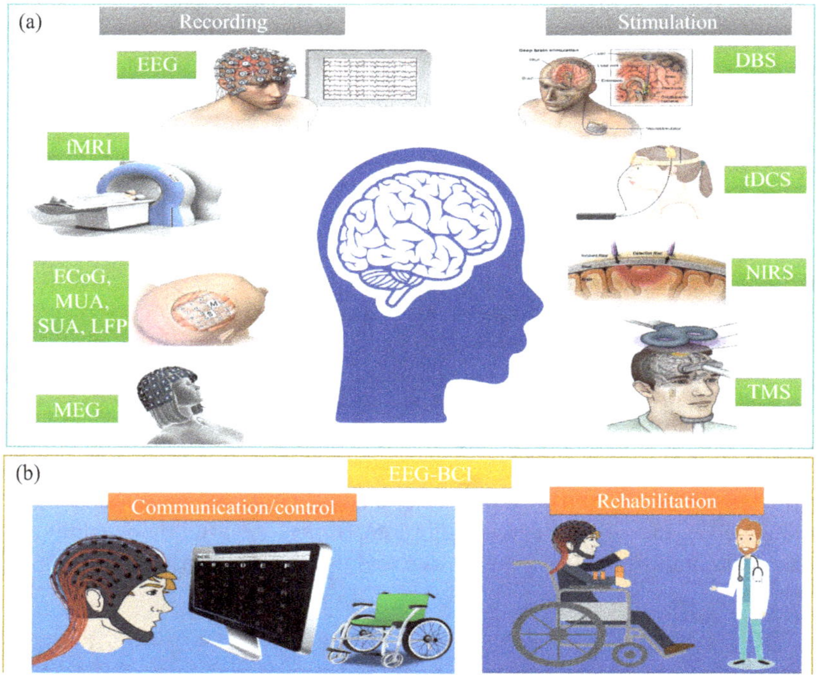

Figure 2.1 The different types of neuroimaging systems for recording brain activity, i.e., ECoG, EEG, fMRI, MEG, or stimulating the brain (DBS, NIRS, TMS, and tDCS), widely used in BCI research field (A) and the potential of deploying the aforementioned systems for communication and control as well as rehabilitation (B)

steady-state visual evoked potentials (SSVEPs). Finally, different BCI applications will be reviewed and assessed qualitatively and quantitatively using two measures: *Classification Accuracy (CA)*, i.e., the percentage of correctly classified BCI controls when the user operates a BCI system and *Information Transfer Rate (ITR)*, i.e., the information transfer rate, given in bits per selection or bits/min.

2.3 Review question

In this review, we focus on noninvasive BCI applications geared toward alternative communication and restoration of movement to paralyzed patients [35]. For the purposes of this systematic review, we use the following study frameworks (Figure 2.2):

- **Population**: People with motor disabilities such as ALS, SCI, PD and MS.
- **Intervention**: EEG-based BCIs using different modalities such as P300, SMR and SCPs
- **Outcomes**: Communication and social integration, environment/domestic appliances

Therefore, we have included several milestone studies focusing on EEG-based BCIs for either assisting people with motor impairment in their everyday life or

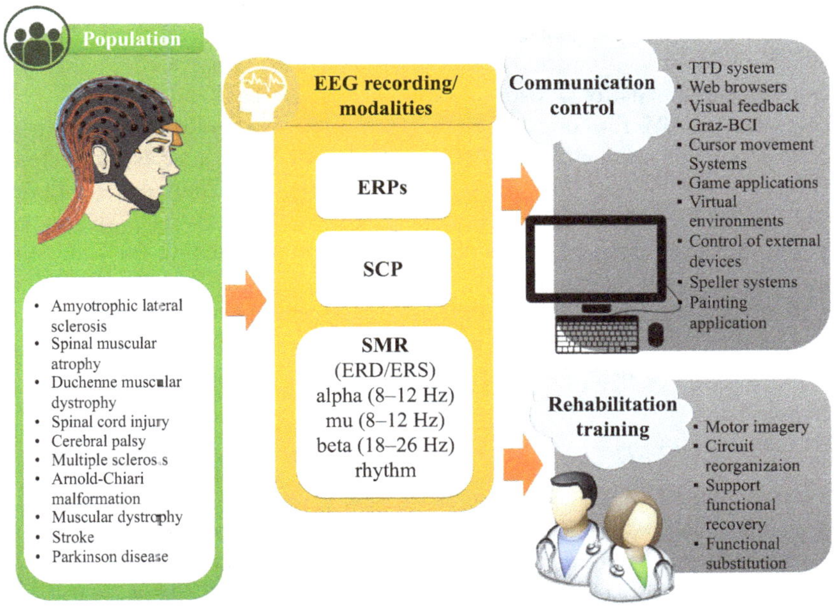

Figure 2.2 EEG-based BCI for communication-control and rehabilitation

supporting them through rehabilitation strategies. We review EEG-based BCI technologies for communication and control based on three different EEG signals (SCP, SMR and P300), and discuss their limitations and advantages. We did not include SSVEP since this type of BCI depends on attentional capacity and vision to be intact, and both are often compromised in patients with advanced and severe neurological disease [21]. Therefore, SSVEPs are not suitable for patients in advanced stages of amyotrophic lateral sclerosis (ALS-LIS) or with uncontrollable eye movements [36]. A number of SSVEP-based BCI systems have been developed and applied to operating a prosthesis [37] or controlling an avatar in a virtual reality environment [38] by healthy individuals. The conventional SSVEP-based BCI systems described above commonly require the basic assumption that the users have a normal oculomotor function and are thus able to maintain gaze at a given visual stimulus consistently [39]. In addition, we examine and analyze the BCI methods for inducing brain plasticity and restoring functions in impaired patients.

2.4 Methods

2.4.1 Search strategy

The following electronic databases have been searched for relevant studies: MEDLINE/PubMed, EMBASE, Scopus, IEEExplore, ResearchGate, Google Scholar and manual search conducted in *Journal of Neuroengineering and Rehabilitation*, *Journal of Clinical Neurophysiology* and *Journal of Medical Internet Research*. We included only English-language journal articles that directly evaluated EEG-based BCI technology on participants with motor disabilities. Our search identified 650 potentially relevant citations. Also, we examined the reference and citation lists of the retrieved articles. We further examined if there were any duplicates of the retrieved articles (35 duplicates identified and removed). A total of 615 articles were screened, based on title and abstract, to only include studies involving individuals with motor disabilities and EEG–BCI systems. Screening of titles, abstracts and full texts yielded 92 publications that met the inclusion criteria. Further screening for EEG studies focused on restoring communication, environmental interaction and rehabilitation after training in the target population reduced the sample to 33 articles. Additional hand-searching of the references of these publications identified another 12 relevant papers as listed in Figure 2.3.

2.4.2 Types of participants and model systems

The selected studies included participants with motor disabilities. Some studies involved both able-bodied and individuals with motor disabilities. Participants' group included: ALS, spinal muscular atrophy type II (SMA II), Duchenne muscular dystrophy (DMD), spinal cord injury (SCI), (Spastic) cerebral palsy (CP), multiple sclerosis (MS), Arnold-Chiari malformation (A-CM), stroke, motor disability and tetraparesis, postanoxic encephalopathy, extrapyramidal syndrome and Parkinsonism.

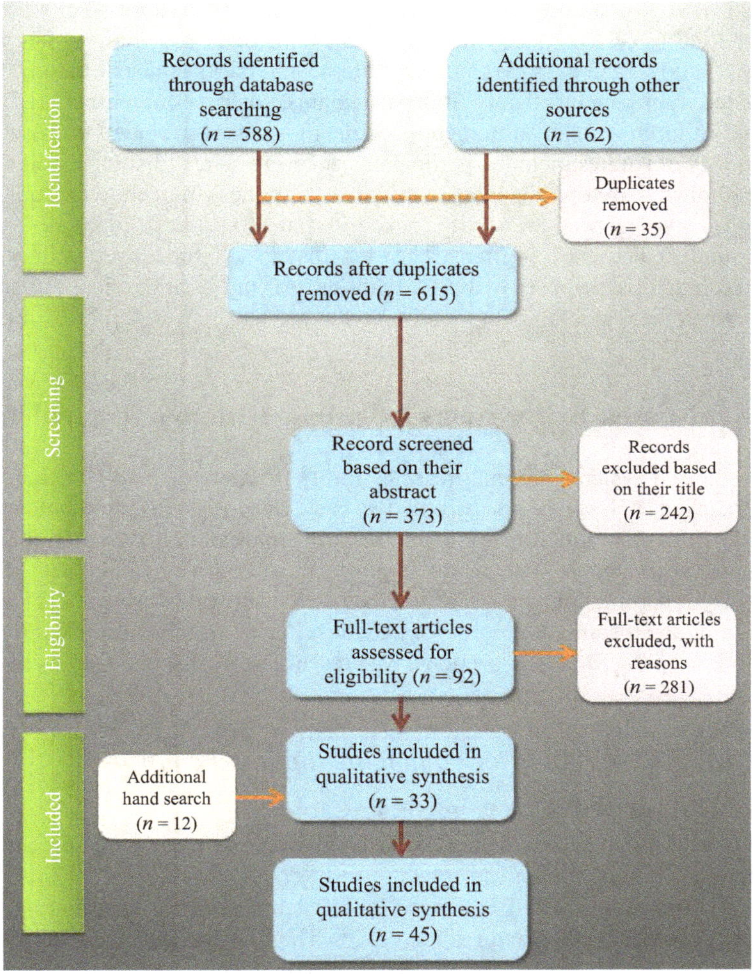

Figure 2.3 PRISMA flow diagram of the article screening and selection process. Article selection was conducted in accordance with PRISMA guidelines for reporting systematic reviews [40]

2.4.3 Data synthesis – description of studies target population characteristics

The selected studies included 10 single-subject reports, 8 studies with fewer than 5 participants, and 11 studies with 5–28 participants with motor disabilities. A total of 20 studies involved both healthy and individuals with the aforementioned characteristics. The majority of studies included participants with ALS (25 studies). Other conditions reported in studies included different levels of SCI (14 studies), SMA II

(1 study), DMD (6 studies), CP (7 studies), MS (3 studies), traumatic brain injury (TBI) (1 study), A-CM (1 study), tetraparesis and myopathy (3 studies), neurofibromatosis (1 study), and extrapyramidal syndrome and Parkinsonism (1 study). All of the selected studies included only adult participants, whereas studies included adolescents and children with the aforementioned diagnosis were excluded from our study. The majority of studies included participants at severe state (13 studies), others contained motor-impaired participants at minimal (5 studies), moderate (8 studies) and advanced (2 studies) level of severity, locked-in state (10 studies), complete locked-in state (2 studies), hemiparesis after stroke (4 studies), and tetraparesis (1 study). Also, studies included participants with SCI with level of injury at cervical (12 studies), thoracic (3 studies) and lumbar (1 study) vertebrae.

2.5 EEG-based BCI systems for people with motor impairment

EEG-based BCI systems present great possibility for a widespread clinical use. The reason for using BCIs can be organized in two basic categories: (a) communication and control and (b) rehabilitation. By communication and control, we mean the ability of BCIs to assist people with motor impairment in communication with multiple devices. By rehabilitation, we refer to rehabilitation strategies that use BCI systems for regaining the lost motor control for people suffering from neurological disorders (e.g., stroke). In the following sections, we present briefly this two-fold use of EEG-based BCIs.

2.5.1 EEG-based BCIs for communication and control

2.5.1.1 EEG-based BCIs using SCP

An EEG-based BCI can rely on SCPs, which allow anatomically specific voluntary activation of different brain areas.

The TTD system. SCP–EEG-based BCI systems require users' training to shift the polarity (positive or negative) of their SCPs. The first seminal study on EEG–BCI with two patients in LIS-ALS using SCP was conducted almost two decades ago [41]. The patients were trained to voluntarily generate SCPs and during the response period, the subjects were required to produce either negativity or positivity greater than specific amplitude. In a subsequent work [3], a TTD was developed. TTD trains participants at locked-in state to self-regulate their SCPs in order to select letters, words or pictograms in a computerized language support program. Kubler *et al.* [42] expanded the previous TTD system by introducing for the first time "feedback training" by testing two participants at late ALS stage. Moreover, in another work [33], the research team trained an ALS patient to use TTD for over 1 year so as to use this device in order to communicate by spelling words. Also, they compared different EEG classification algorithms of the three phases of training in two sets (Set 1: "feedback training" phase by rewarding the patient for producing cortical shifts in a requested direction, Set 2: "copy-spelling mode" by requiring the participant to copy a text, and "free-spelling mode" by self-selecting letters and words). The patient's average correct response rates in the online training were 83% and 72% for Set 1 and Set 2, respectively.

In another work, five ALS participants were trained on how to use their SCPs for effective communication. Particularly, the procedure included a training phase, while during the copy spelling and free spelling mode, participants had to select a letter by producing positive SCP amplitude shifts or to reject a letter by producing negative SCP amplitude shifts [43]. In essence, it was shown that the performance in SCP self-regulation after several sessions of training can be predicted from an initial performance. Another approach investigated the potential use of SCPs as a control mechanism in order to promote communication of an ALS patient by providing visual feedback of his actual SCP amplitude [44]. After 6 months of training, the participant managed to self-regulate his SCP by producing two different brain responses. In the end he managed to produce 454 words in German language even though the speller yielded one letter per minute.

Web browsers. Another approach to using the TTD system for handling a Web Browser, namely "Descartes," was firstly introduced in Karim *et al.* [6]. The embedded system was tested with only one ALS patient. "Descartes" provided feedback of SCP amplitude in a time-locked manner after several sessions of training. This was the first study to show that an EEG-controlled Web browser, which is based on SCP self-regulation, could be efficiently and successfully operated by a severely paralyzed patient. It was possible to access the Internet via the "Descartes" system but there was great difficulty with respect to selecting an icon or a picture on a Web page or selecting from an alphabetically sorted decision tree. Although the training phase of the aforementioned study may be a proper procedure in order to support patients so as to take advantage of BCIs and foster communication, the participant had to exceed or to remain below a certain threshold (7.7 μV) for operating the system and this containment may not be suitable for daily usage.

2.5.1.2 EEG-based BCIs using SMRs

People with NMDs can also learn to modulate their SMRs generated when a specific movement is executed or simply imagined (MI).

Graz-BCI system. One of the most highlighted SMR-BCI studies, namely the "Graz-BCI," was tested on a severely paralyzed participant with CP, who had lost the ability to communicate completely [45]. This particular study showed that "Graz-BCI" could effectively decode changes in SMR due to MI with a CA of 70% in the task of letter selection. Nevertheless, the spelling speed rate was very low (one letter per minute) and many adaptations of the classifier were necessary. More specifically, the task in this study was a right-hand movement imagination, the so-called Basket-paradigm with feedback, where the participants had to move the ball into the correct "basket" [46]. One aspect that should be highlighted in this study was that the participants achieved high CA and an ITR between 8 and 17 bits/min without participants having any experience with the BCI system before. This proved that the "Graz-BCI" paradigm can be easily learned even by people who are not skilled at these applications.

Cursor movement systems. Similarly, Wolpaw *et al.* [47] developed an EEG–BCI based on SMR regulations. Their results showed that people can learn to use scalp-recorded EEG rhythms to control movement of a cursor in two dimensions, while this control is developing gradually over training sessions. This study demonstrated that performance gradually improved over the training sessions while participants gradually gained better control over the rhythm amplitudes that controlled the cursor. In another study [48], four ALS participants were trained to move the cursor steadily across the screen with its vertical movement controlled by SMR amplitude. Their results showed that over the initial 20 sessions of training, all four participants acquired SMR control, proving that an SMR-based BCI might help participants with ALS to maintain an acceptable quality of life with appropriate communication systems [48].

In exploring the feasibility and accuracy of the SMR-based systems, one recent study [49] applied high-resolution electroencephalographic (HREEG) techniques that estimated cortical activity by using appropriate models of volume conduction and neuroelectrical sources. In this study, the authors examined five able-bodied participants and one with a traumatic stabilized lesion located at the dorsal level. In this study, the lateralization of electrical activity, which is expected to be contralateral to the imaginary movement, is more evident on the estimated cortical current density (CCD) than in the scalp potentials and showed that subjects who underwent training could use voluntary modulation of estimated CCDs for accurate online control of a cursor. Furthermore, McFarland *et al.* [50] showed remarkable results as people with severe motor disabilities could use brain signals for sequential multidimensional movement, selection with two-dimensional cursor movement and target selection through self-regulation of their SMR. This is one of the first and most highlighted studies which showed that people with motor impairment can learn to use scalp-recorded EEG rhythms so as to move a cursor in two dimensions to reach a target and then to select the target.

Game applications. A noteworthy study tried to address the problem of long training periods needed for people with motor impairment to learn how to operate an EEG–BCI system. A game experiment that could achieve a successful BCI operation in less than 30 min was proposed [51]. Their results indicated that three out of six subjects learned to control a BCI after training and hit the target with rates 2.2–3.8 hits/min, with accuracy of 94%, 67%, and 57%, respectively, and an ITR of 8 bits/min. They concluded that subjects could improve their performance after more training since they could learn to produce more distinctive brain activations during the attempted movements.

In this common vein, Bai *et al.* [52] investigated the role of extensive training in the ability to control SMR. This is one of the few studies where people with motor impairment showed equal performance, in terms of CA, with able-bodied ones. In particular, two motor-impaired participants (one with stroke and one with ALS) and nine healthy ones participated and it was found that by using MI, subjects were able to operate the proposed system with good CA and with fast transfer rates (10–12 bits/min), while in terms of CA, healthy, stroke and ALS participants had similar performance. Another study [53] developed a different approach for self-control of SMR rhythms throughout a game, namely the "Connect-Four." The proposed system showed low effectiveness, as compared with effectiveness achieved by end users in other SMR-based BCI systems (70%–100%) [51].

Although some of the aforementioned studies tried to address SMR-based BCI systems by reducing the time needed for training, their results revealed low ITR 2.2–3.8 hits/min and showed that participants' performance improved with longer training periods [51,54].

Virtual environments. A milestone approach investigated how a tetraplegic subject could control his movements in a virtual environment (VE) with a self-paced (asynchronous) BCI system [11]. After four runs, the subject was able to reach accuracy of 100%. The fact that after only four runs the subject reached a successful performance of 100% shows that motivation induced by responses of the avatars and more realistic conditions enable the participant to achieve better performance.

Control of external devices. There are plenty of studies that have highlighted the importance of EEG-based BCI for controlling external devices. In Cincotti *et al.* [55], it was found that people with severe NMDs can acquire and maintain control over detectable patterns of brain signals and use this in order to control devices. This particular study showed for the first time that an EEG-based BCI system can be integrated into an environmental control system. Over the 10 sessions of training, subjects acquired brain control with an average CA higher than 75% in a binary selection task. However, it should be underlined that only four out of 14 participants managed to achieve the aforementioned CA. In another study, the potential use of SMR in control of external domestic devices was investigated. During the training sessions, subjects were asked to imagine the same kinesthetic movement during the visual session. Four out of six participants with DMD were able to control several electronic devices in the domestic context with the BCI system with a percentage of correct responses averaging over 63%, whereas healthy subjects achieved rates of 70%–80%. Furthermore, another study [56] introduced the "one-dimensional feedback task," i.e., moving a cursor from the center of a monitor to a randomly indicated horizontal direction. Four out of seven subjects were able to operate the BCI system via attempted (not imagined) movements with their impaired limbs (both foot and hand) with up to 84% CA.

Additionally, a recent approach investigated the potential use of a telepresence robot, which is remotely controlled by a BCI system during a navigation task. In this task, the participants with motor impairment achieved similar CA to ten able-bodied participants who were already familiar with the environment [57]. Some end users were able to press specific buttons on a modified keyboard, while others were using "head switches" by imagining left hand, right hand and feet movements during calibration recordings. Nevertheless, people with motor impairment needed more time as compared with healthy participants to complete the task.

2.5.1.3 EEG-based BCIs using P300

The widely known P300 component is an EEG modality that is very frequently used for operating a BCI system, since it is a late positive component evoked in response to an external task-relevant stimulus. In particular, the P300 component has been used in order to control devices, such as wheelchairs, operate real and VEs, and allows the user to use paint interfaces or access the Internet. The following categories have been examined so far with EEG–BCI systems using P300 as modality:

Speller systems. Speller systems are the main BCI applications that use the P300 modality. One seminal work [58] investigated whether P300–BCI could be used as an alternative EEG-based BCI modality for communication in ALS population. In terms of CA, groups' performances were similar, suggesting that a P300-based BCI can be considered as a powerful and cost-effective, nonmuscular communication tool both for ALS and non-ALS, able-bodied users. However, the ITR (bits/selection) in this study is considered rather low as compared with other similar, more recent P300-based BCI studies [7]. In addition, the most severely impaired ALS participants had the worse CA as compared with other participants, which reflects also the incapability of P300 to be an effective communication solution for more severe impairments. In this common line, another study [59] developed a P300-based BCI system tested on five people with motor impairment with separate pathologies and four able-bodied subjects. Four out of the five disabled participants and two out of the four healthy subjects achieved 100% offline accuracy. Similarly, Nijboer *et al.* [60] examined severely disabled ALS individuals on a P300-based BCI system for writing text by using a two-phase experimental procedure. The research team found that the ERP response remained stable for several months and that CA improved during free-spelling phase.

In another study [61], four ALS participants who had already been involved in similar studies operated a 5×5 spelling matrix (all letters of the alphabet, except for letter Z) with auditory feedback. This study showed extremely low selection and CA, which demonstrates that it is exceptionally difficult for participants to focus and maintain their attention on the numbers. Another study [62] used P300 Speller Paradigm with 7×7 matrix of alphanumeric characters, where two types of stimuli are presented with different probability (infrequent target stimuli and frequent nontarget stimuli). People with motor impairment showed lower CA with regard to the able-bodied group, while the ITR was also lower for the people with motor impairment. A more recent approach [63] compared the "Classic Flashing (CF)" P300 Speller paradigm, in which rows and columns are highlighted randomly (as in common P300 Spellers), and the "Face Flashing (FF)," in which characters (letter and numbers) are overlaid with faces, so as to investigate effects of face familiarity on spelling accuracy. Two motor-impaired participants were not able to communicate with more than 40% CA with CF, whereas with FF the same participants spelled with an average accuracy of approximately 82%. On the contrary, low ITR and CA reported in a study that examined 10 SCI participants with a P300-based BCI system. In this study, participants had to spell the 5-character word "SPINE" by using a 36-character matrix (6×6 matrix layout) after 10 min of training [54]. A recent study [64] introduced a thought-based, row-column filtering correspondence board, emulating user-centered configuration standard for individuals with CP. They found that the participants became gradually more capable to communicate by using the proposed system.

In this common line, another study evaluated the possibility and usability of an assistive model operated by a P300-based BCI system in order to promote communication and environmental control of domestic appliances and applications to users with ALS [65]. Their results of the three experimental conditions showed that the effectiveness, adequacy and the end user's fulfillment did not vary within those three experiments. However, in terms of CA, the participants' performance was not successful. In this vein, another study [66] evaluated the impact of a hybrid control of a P300-based BCI technology (using both EEG and Electromyography) that was developed to operate an assistive technology software. It was found that hybrid BCI might enable end users to take advantage of some remaining muscular activity, which may not be fully reliable for properly controlling an assistive technology device. Moreover, [67] introduced also a 6 × 6 matrix where the participant had to "copy-spell" (35 letters) or "texttospell." It is noteworthy that this study concluded that there is no significant correlation between level of severity and CA. Finally, a combined approach tested a heterogeneous group of ALS participants in a study of a P300–BCI with an MI task and showed for the first time that the quality of the control signals depends on the cognitive function of the participants and that behavioral dysfunction negatively affects P300 speller performance [68].

Web browsers. Another approach refers to BCI systems for Internet access. More recently, a new solution of Internet access, the "true web access," was proposed by Mugler *et al.* [7] where Internet surfing could be successfully executed through a P300–BCI browser. It was the first study to use real-life scenario for Internet access by using the open-source technology of Mozilla's Firefox. Their results showed that participants with ALS achieved a CA of 73%, while healthy subjects achieved a CA of 90%, and ITRs achieved a CA of 8.6 and 14.3, respectively. However, the positive response and acceptance of this useful tool were outlined by the ALS participants. Moreover, a recent work assessed the effectiveness, efficiency and user satisfaction in two spelling tasks, an email sending and an Internet browsing task [69] by testing a commercial "AT-software QualiWORLD" (QW) controlled by the P300–BCI with four end users and three AT experts. The performance was high in all tasks and always above 70% CA.

Paint application. A different direction of a P300–BCI system is toward supporting users to paint [70], such as the "Brain Painting." Despite the low ITR (ALS: 5.8, able-bodied: 8.57), the CA of both people with motor impairment and able-bodied was high (ALS: 79%, able-bodied: 92%). Moreover, the patients found the application extremely useful. However, this study showed that P300 amplitudes may be affected by illness severity as has been already mentioned in previous studies. Additionally, another P300-based BCI system used a "paint" application, achieving high-performance levels (80% accuracy) in both free painting and copy painting conditions, whereas ITRs were rather low (4.47–6.65 bits/min) as compared with other P300 applications. In general, P300 Brain Painting application was effective and the end users with severe motor paralysis declared that they might use the suggested application in their everyday routine [71]. A recent study [72] reported the use of a Brain Painting 6 × 8 matrix including 48 tools for painting combined with a home-use P300–BCI application. After approximately 2 years of training, a high accuracy of 70%–90% was accomplished, while, most importantly, the end users were highly satisfied with the BCI–Brain Painting system. This study highlighted the potential of using a BCI application independently by the users while promoting satisfaction and enjoyment to people with motor impairment.

Control of external devices. Another communication mode of P300-based systems is to control external devices. A recent study [73] developed a BCI system that was tested by two groups of participants, where they had to select a specific path for a virtual object to reach the goal-point by using the P300 activity. It is worth noting that this is one of the few studies which have deployed such a system to people with motor impairment other than the common diseases (e.g., ALS and stroke) by examining five paralyzed participants with various impairments. This study also highlights the fact that correct response without training by using P300 modulation is feasible as an endogenous response to a stimulus. Furthermore, Mauro *et al.* [74] compared two interfaces for controlling the movement of a virtual cursor on a monitor and found that online CA was more than 70% both for able-bodied and ALS participants. Their results showed that ALS participants with residual motor abilities were able to maintain their attention by controlling the movement of a virtual cursor on a monitor, similar to the able-bodied ones. Another approach evaluated different visual P300 BCI systems on subjects with severe disabilities and their results revealed that P300-based BCI operation may be affected by the severity of the disease, whereas long periods of BCI execution revealed tiredness symptoms and a decrease in performance rates [75].

Furthermore, a recent work on this field tested a P300-based BCI system on participants with different motor and cognitive limitations, aiming at managing eight real domestic devices by means of 113 control commands [32]. Ten out of 15 participants were able to properly execute the suggested apparatus for precision higher than 75%. Eight of them reached accuracy above 95%. Moreover, high ITRs, up to 20.1 bit/min, were reached. In another study, a BCI system decodes the user's intentions and facilitates navigation, exploration and bidirectional communication with a robotic system that is remotely controlled [76]. More specifically, the system is composed of a user station (patient environment) and a robot station (placed anywhere in the world), both remotely located and through the Internet to promote communication. Although the ITR was significantly low (7 bits/min) the CA was interestingly high (90%–100%).

2.5.2 EEG-based BCIs for rehabilitation and training

One of the most innovative applications for BCI technology concerns rehabilitation systems that aim to support people with motor impairments to regain the lost motor control. Many clinical BCI studies showed evidence for the feasibility and positive effect of MI-based BCI systems in combination with physiotherapy and robotic assistive orthotic devices for motor poststroke recovery [16,77–80]. It is conjectured [21,77,78,80,81] that this efficacy of BCI systems on motor rehabilitation is due to the underlying mechanism of synaptic plasticity [82]. Various recent studies have shed light in the BCI-based rehabilitation training and motor retraining through real, virtual [11] and augmented approaches and have managed to identify the characteristics of a MI-based BCI system for rehabilitation.

In this vein, it is suggested that there are two ways for paralyzed people to regain motor abilities by using BCI systems: (i) train patients to produce more reliable motor brain signals and (ii) train patients to activate a device that assists movement by improving the motor function [10]. Even though people with acquired motor impairment often exhibit damaged cortex or disrupted motor connection integrity, EEG-based BCI methods are still capable of identifying meaningful improvement and gradual change. The combined approach of BCI systems together with traditional physiotherapy appears to be very effective. In another work, a tetraplegic participant gained control of a hand orthosis in order to improve his functionality of residual muscle activity and restore "hand grasp function" of the upper limb. This was achieved by regulating his SMR through motor imaging of his foot movement [16]. Following the training procedure, the participant was capable of a successful operation of the orthosis by closing the hand orthosis while imaging both-feet-movement and by opening the orthosis by imaging right-hand-movement, nearly error-free with CA close to 100%. Since a stable performance of above 90% correct responses was reached, the patient started to practically use the orthosis to lift light-weighted objects.

Another similar study evaluated the results of daily BCI training to beneficial effects of physiotherapy in patients with severe paresis [81]. Successful SMR control resulted in concurrent movements of the arm and hand orthoses in the experimental group, while in the control group participants sham feedback was received, i.e., random movements of the robotic orthoses (not related to the ipsilesional SMR oscillations). This was the first study to present SMR–BCI intervention as a rehabilitative approach for a relatively large number ($N = 32$) of stroke survivors. A recent study [83] highlighted the importance of the currently used BCI technology in assisting MI practice, which notably contributes to the improvement of motor functionality in subacute stroke patients with severe motor disabilities. More specifically, they used BCIs which can support rapid measure of brain activity generated by MI. The importance of this study lies in the fact that it was a randomized controlled trial and the first to demonstrate a clinical, pre-post improvement of the subacute stroke patients with detailed reports about the underlying neurobiology.

Moreover, a recent study suggested enhancing motor recovery in people who survived a stroke through passive movement (PM) with a haptic robot and MI [79]. These results revealed better CA of stroke participants (75%) than observed by able-bodied (67.7%). Finally, another study showed the potential application of an MI task as a mechanism for stroke rehabilitation. They classified the participants into three subgroups based on the different lesion locations in order to perform three different motor tasks (MI, passive motion and active motion). They found different β band EEG patterns in each patient group, while two groups showed positive laterality coefficient (LC) values (LC of the ERD/ERS power of stroke patients is affected by brain damage) in the active and MI tasks.

Finally, a recent study [84] explored the EEG activity from six NMD participants and six able-bodied individuals during two randomly alternating, externally cued, MI tasks (clenching either left or right fist) and a rest condition. In detail, the participants had to imagine the movement of their left or right hand. The cue for the initiation of movement imagination was given by a red arrow (onset), appearing either on the left or right side of the screen, pointing in the same direction and indicating the corresponding imagery movement. Their results revealed increased phase synchrony and richer network organization in NMD patients since they appear to possess an inherent advantage, over able-bodied participants, in the use of phase-synchrony-related MI–BCIs.

2.6 Discussion

The goal of this systematic review was to identify and synthesize findings on the grounds of noninvasive EEG-based BCI systems. We presented published studies which promote communication and control of appliances for people with motor impairment and approaches which applied adaptable rehabilitation strategies concerning the modern restorative physiotherapy. This chapter reviewed BCI systems of 45 published studies that were generated from the year 2000 till 2016, discussing the added value of this novel technology and highlighting the important role of BCIs in motor-impaired people's life. These studies involved applications of BCI systems in fields such as medical and clinical applications, control of wheelchair, games and entertainment (e.g., painting), communication, rehabilitation and environmental control. The majority of the studies included participants with adult-onset ALS, while most of them were at severe level of paralysis of locked-in state. In this common line, in studies focusing on rehabilitation, the majority of participants were tetraplegic with SCI at cervical vertebrae. This contributes to the line of research that tackles the importance of investigating new solutions for people who are severely paralyzed and concentration in BCI research. It also underlines the critical need of severely paralyzed motor-impaired people for communication even at late stages. Moreover, different control mechanisms have been used to assess the CA and ITR of the system and the ability of the user to modulate brain patterns. From the reviewed studies, it is evident that each one of the three EEG-based modalities (i.e., SCP, SMR and P300), used in noninvasive BCIs, comprises a promising solution for EEG–BCI system realization.

Moreover, SMR-based games [51,52] indicated that people with motor impairment show reliable performance and a successful BCI operation. In Kauhanen *et al.* [51], three out of six subjects learned to control a BCI after training, where a great advantage was that participants received feedback and could change their strategy in response to the feedback. In Bai *et al.* [52], both patients were able to use the proposed SMR–BCI game system with high performance as well. Both studies are highlighting the adaptability of the systems and the general acceptance of game applications by people with motor impairment. However in SMR–BCI applications, the main disadvantage is that although ERD/ERS is observed in the majority of participants, some subjects (even able-bodied) may have no detectable ERD/ERS components (on the contrary, P300 component is always observed). Initially, BCI approaches for entertainment were typically not at high priority in the field of BCI research due to the fact that BCI research has mainly focused on applications to address communication and independency through assistive technological ways (i.e., spelling devices and control of external devices). Nevertheless, game-oriented solutions seem to be really promising since they use additional assistive tools while enhancing participant's motivation. Furthermore, reported BCI studies involving both able-bodied individuals and those with severe disabilities have pointed delays in reaction time, low ITR and worse CA of people with motor impairment [7,52–54,62,75,76]. Moreover, low ITRs of current

BCI systems do not allow for general conclusions regarding the effective use of BCI on a daily basis. Also, except for few cases [85], the majority of BCI systems and applications are mainly used in a research environment (research laboratory, etc.) and have not been deployed yet in patient's homes for continuous and everyday use, as they would need adaptation and fixation during the operation.

P300 shows higher ITRs and does not need training but is greatly affected by the level of severity of the disease. Nevertheless, many studies have also shown that even patients in the LIS can use a P300–BCI for longterm periods [72,86]. However, in terms of ITR, able-bodied group reached higher maximum bit rates than disabled subjects in almost all studies of P300–BCI [7,62,74–76,87]. Moreover, in some studies, the patients did not even complete the experimental process [74]. In addition, in most cases the most severely paralyzed participants seem not to be able to operate successfully the EEG-based BCI system [75,88]. These results indicate that (i) P300 may be affected by the level of severity and (ii) participants have worse performance during the sessions due to a "habitual effect" [89–91].

The reviewed articles largely focused on individuals with adult-onset disabilities. It is unclear if the findings of these studies could be generalized to individuals with congenital disabilities, who often have never experienced any terms of communication or motion. For example, to the best of our knowledge, BCI systems that use SMR for system operation rely on MI of upper and lower limbs and have not been tested with individuals who never experienced voluntary control of their movements [92].

Albeit slow, in the majority of the studies SCP speller (namely the TTD) was around one letter per minute and satisfied the requirements for a successful BCI system [3,33]. Although the course of the SCP shifts of participants who used the TTD remained stable over time, a huge disadvantage of BCIs that demand "self-control" of an EEG component is that the user must undergo long-term preparation and extensive training for several weeks so as to gain the level of CA needed to use, e.g., "brain-controlled cursor movement" for communication. Moreover, BCI technology that does not replace but complement existing therapies is a novel and promising field. Studies in stroke patients have shown that, with a motor relearning intervention, EEG features change in parallel with improvement in motor function and that sensorimotor rehabilitation using BCI training and MI may improve motor function after CNS injury. Taking into account all the aforementioned pieces of evidence, there is a strong indication that BCIs can eventually promote independence through novel communication techniques and promote motor rehabilitation of patients with NMD, SCI and stroke.

2.7 Summary

The development of novel BCIs raises new hopes for the communication and control as well as the motor rehabilitation of people with motor impairment. However, the majority of current published works are basically proof of concept studies with no clinical-based evidence of daily use by people with motor impairment. Research interest in the field of BCI systems is expected to increase and BCI design and development and will most probably continue to bring benefits to the daily lives of people with motor impairment. Moreover, to address the need for extensive training for self-regulation of SMR, and considering the effect of motivation in the BCI control performance, more enjoyable solutions such as Virtual Reality or Gaming/Painting could be used. These approaches re-enable patients to be creatively active and consequently promote feelings of happiness, self-esteem and well-being, and promote better quality-of-life. Also, as the goal of future studies should be the demonstration of a long-term beneficial impact of BCI technology on functional recovery and motor rehabilitation, extensive randomized controlled trials are required.

References

[1] Vaughan TM, McFarland DJ, Schalk G, Sarnacki WA. EEG-based brain–computer interface: development of a speller. In: Neurocience Meeting Planner. 2001.

[2] Kübler A, Kotchoubey B, Kaiser J, Wolpaw JR, Birbaumer N. Brain–computer communication: Unlocking the locked in. Vol. 127, Psychological Bulletin. US: American Psychological Association; 2001. p. 358–75.

[3] Birbaumer N, Kübler A, Ghanayim N, *et al.* The thought translation device (TTD) for completely paralyzed patients. IEEE Trans Rehabil Eng. 2000;8(2):190–3.

[4] Li W, Sperry JB, Crowe A, Trojanowski JQ, Iii ABS, Lee VM. NIH public access. Evolution (N Y). 2009;110(4):1339–51.

[5] Blankertz B, Krauledat M, Dornhege G, Williamson J, Murray-Smith R, Müller K-R. A note on brain actuated spelling with the Berlin Brain–Computer Interface. Univers Access Human–Computer Interact Ambient Interact [Internet]. 2007;4555/2007:759–68. Available from: http://dx.doi.org/10.1007/978-3-540-73281-5_83%5Cn http://www.springerlink.com/content/03hr3j4812561652/

[6] Karim A, Hinterberger T, Richter J, *et al.* Neural internet: Web surfing with brain potentials for the completely paralyzed. Neurorehabil Neural Repair [Internet]. 2006;20(4):508–15. Available from: http://www.ncbi.nlm.nih.gov/pubmed/17082507

[7] Mugler EM, Ruf CA, Halder S, Bensch M, Kubler A. Design and IMPLEMENTATION OF A P300-based brain–computer interface for controlling an internet browser. Vol. 18, IEEE Transactions on Neural Systems and Rehabilitation Engineering. 2010. p. 599–609.

[8] Sherif T, Kassis N, Rousseau M-É, Adalat R, Evans AC. BrainBrowser: distributed, web-based neurological data visualization. Front Neuroinform [Internet]. 2014;8:89. Available from: http://journal.frontiersin.org/Article/10.3389/fninf.2014.00089/abstract

[9] Karim AA, Hinterberger T, Richter J, *et al.* Neural internet: Web surfing with brain potentials for the completely paralyzed. Neurorehabil Neural Repair. 2006;20(4):508–15.

[10] Remsik A, Young B, Vermilyea R, *et al.* A review of the progression and future implications of brain–computer interface therapies for restoration of distal upper extremity motor function after stroke. Expert Rev Med Devices

[Internet]. 2016;4440:17434440.2016.1174572. Available from: http://www.tandfonline.com/doi/full/10.1080/17434440.2016.1174572

[11] Leeb R, Friedman D, Müller-Putz GR, Scherer R, Slater M, Pfurtscheller G. Self-paced (asynchronous) BCI control of a wheelchair in virtual environments: a case study with a tetraplegic. Comput Intell Neurosci. 2007;2007.

[12] Daly JJ, Wolpaw JR. Brain–computer interfaces in neurological rehabilitation. Lancet Neurol. 2008;7(11):1032–43.

[13] Neuper C, Pfurtscheller G. Neurofeedback training for BCI control [Internet]. Brain–Computer Interfaces, The Frontiers Collection. 2010. 65–78 p. Available from: http://link.springer.com/10.1007/978-1-84996-272-8

[14] Takeuchi N, Mori T, Nishijima K, Kondo T, Izumi S-I. Inhibitory transcranial direct current stimulation enhances weak beta event-related synchronization after foot motor imagery in patients with lower limb amputation. J Clin Neurophysiol. 2015;32(1):44–50.

[15] Birbaumer N, Cohen LG. Brain–computer interfaces: communication and restoration of movement in paralysis. J Physiol [Internet]. 2007;579(Pt 3): 621–36. Available from: http://www.pubmedcentral.nih.gov/articlerender.fcgi?artid=2151357&tool=pmcentrez&rendertype=abstract

[16] Pfurtscheller G, Guger C, Müller G, Krausz G, Neuper C. Brain oscillations control hand orthosis in a tetraplegic. Neurosci Lett. 2000;292(3):211–14.

[17] Mackenzie L, Bhuta P, Rusten K, Devine J, Love A, Waterson P. Communications Technology and motor neuron disease: an Australian survey of people with motor neuron disease. JMIR Rehabil Assist Technol [Internet]. 2016;3(1):e2. Available from: http://rehab.jmir.org/2016/1/e2/

[18] Baram Y, Miller A. Virtual reality cues for improvement of gait in patients with multiple sclerosis. Neurology. 2006;66(2):178–81.

[19] Fulk GD. Locomotor training and virtual reality-based balance training for an individual with multiple sclerosis: a case report. J Neurol Phys Ther. 2005;29(1):34–42.

[20] Lozano-Quilis J-A, Gil-Gomez H, Gil-Gomez J-A, *et al.* Virtual rehabilitation for multiple sclerosis using a Kinect-based system: randomized controlled trial. J Med Internet Res. 2014;16(11):1–8.

[21] Chaudhary U, Birbaumer N, Ramos-murguialday A. Brain–computer interfaces for communication and rehabilitation. Nat Rev Neurol. 2016;12(9).

[22] Bowsher K, Civillico EF, Coburn J, *et al.* Brain–computer interface devices for patients with paralysis and amputation: a meeting report. J Neural Eng [Internet]. 2016;13(2):023001. Available from: http://stacks.iop.org/1741-2552/13/i=2/a=023001?key=crossref.7c5c3fcf0a5aace1b2e903542a933746

[23] Bleichner MG, Freudenburg Z V, Jansma JM, Aarnoutse EJ, Vansteensel MJ, Ramsey NF. Give me a sign: decoding four complex hand gestures based on high-density ECoG. Brain Struct Funct. 2016;221(1):203–16.

[24] Watanabe K. Systems neuroscience and rehabilitation. Neuroscience [Internet]. 2011;3:117–29. Available from: http://www.springerlink.com/index/10.1007/978-4-431-54008-3

[25] Jerbi K, Freyermuth S, Minotti L, Kahane P, Berthoz A, Lachaux J-P. Watching brain TV and playing brain ball exploring novel BCI strategies using real-time analysis of human intracranial data. Int Rev Neurobiol. 2009;86:159–68.

[26] Ahn M, Ahn S, Hong JH, *et al.* Gamma band activity associated with BCI performance: simultaneous MEG/EEG study. Front Hum Neurosci. 2013; 7:848.

[27] Zich C, Debener S, Kranczioch C, Bleichner MG, Gutberlet I, De Vos M. Real-time EEG feedback during simultaneous EEG-fMRI identifies the cortical signature of motor imagery. Neuroimage. 2015;114:438–47.

[28] Hosseini SMH, Pritchard-Berman M, Sosa N, Ceja A, Kesler SR. Task-based neurofeedback training: A novel approach Toward training executive functions. Neuroimage [Internet]. 2016;134:153–9. Available from: http://www.sciencedirect.com/science/article/pii/S1053811916002408

[29] Hanselmann S, Schneiders M, Weidner N, Rupp R. Transcranial magnetic stimulation for individual identification of the best electrode position for a motor imagery-based brain–computer interface. J Neuroeng Rehabil. 2015;12:71.

[30] Ang KK, Guan C, Phua KS, *et al.* Facilitating effects of transcranial direct current stimulation on motor imagery brain–computer interface with robotic feedback for stroke rehabilitation. Arch Phys Med Rehabil. 2015; 96(3 Suppl):S79-87.

[31] Daly JJ, Cheng R, Rogers J, Litinas K, Hrovat K, Dohring M. Feasibility of a new application of noninvasive brain–computer interface (BCI): a case study of training for recovery of volitional motor control after stroke. J Neurol Phys Ther. 2009;33(4):203–11.

[32] Corralejo R, Nicolas-Alonso LF, Alvarez D, Hornero R. A P300-based brain–computer interface aimed at operating electronic devices at home for severely disabled people. Med Biol Eng Comput. 2014;52(10):861–72.

[33] Hinterberger T, Kübler A, Kaiser J, Neumann N, Birbaumer N. A brain–computer interface (BCI) for the locked-in: Comparison of different EEG classifications for the thought translation device. Clin Neurophysiol. 2003;114(3):416–25.

[34] Klobassa DS, Vaughan TM, Brunner P, *et al.* Toward a high-throughput auditory P300-based brain–computer interface. Clin Neurophysiol [Internet]. 2009;120(7):1252–61. Available from: http://dx.doi.org/10.1016/j.clinph.2009.04.019

[35] Lazarou I, Nikolopoulos S, Petrantonakis PC, Kompatsiaris I, Tsolaki M. EEG-Based Brain–Computer Interfaces for Communication and Rehabilitation of People with Motor Impairment: A Novel Approach of the 21st Century. Front Hum Neurosci [Internet]. 2018b;12(January):1–18. Available from: http://journal.frontiersin.org/article/10.3389/fnhum.2018.00014/full

[36] Nicolas-Alonso LF, Gomez-Gil J. Brain–computer interfaces, a review. Sensors. 2012;12(2):1211–79.

[37] Muller-Putz GR, Pfurtscheller G. Control of an electrical prosthesis with an SSVEP-based BCI. IEEE Trans Biomed Eng. 2008;55(1):361–4.

[38] Faller J, Müller-Putz G, Schmalstieg D, Pfurtscheller G. An application framework for controlling an avatar in a desktop-based virtual environment via a software SSVEP brain–computer interface. Presence Teleoperators Virtual Environ. 2010;19(1):25–34.

[39] Lim J-H, Hwang H-J, Han C-H, Jung K-Y, Im C-H. Classification of binary intentions for individuals with impaired oculomotor function: "eyes-closed" SSVEP-based brain–computer interface (BCI). J Neural Eng [Internet]. 2013;10(2):026021. Available from: http://www.ncbi.nlm.nih.gov/pubmed/23528484

[40] D. Moher, A. Liberate, J. Tetzlaff, D. Altman, PRISMA Group, Preferred reporting items for systematic reviews and meta-analyses: the PRISMA statement, PLoS Med. 6 (7) (2009) pp. e1000097

[41] Birbaumer N, Ghanayim N, Hinterberger T, *et al.* A spelling device for the paralysed. Nature [Internet]. 1999;398(6725):297–8. Available from: http://www.nature.com/doifinder/10.1038/18581

[42] Kubler A, Neumann N, Kaiser J, Kotchoubey B, Hinterberger T, Birbaumer NP. Brain–computer communication: self-regulation of slow cortical potentials for verbal communication. Arch Phys Med Rehabil. 2001;82(11):1533–9.

[43] Neumann N, Birbaumer N. Predictors of successful self control during brain–computer communication. J Neurol Neurosurg Psychiatry. 2003;74:1117–21.

[44] Neumann N, Kuebler A, Kaiser J, Hinterberger T, Birbaumer N. Conscious perception of brain states: mental strategies for brain–computer communication Bewusste Wahrnehmung von Hirnzustaenden: Mentale Strategien fuer die Gehirn–Computer-Kommunikation. Neuropsychologia. 2003;41:1028-1036 URLJ: http://www.elsevier.nl/inca/public

[45] Neuper C, Müller GR, Kübler A, Birbaumer N, Pfurtscheller G. Clinical application of an EEG-based brain–computer interface: a case study in a patient with severe motor impairment. Clin Neurophysiol. 2003;114(3):399–409.

[46] Krausz G, Scherer R, Korisek G, Pfurtscheller G. Critical decision speed and information transfer in the "Graz-brain–computer interface." Appl Psychophysiol Biofeedback. 2003;28(3):223–40.

[47] Wolpaw JR, McFarland DJ. Control of a two-dimensional movement signal by a noninvasive brain–computer interface in humans. Proc Natl Acad Sci USA [Internet]. 2004;101(51):17849–54. Available from: http://www.ncbi.nlm.nih.gov/pubmed/15585584

[48] Kübler A, Nijboer F, Mellinger J, *et al.* Patients with ALS can use sensorimotor rhythms to operate a brain–computer interface. Neurology. 2005;64(10):1775–7.

[49] Cincotti F, Mattia D, Aloise F, *et al.* High-resolution EEG techniques for brain–computer interface applications. J Neurosci Methods. 2008;167(1):31–42.

[50] McFarland DJ, Krusienski DJ, Sarnacki W a, Wolpaw JR. Emulation of computer mouse control with a noninvasive brain–computer interface. J Neural Eng. 2008;5(2):101–10.

[51] Kauhanen L, Jylänki P, Lehtonen J, Rantanen P, Alaranta H, Sams M. EEG-based brain–computer interface for tetraplegics. Comput Intell Neurosci. 2007;2007.

[52] Bai O, Lin P, Vorbach S, Floeter MK, Hattori N, Hallett M. A high performance sensorimotor beta rhythm-based brain–computer interface associated with human natural motor behavior. J Neural Eng [Internet]. 2008;5(1):24–35.

Available from: http://dx.doi.org/10.1088/1741-2560/5/1/003%5Cn http://iopscience.iop.org/article/10.1088/1741-2560/5/1/003/pdf

[53] Holz EM, Höhne J, Staiger-Sälzer P, Tangermann M, Kübler A. Brain–computer interface controlled gaming: evaluation of usability by severely motor restricted end-users. Artif Intell Med [Internet]. 2013;59(2):111–20. Available from: http://dx.doi.org/10.1016/j.artmed.2013.08.001

[54] Jeon H, Shin DA. Experimental set up of P300 based brain–computer interface using a bioamplifier and BCI2000 system for patients with spinal cord injury. Korean J Spine. 2015;12(3):119–23.

[55] Cincotti F, Mattia D, Aloise F, *et al.* Non-invasive brain–computer interface system: towards its application as assistive technology. Brain Res Bull. 2008;75(6):796–803.

[56] Conradi J, Blankertz B, Tangermann M, Kunzmann V, Curio G. Brain–computer interfacing in tetraplegic patients with high spinal cord injury. Intell Data Anal [Internet]. 2009;11(2):1–4. Available from: http://eprints.pascal-network.org/archive/00006452/

[57] Leeb R, Tonin L, Rohm M, Desideri L, Carlson T, Millan J del R. Towards independence: a BCI telepresence robot for people with severe motor disabilities. Proc IEEE. 2015;103(6):1–12.

[58] Sellers EW, Donchin E. A P300-based brain–computer interface: initial tests by ALS patients. Clin Neurophysiol. 2006;117(3):538–48.

[59] Hoffmann U, Vesin JM, Ebrahimi T, Diserens K. An efficient P300-based brain–computer interface for disabled subjects. J Neurosci Methods. 2008;167(1):115–25.

[60] Nijboer F, Sellers EW, Mellinger J, *et al.* A P300-based brain–computer interface for people with amyotrophic lateral sclerosis. Clin Neurophysiol. 2008;119(8):1909–16.

[61] Kübler A, Furdea A, Halder S, Hammer EM, Nijboer F, Kotchoubey B. A brain–computer interface controlled auditory event-related potential (p300) spelling system for locked-in patients. Ann N Y Acad Sci. 2009;1157: 90–100.

[62] Nam CS, Woo J, Bahn S. Severe motor disability affects functional cortical integration in the context of brain–computer interface (BCI) use. Ergonomics [Internet]. 2012;55(5):581–91. Available from: http://www.ncbi.nlm.nih.gov/pubmed/22435802

[63] Kaufmann T, Schulz SM, Köblitz A, Renner G, Wessig C, Kübler A. Face stimuli effectively prevent brain–computer interface inefficiency in patients with neurodegenerative disease. Clin Neurophysiol [Internet]. 2013;124(5): 893–900. Available from: http://dx.doi.org/10.1016/j.clinph.2012.11.006

[64] Scherer R, Billinger M, Wagner J, *et al.* Thought-based row-column scanning communication board for individuals with cerebral palsy. Ann Phys Rehabil Med. 2015;58(1):14–22.

[65] Schettini F, Riccio A, Simione L, *et al.* Assistive device with conventional, alternative, and brain–computer interface inputs to enhance interaction with the environment for people with amyotrophic lateral sclerosis: a feasibility and usability study. Arch Phys Med Rehabil. 2015;96(3 Suppl):S46–53.

[66] Riccio A, Holz EM, Arico P, *et al.* Hybrid P300-based brain–computer interface to improve usability for people with severe motor disability: electromyographic signals for error correction during a spelling task. Arch Phys Med Rehabil. 2015;96(3 Suppl):S54–61.

[67] McCane LM, Heckman SM, McFarland DJ, *et al.* P300-based brain–computer interface (BCI) event-related potentials (ERPs): People with amyotrophic lateral sclerosis (ALS) vs. age-matched controls. Clin Neurophysiol. 2015;126(11):2124–31.

[68] Geronimo A, Simmons Z, Schiff SJ. Performance predictors of brain–computer interfaces in patients with amyotrophic lateral sclerosis. J Neural Eng. 2016;13(2):26002.

[69] Zickler C, Riccio A, Leotta F, *et al.* A brain–computer interface as input channel for a standard assistive technology software. Clin EEG Neurosci. 2011;42(4):236–44.

[70] Münßinger JI, Halder S, Kleih SC, *et al.* Brain painting: First evaluation of a new brain–computer interface application with ALS-patients and healthy volunteers. Front Neurosci. 2010;4:1–11.

[71] Zickler C, Halder S, Kleih SC, Herbert C, Kübler A. Brain painting: Usability testing according to the user-centered design in end users with severe motor paralysis. Artif Intell Med [Internet]. 2013;59(2):99–110. Available from: http://dx.doi.org/10.1016/j.artmed.2013.08.003

[72] Holz EM, Botrel L, Kübler A. Independent home use of brain painting improves quality of life of two artists in the locked-in state diagnosed with amyotrophic lateral sclerosis. Brain–Computer Interfaces. 2015;2(2–3):117–34.

[73] Piccione F, Giorgi F, Tonin P, *et al.* P300-based brain–computer interface: reliability and performance in healthy and paralysed participants. Clin Neurophysiol. 2006;117(3):531–7.

[74] Mauro M, Francesco P, Stefano S, Luciano G, Konstantinos P. Spatial attention orienting to improve the efficacy of a brain–computer interface for communication. ACM Int Conf Proceeding Ser [Internet]. 2011;114–17. Available from: http://dx.doi.org/10.1145/2037296.2037325

[75] Pires G, Nunes U, Castelo-Branco M. Evaluation of brain–computer interfaces in accessing computer and other devices by people with severe motor impairments. Procedia Comput Sci. 2012;14:283–92.

[76] Escolano C, Murguialday AR, Matuz T, Birbaumer N, Minguez J. A telepresence robotic system operated with a P300-based brain–computer interface: initial tests with ALS patients. 2010 Annual International Conference of the IEEE Engineering in Medicine and Biology. 2010. p. 4476–80.

[77] Grosse-Wentrup M, Mattia D, Oweiss K. Using brain–computer interfaces to induce neural plasticity and restore function. J Neural Eng [Internet]. 2011;8(2):025004. Available from: http://www.pubmedcentral.nih.gov/articlerender.fcgi?artid=4515347&tool=pmcentrez&rendertype=abstract

[78] Nicolas-Alonso LF, Corralejo R, Gomez-Pilar J, *et al.* The feasibility of a brain–computer interface functional electrical stimulation system for the restoration of overground walking after paraplegia. J Neuroeng Rehabil. 2015; 12(4):80.

[79] Arvaneh M, Guan C, Ang KK, *et al.* Facilitating motor imagery-based brain–computer interface for stroke patients using passive movement. Neural Comput Appl [Internet]. 2016. Available from: http://link.springer.com/10.1007/s00521-016-2234-7

[80] Kasashima-Shindo Y, Fujiwara T, Ushiba J, *et al.* Brain–computer interface training combined with transcranial direct current stimulation in patients with chronic severe hemiparesis: Proof of concept study. J Rehabil Med. 2015; 47(4):318–24.

[81] Ramos-Murguialday A, Broetz D, Rea M, *et al.* Brain–machine-interface in chronic stroke rehabilitation: a controlled study. Ann Neurol [Internet]. 2013;74(1):100–8. Available from: http://www.ncbi.nlm.nih.gov/pubmed/23494615

[82] Morris RG. D.O. Hebb: The Organization of Behavior, Wiley: New York; 1949. Brain Res Bull. 1999;50(5–6):437.

[83] Pichiorri F, Morone G, Petti M, *et al.* Brain–computer interface boosts motor imagery practice during stroke recovery. Ann Neurol. 2015;77(5):851–65.

[84] Georgiadis K, Laskaris N, Nikolopoulos S, Kompatsiaris I. Exploiting the heightened phase synchrony in patients with neuromuscular disease for the establishment of efficient motor imagery BCIs. J Neuroeng Rehabil. 2018;15(1):1–18.

[85] Vansteensel MJ, Pels EGM, Bleichner MG, *et al.* Fully implanted brain–computer interface in a locked-in patient with ALS. N Engl J Med [Internet]. 2016;NEJMoa1608085. Available from: http://www.nejm.org/doi/10.1056/NEJMoa1608085

[86] Sellers EW, Vaughan TM, Wolpaw JR. A brain–computer interface for long-term independent home use. Amyotroph Lateral Scler. 2010;11(5):449–55.

[87] Vaughan TM, Wolpaw JR, Sellers EW, Marie SS. A novel P300-based brain–computer interface stimulus presentation paradigm: moving beyond rows and columns. 2011;121(7):1109–20.

[88] Pasqualotto E, Matuz T, Federici S, *et al.* Usability and workload of access technology for people with severe motor impairment: a comparison of brain–computer interfacing and eye tracking. Neurorehabil Neural Repair [Internet]. 2015;29(10):950–7. Available from: http://www.ncbi.nlm.nih.gov/pubmed/25753951

[89] Ravden D, Polich J. Habituation of P300 from visual stimuli. Int J Psychophysiol. 1998;30(3):359–65.

[90] Ravden D, Polich J. On P300 measurement stability: habituation, intra-trial block variation, and ultradian rhythms. Biol Psychol. 1999;51(1):59–76.

[91] Kececi H, Degirmenci Y, Atakay S. Habituation and dishabituation of P300. Cog Behav Neurol. 2006;19:130–4.

[92] Moghimi S, Kushki A, Marie Guerguerian A, Chau T. A review of EEG-based brain–computer interfaces as access pathways for individuals with severe disabilities. Assist Technol [Internet]. 2013;25(2):99–110. Available from: http://www.tandfonline.com/doi/abs/10.1080/10400435.2012.723298

Chapter 3

Brain–computer interfaces in a home environment for patients with motor impairment—the MAMEM use case

Sevasti Bostantjopoulou[1], Zoe Katsarou[2], and Ioannis Dagklis[1]

Individuals with motor disabilities are marginalized and unable to keep up with the rest of the society in a digitized world with little opportunity for social inclusion. Specially designed electronic devices are required so as to enable patients to overcome their handicap and bypass the loss of their hand motor dexterity, which constitutes computer use impossible. The MAMEM's ultimate goal is to deliver technology in order to enable people with motor disabilities to operate the computer using interface channels that can be controlled through eye-movements and mental commands. Three groups of 10 patients with motor disabilities each were recruited to try the MAMEM platform at their home: patients diagnosed with high spinal cord injuries, patients with Parkinson's disease and patients with neuromuscular diseases. Patients had the MAMEM platform—including a built-in monitoring mechanism—at home for 1 month. Some of the participants used the platform extensively participating in social networks, while others did not use it that much. In general, patients with motor disabilities perceived the platform as a useful and satisfactory assistive device that enabled computer use and digital social activities.

3.1 Introduction

People with disabilities meet multiple barriers that have an impact on their quality of life. The World Health Organization (WHO) [1] defines barriers as factors in a person's environment that through their absence or presence cause limitation of daily functioning (an umbrella term that covers body functions, body structures, activities and participation) and create disability. Furthermore, according to the International Classification of Functioning, Disability and Health [1], disability refers to impairments, such as loss or abnormality in body structure or physiological function, limited

[1]3rd University Department of Neurology, Aristotle University of Thessaloniki, Thessaloniki, Greece
[2]Department of Neurology, Hippokration General Hospital, Thessaloniki, Greece

physical activity and restricted participation in daily activities. People with motor disabilities have limited capabilities in moving, performing manual tasks, gaining employment, participating in recreational everyday activities and organizing physical meeting or having an effective communication due to physical, communication, attitudinal and social barriers.

Nowadays, technology plays a pivotal role in several aspects of life and people with disabilities will gain the most from the new technologies lowering the multiple barriers they meet. Brain–computer interface (BCI) technology translates signals recorded from the brain into outputs that enable people with disabilities to communicate and control applications without the participation of peripheral nerves and muscles [2–4]. The area of BCI technology will help people with nervous system diseases and disabilities to improve both their communication and mobility [5–7]. Furthermore, the use of computer with Internet connection has changed our lives. Computers are important for accessing global information, communicating with others, collaborating with people worldwide and increasing both creativity and self-expression, thus helping people to become more social and at the same time preserve their independence. However, the needs and concerns of older people are different from those of younger people. Older people meet a variety of barriers that hold them from using the computer such as motor and sensory changes, cognitive changes as well anxiety toward computers [8–10]. Furthermore, people with motor disabilities face multiple barriers for computer accessibility such as difficulties handling standard input devices (keyboard and mouse), navigation problems, etc. Moreover, patients with neurological diseases causing motor impairment face specifically many problems and difficulties when using a computer.

3.1.1 Parkinson's disease

Parkinson's disease (PD) is the second most common neurodegenerative disorder affecting about 1% of the population over 60 years and the prevalence of the disease is increasing with age from 107/100,000 persons between ages 50 and 59 years to 1087/100,000 persons between 70 and 79 years [11,12]. PD is characterized by motor symptomatology including tremor at rest, rigidity, bradykinesia, postural instability and freezing episodes [13,14]. However, the clinical spectrum of PD is more extensive covering a wide range of non-motor symptoms (depression, apathy, sleep disorders, autonomic dysfunction, cognitive impairment, etc.) [15,16].

Despite significant advances in the pharmacotherapy of the disease, drug treatment remains only symptomatic and slowing disease progression still remains an unmet need [17]. Therefore as the disease advances motor, symptoms become progressively worse, leading to difficulties of daily activities and functional impairment. Furthermore, after years of treatment with L-Dopa patients develop motor complications (fluctuations and dyskinesias) [18]. Patients' quality of life deteriorates due to the significant motor disability, communication difficulties, loss of employment, social embarrassment and isolation [19,20].

Patients with PD want to continue living a normal life and technology will help them to compensate their disability and have social participation. Little is known about the needs and troubles of patients with PD in relation to computer usage. Cunningham *et al.* [21] reported that the loss of high level of finger dexterity makes difficult the use of a keyboard or mouse. During computer use the issues that emerged were keeping the hand steady when navigating, losing the cursor, moving in the wrong direction, slipping off menus, running out of room on the mouse mat and the mouse ball getting stuck [21]. In the study of Begnum [22], nearly 80% of patients with PD had severe challenges using the computer mainly linked to the motor symptoms (stiffness, bradykinesia and tremor), inertia, pain and fatigue. Main problems arise in the use of standard input devices such as computer mouse and keyboard. Furthermore, Begnum and Begnum [23] evaluated the usefulness of different adaptations according to the patient difficulty such as mouse adaptations, keyboard adaptation and ergonomic adaptations. Although potential solutions with different adaptations have been found for individual cases no single solution was found for all patients with PD.

3.1.2 Patients with cervical spinal cord injury

Spinal cord injury (SCI) and especially of the cervical region are devastating conditions with enormous physical and psychosocial burden. Spinal cord lesions can be divided into two categories: traumatic and nontraumatic. The annual worldwide crude incidence of traumatic spinal cord injuries ranges from 12.1 to 57.8 cases per million people [24]. The traumatic SCIs have a bimodal age distribution: the first peak in young adults and the second peak in older adults [24]. Patients with cervical lesions show tetraparesis or tetraplegia together with sensory loss, pain, spasticity and autonomic dysfunction. SCI can be complete or incomplete and the disability it causes is related to the level and extent of the damage. Complete high cervical cord lesions are the most severe, rendering the patient totally unable to move all extremities (tetraplegia) and torso [25,26]. They may also cause respiratory problems. Complete lower cervical cord lesions also cause tetraplegia, but may allow some mobility of the upper extremities. Incomplete cervical cord lesions may cause less severe muscular weakness in upper and lower extremities (tetraparesis). However, even in this category of patients finger motor dexterity is compromised. Spinal cord lesions are accompanied by other neurological symptoms such as sensory loss, pain, spasticity and autonomic dysfunction that increase the patients' discomfort and disability. The functional, psychological and financial impacts of traumatic spinal cord injury (TSCI) are broad. There are different scales that assess independence and disability in patients with SCI as well as the impairments of the arm and hand that determined the level of functioning of these patients [27]. For patients with SCI, a variety of computer interface devices have been created in order to help them in their everyday living. These devices are classified according to the handling methods that are operated with the mouth and that sense the movement of a specific part of the body [28]. According to Goodman *et al.* [29], 69.2% of patients with SCI used a computer with 19.1% of them having an assistive device. A proportion of 94.2% of the computer users accessed the Internet for e-mail, shopping sites and health sites [29]. Patients before the age of 18 years had the highest computer use. Furthermore, the motor and functional impairments of patients with SCI have a negative impact on their quality of life and the assistive devices for computer use present a possibility for access to information, social networks, work and leisure activities [30]. Baldassin *et al.* [30] in a systematic review about computer assistive technology and associations with quality of life for patients with SCI conclude that "despite the scarcity of studies and their methodological limitations, there is evidence that assistive technology for computer access favors the quality of life of people with tetraplegia due to SCI, since it improves participation, independence, and self-esteem."

3.1.3 Patients with neuromuscular diseases

Neuromuscular diseases (NMDs) are a heterogeneous group of diseases that are inherited or acquired and include motor neuron diseases, neuropathies, neuromuscular junction diseases and muscular diseases [31,32]. Loss of muscle strength is the main problem of the NMD. However, muscle weakness differs in localization and type of progression between the different NMDs. Furthermore, neuromuscular disorders are associated with disability in body structure and function that have an impact on their activities and participation [31].

Concerning body structure and function patients may present with loss of strength, atrophy, contractures, pain and cardiopulmonary symptoms [31]. They also have problems in activities (fatigue, exercise intolerance, walking and mobility problems and psychological problems) as well as participation problems [31]. As the disease progresses, functional problems and limitations in activities become worse and there is an augmentation of the psychological problems. Patients with NMD will benefit from the use of assistive technology that facilitates communication, house activities and mobility. Few studies analyzed computer task performance in patients with Duchenne muscular dystrophy (DMD). Vilozni *et al.* [33] used computer games to encourage their respiratory efforts. Recently, Malheiros *et al.* [34] determined the computing task performance in patients with DMD and they reported that patients with DMD improved their functionality after practicing a computational task. Climans *et al.* [35] conducted a survey in Canada about myotonic dystrophy patients' interaction with technology. According to their study the majority of the patients used computer and smartphones. They used technology for e-mails, social media and to obtain information about their disease. Those who did not use technology found it too complicated and too expensive. For patients with amyotrophic lateral sclerosis, high-tech augmentative and alternative communication technologies such as eye-tracking-based computer devices and BCIs have a high potential for improving communication and environmental control [36,37]. Recently, Wolpaw *et al.* [38] reported that the benefit exceeded the burden of communication for patients with amyotrophic lateral sclerosis that used independently at home the Wadsworth electroencephalogram (EEG)-based BCI.

The goal of the MAMEM project was to provide a tool for disabled people (patients with PD, SCI and NMDs) that can enable them to integrate back into society, by allowing them a better use of computers and thus a better option to participate in social networks [39]. Therefore, a novel way to control computers for multimedia authoring and management using primarily their eye-gaze and mental commands were developed—the MAMEM platform.

This chapter initially tackles the habits and difficulties that patients with PD, SCI and NMD face when using a computer. Subsequently the clinical trials for the use of the MAMEM platform at home are being analytically presented.

3.2 Computer habits and difficulties in computer use

A set of questionnaires was developed concerning computer use habits, difficulties and needs for patients with PD, SCI and NMD in order to design the MAMEM platform. Questions for computer use habits included type of computer system used, computer using hours per day, years of experience and main uses of computer (social participation, communication, productive activities, recreation, etc.). The questions about difficulties using a computer referred to problems with the cursor, the keyboard, browsing/navigating the Internet, etc. Furthermore the patients reported how much their disease affected computer use aspects (comfort, independence, speed and accuracy of operation, pain, fatigue, effectiveness, etc.). The results of the questionnaires helped defining the clinical requirements for operating the MAMEM platform [40–42].

3.2.1 Patients with PD

Most of the patients used a desktop (78.9%) and a laptop (57.9%) and they had 4.34 ± 2.8 h of computer use every day. Their most important uses of computer system were communication (e-mail, Skype, etc.) (73.7%) and information (Wikipedia, news, etc.) (63.2%). According to them, the three most important aspects of computer contribution to their life were emotional well-being, educational attainment and interpersonal interactions and relationships. Their main difficulties using the computer were double-clicking with the cursor (63.2%), moving the cursor on the screen (63.2%), identifying the cursor on the screen (36.8%) and using two keys at the same time (36.8%).

3.2.2 Patients with cervical spinal cord injuries

Most of the patients used a laptop (73.3%), smartphone (66.7%) and a tablet (60%) and they had 5.4 ± 3.13 h of computer use every day. Their most important uses of computer system were communication (e-mail, Skype, etc.) (45.5%) and productive (writing, editing, etc.) activities (45.5%). According to them, the three most important aspects of computer contribution to their life were educational attainment, interpersonal interactions and relationships and work and employment status. Their main difficulties using the computer were using two keys at the same time (70%), typing with the keyboard (60%) and zooming (50%).

3.2.3 Patients with NMDs

Most of the patients used a laptop (78.9%) and equally a smartphone and a desktop (63.2%) and they had 6.05 ± 2.99 h of computer use every day. Their most important uses of computer system were social participation (Facebook, forums, etc.), communication and recreation (movies, music etc.). According to them, the three most important aspects of computer contribution to their life were interpersonal interactions and relationships, educational attainment and work and employment status. Their main difficulties using the computer were using the keyboard (57.9%), identifying the letters on the keyboard (57.9%) and using two keys at the same time (52.6%).

In conclusion, patients used computers extensively despite their disability. They all agreed that computer use has a major contribution in their lives especially for interpersonal interactions and relationships and educational attainment. Therefore, every improvement that will facilitate computer use will improve their quality of life and socialization.

3.3 MAMEM platform use in home environment

The clinical trials for the use of the MAMEM platform at home, that is a novel way of using computer in real-world conditions, had two objectives: to assess the feasibility and usability of the system in home environment by patients with disabilities and to test the ability of the platform to enhance the social communication activities of the patients in real-world conditions. Therefore the purpose of the clinical trials was to determine whether the MAMEM platform can indeed provide a better computer operating solution by a sample of potential users.

3.3.1 Subjects selection

Three different cohorts of patients were selected fulfilling the inclusion/exclusion criteria: patients with PD, SCI and NMD (muscular dystrophies and spinal muscular atrophy II). There was a pretest trial in the hospital involving 20 patients from each disease type in order to test patients' ability to operate the eye-tracker and the platform. Thus, ten patients from each cohort with the best performance were selected for the home trial. In the PD group, there were six men and four women with an average age of 55.6 years and 10 years of disease duration. They were in an average 2.1 stage of the disease according to the Hoehn and Yahr [43] scale. All of them had bilateral arm bradykinesia and rigidity, four had bilateral arm tremor while all had dyskinesias. The majority of them (90%) reported of a slight to medium interference of their clinical condition on the computer use. They use mainly laptop and desktop computer. Their most important computer uses were information, communication and study (online courses, articles, etc.). In the SCI group (with complete or incomplete injury above C5), all patients were male with an average age of 38.1 years. They had partial or complete tetraparesis confined to a wheelchair. Most of them (62.5%) reported large interference of their medical condition on computer use. Their most important computer uses were recreation activities and information. In the NMD group, there were six men and four women with an average age of 31.5 years. They reported slight (50% of them) and fairly much (20% of them) interference of their medical condition on their computer use. The most important computer uses were social participation and communication.

3.3.2 Method

3.3.2.1 Apparatus

The apparatus included a standard laptop with "GazeTheWeb" installed on. The laptops were relatively new with i5 6th-generation Intel processors, 4GB RAM and 240GB SSD hard drives. For the gaze behavior analysis, the MAMEM apparatus also included the myGaze [44] eyetracking system. The MAMEM platform included the final version of "GazeTheWeb" (the tool that was developed within the MAMEM platform that enables surfing the Internet with the use of the eyes) on each computer, in addition to supporting software for the trials. This supporting software included the TeamViewer application [45] which was for remote technical support if needed. In addition, the MAMEM platform included a built-in monitoring mechanism that recorded every action that the user performed with the system. This monitoring mechanism had a temporary "turn-off" option for privacy reasons. As the default page of GazeTheWeb, the MAMEM dashboard/Homepage was used so as to inform the participants for their digital indicators of social integration. Another monitoring mechanism was the social tracker application which monitored the public activities of the participants in online social networks. Moreover, in selected subjects (two participants from the PD cohort and three participants from the NMD cohort) the apparatus also included the ENOBIO 8 EEG [46] device and the Shimmer GSR sensor [47] that were set up (additionally to the eye-tracker) in collaboration with the experimenter so as to test MAMEM's multimodal interfaces including the error-aware gaze-based keyboard and the hands-free version of Tetris (MM-Tetris).

3.3.2.2 Procedure

The laptop with the MAMEM platform installed on it, as well as an eye-tracker was given to the patients for a period of 1 month to use at their homes. Participation in the study was performed in several stages. Initially, 1 month before the actual trial, there was a phone call that started a social monitoring mechanism by asking the "Facebook," "Twitter" and "Google Plus" usernames from the participants, assuming that they had one, and entering them into a social tracker application that was created for the MAMEM project. In the next stage, the first visit of the trial in each participant's home took place. In case the participant had a Facebook account the username was entered into the MAMEM Facebook developer's application as an additional social monitoring mechanism. Then the laptop was located in an appropriate operation station with a certain height and angle in order to enable the proper use of the eye-tracker. After the installation of the platform, including connecting all the devices as well as connecting the laptops to the local WIFI Internet network each participant was given an oral and written explanation on how to operate the system (how to turn it on and off, operate the "GazeTheWeb" interface using their eyes, open pages, scroll in them and save them as bookmarks or how to use the "GazeTheWeb" keyboard, etc.). In addition, the participants were given a full and profound explanation about the GazeTheWeb built-in monitoring mechanism that recorded each of their action, and how to turn it off when they desire. Furthermore, participants were notified about sites that promote social inclusion; they were suggested to visit them and were taught how to view their social activities on the "GazeTheWeb" dashboard (i.e., MAMEM dashboard/Home page). This dashboard was designed to provide the participants with feedback about their progress in accessing social sites from different categories as well as with information about their progress in the training games according to the persuasive design principles and user models that were formulated within the MAMEM project for this purpose. Two weeks after the installation visit, an experimenter performed a monitoring phone call to the participants asking about the experience they had with the MAMEM platform and the technical difficulties they were confronted. The final visit was carried out 1 month after the first visit and the computer was removed from the patient's home. In selected subjects, during the first visit the ENOBIO 8 EEG device and the Shimmer GSR sensor were installed (additionally to the eye-tracker) in collaboration with the experimenter and a few hours were spent on testing MAMEM's multimodal interfaces including the error-aware gaze-based keyboard and the hands-free version of Tetris (MM-Tetris). All patients signed an informed consent form for all the stages of the trial.

3.3.2.3 Evaluation

The primary outcomes of the trials were the impact of various aspects of the MAMEM platform on the computer use habits and the social lives of the participants. For the computer use habits outcome, we extracted five measures of usage: (i) active hours of usage, (ii) unique sites that the user visited, (iii) keystrokes that were made in the keyboard, (iv) clicks that were made on the screen and (v) typing speed (calculated as seconds per character). For the social life outcome, we chose the five most popular social sites and extracted three measures of usage in them: (i) number of sessions, (ii) total time spent in the site and (iii) number of keystrokes that were made in the site.

The secondary outcomes comprised user satisfaction and perceived usability of the system, as measured by the QUEST 2.0 (Quebec User Evaluation of Satisfaction with assistive Technology) [48] and the System Usability Scale (SUS) [49,50] questionnaires. The QUEST 2.0 item scores are averaged and the final score ranges between 1 and 5 (not satisfied at all—highly satisfied). The QUEST 2.0 scores were calculated by averaging the first part of the questionnaire that concerns the different physical and usability aspects of the assistive system. The SUS scores ranged between 0 and 100, and a SUS score above a 68 would be considered above average and anything below 68 is below average.

Two weeks after the installation visit, the patients answered a structured questionnaire on the phone. The questionnaire included three Likert style questions regarding the experience with the platform [on a scale of 10 to 1 (10—very satisfied and 1—not at all satisfied)] how satisfied are you by using the MAMEM up to this point? in comparison to the previous digital device how satisfied are you with MAMEM on a scale of 1–5 (5—by comparison more satisfied and 1—by comparison not at all satisfied)? and now that you have tried MAMEM for 2 weeks how probable is it that you could recommend it to another person with the same disability (a scale of 10 would definitely recommend to 1 would not at all recommend)?

The qualitative outcome included patients' testimonials, technical problem, experimenter's impression regarding the participants, case study analysis and multimodal interface experience. The main areas that the case study analyses focused were demographics, mobility status, history with digital devices, ability of learning how to use the device, the experience of MAMEM over time, the range of MAMEM usage, some critical satisfaction and dissatisfaction factors, core learnings and future perspectives.

3.3.3 Results

All patients completed the 1-month trial using the MAMEM platform at home.

Primary outcome: In the three groups of patients, nine participants used the platform for a short duration, fifteen participants made a moderate use and six participants were frequent users. Active usage hours per day in the three cohorts are shown in Figure 3.1. Patients with NMD tended to use the platform more frequently, while patients with PD used it less and SCI very little. In detail, the usage outcome for the participants with moderate and frequent usage was as follows: keystrokes per day ranged from 0.27 to 277.47 (mean ± standard deviation [SD] = 56.15 ± 91.3), click per day ranged from 0.4 to 59.44 (mean ± SD = 22.04 ± 27.23) and typing speed per day (seconds per character) ranged from 0.23 to 3.87 (mean ± SD = 1.43 ± 1.2). Concerning social activity primary outcomes patients hardly used social media sites except for YouTube and Facebook (Figures 3.2 and 3.3). The most popular websites were search engine sites.

Secondary outcome: All three cohorts reported an above-average score to the interface design of the system measured by the SUS questionnaire (the SUS score over 68 is considered above average). PD cohort had a SUS score of mean ± SD = 75.5 ± 13 [51], NMD cohort had a score of mean ± SD = 70 ± 17 and the score for the SCI cohort was mean ± SD = 73.33 ± 15.81. For the physical attributes of the system measured by the QUEST 2.0, the scores were mean ± SD = 4.2 ± 0.5 for the PD group [51], mean ± SD = 3.8 ± 0.68 for the NMD group and mean ± SD = 4.33 ± 0.48 for the SCI group. The results were similar across cohorts and generally favorable toward the MAMEM platform. The MAMEM platform was perceived by the patients as a useful, usable and a satisfactory assistive device that enables computer usage and digital social activities.

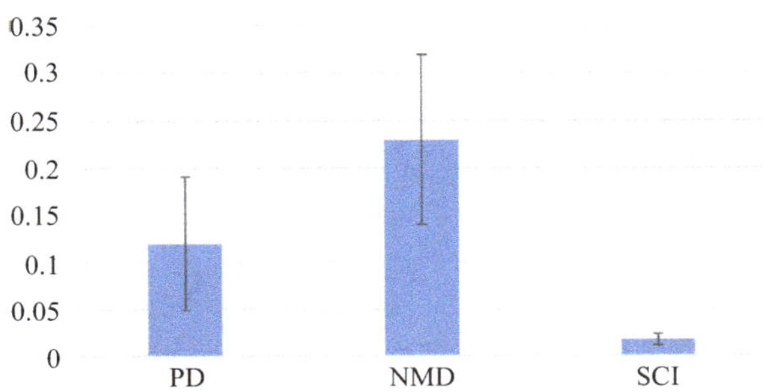

Figure 3.1 Average active hours using the MAMEM platform in the three cohorts

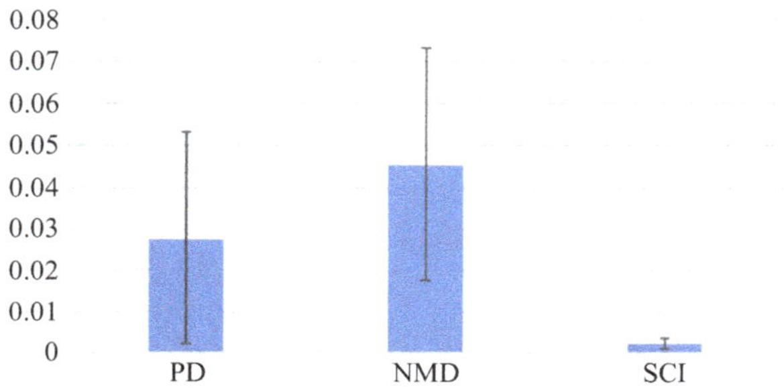

Figure 3.2 Average Facebook active hours using the MAMEM platform in the three cohorts

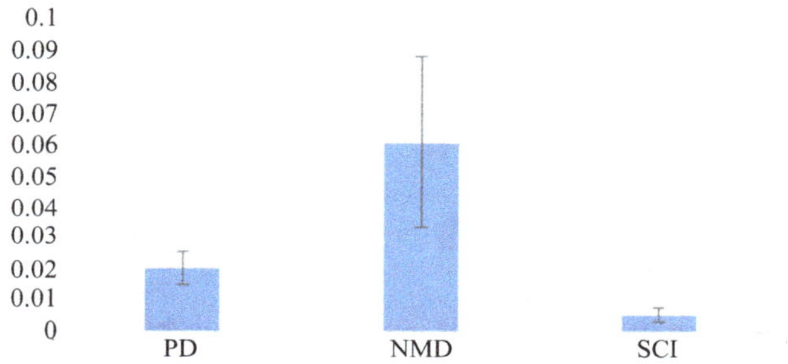

Figure 3.3 Average YouTube active hours using the MAMEM platform in the three cohorts

Qualitative outcome: Most of the patients found the platform useful, easy to learn and manageable. They enjoyed it and they reported that usage was getting easier and faster as time went up. They concluded that it is a great tool for people with disabilities. However, some patients complained of technical problems such as need to restart the program due to frequent crashes and recalibration of the eye-tracker.

PD patients' testimonials were as follows: "I was familiarized with the system very quickly and I used it almost day long but I faced problems focusing the keyboard after using it for significant time," "I used the system easily," "I had difficulty to use it since I am not familiar with computers and I asked the help of my daughter," "I found MAMEM very useful and easy use but faced problems with the eye-tracker (unplug and plug the USB cable)," "I found the use of the system straightforward and useful but the recalibration when leaving the laptop was tiring," "after some delay I was thrilled with the use of the platform," "I found difficulty using the platform and the eye-tracker crashed several times," "the more time I spent with the system the easier I got to use it but problems with the eye-tracker," "I found the system very easy and I used it as part of my occupation."

SCI patients' testimonials were as follows: "MAMEM tool is a great tool for people with disabilities and useful for my needs," "I had many problems with the eye-tracker and I am not so sure that I can use it," "many problems with the eye-tracker and I am not very positive," "positive because I found a better way to perform activities for social participation that did not have before," "the MAMEM platform must provide a better solution for me in order to leave my laptop," "the MAMEM platform must provide a better solution for me in order to leave my smartphone," "I like the system and I think it can be very beneficial for people with disabilities."

NMD patients' testimonials were as follows: "the platform had no flash support and needed to be re-calibrated and pull out and put the USB," "nice, useful and enjoyed it, very useful for many people, eye-tracker problems," "difficulty writing using my eyes but if you are familiar with the platform it can be useful for many people," "I used the platform 2–5 h per day for many social activities, usage was getting easier and faster as time went by, I organized a trip through the MAMEM use," "easy learning of the system from the first usage time, several times program crashed, after stopping the use need of calibration again, the eyes got tired quickly and cannot used it for hours so I prefer my own computer mainly because of fatigue with the MAMEM platform," "if you are familiar with the system it is easier for you, affects positively job opportunities," "several times the program crashed and needed to restart the system," "the system worked well and is easily manageable, many times needed to restart and calibration again after stopping using it for a while."

The 2-week phone calls revealed that all patients were satisfied with the platform. MAMEM platform was found a bit better than the one they were currently using. They would recommend it to someone with the same disability. Concerning the structural questionnaire of patients with PD had an average score of mean $\pm$ SD $= 8 \pm 1.6$ for the first question, an average score of mean $\pm$ SD $= 3.5 \pm 1.06$ for the second one and an average score of mean $\pm$ SD $= 9.25 \pm 1.03$ for the third question. SCI patients' scores were mean $\pm$ SD $= 6 \pm 1.8$ for the first question, mean $\pm$ SD $= 3.25 \pm 1.3$ for the second and mean $\pm$ SD $= 6.9 \pm 2.2$ for the third question. Finally, patients with NMD had mean $\pm$ SD $= 6.5 \pm 2.8$ for the first question, mean $\pm$ SD $= 3 \pm 1.7$ for the second question and mean $\pm$ SD $= 7.6 \pm 2.84$ for the third question.

Concerning case study analysis we report the comments of three patients, one of each category, which used the system frequently and reported high satisfaction with MAMEM:

> **Case 1:** Patient with PD—he expressed high satisfaction with MAMEM because it facilitated greatly his use of the computer, allowing him to reexperience what is like to be using the computer effortlessly. He was able to carry out mostly all his daily activities and obligations. He mentioned that the difference MAMEM made in the way he uses the computer was 2-fold: (1) carrying his regular tasks with more ease and with much less frustration caused by slow movements and (2) using the computer more effortlessly and faster so that for the amount of time he spent on the computer he was able to make better use of his time and achieve more.

> **Case 2:** Patient with NMD—the patient started with MAMEM from not using the computer before to use the computer 2–3 h every day. She mentioned that MAMEM had a tremendous impact in her life as it opened up a wide range of opportunities to learn, to connect and to entertain. The patient mentioned a radical

difference in her life in the sense that it actually gave her access to the Internet. The critical satisfaction factor of the MAMEM use was the gift of independent use of the Internet. When she had to give back the device at the end of the month she said that she would miss all the activities that she was able to carry out with it.

Case 3: Patient with SCI—he felt that as time went by he has gotten used to operating the platform and the experience of the system has gotten better over the time. He used the platform to go to social network sites and searching various things on the Internet. The critical satisfaction factors for this patient were the ease of operation and the usefulness of the assistive device. He was satisfied with the MAMEM platform and he learned that technology gets better over time.

3.4 Summary

The clinical trials were designed to evaluate the use of the MAMEM platform for multimedia authoring among patients with motor disabilities. The platform was used by three groups of patients at home, which is an uncontrolled environment without the helpful presence of experimenters. The novelty of the MAMEM platform as well as its usefulness and usability motivated the patients to use the platform.

The use of the platform at home provided positive indications for MAMEM as an assistive device that enables computer usage and digital social activities. However the usage pattern varied between patients. Some patients used the platform extensively during the 1 month that they had the platform, while others hardly used it. Those who used the system extensively used it for various activities especially for participation in social networks, thus promoting social inclusion. Several patients reported that they could do various sophisticated Web activities that they could not do before. The reason why some patients used the platform a lot and some not as much is unclear. Various subjective measures of user impairment stage, age, preferences, prior interaction experience, performance and accuracy are implicated. Furthermore, some users that rarely used the platform they were already using computer by another assistive technology (e.g., mouse/switch) and they were reluctant to use a novel eye-tracking technology.

In general, the MAMEM platform was perceived as a useful, usable and satisfactory assistive device for surfing the Internet and social networks participation among some patients with motor disabilities. Certain technical problems arose that had to do with the hardware technology that is the eye-tracker functionality and usability, the calibration, the restart requirements and difficulties installing the platform.

BCI systems are a fast-growing technology that has a wide array of potential clinical application-restoring communication and motor function. These systems should be advanced to become portable, more convenient, easy to use and able to function in different environments. Therefore, the MAMEM platform joins the family of assistive devices for computer use for people with motor hand disabilities offering a unique way to surf the web and to participate in online social networks using eye gaze.

The area of interfacing brain with computer for communication and social participation is continuously improving with further improvements in the ease and convenience of their daily use. People with disabilities will gain the most from the new technology. However, in order to offer the best to the patients, this technology must cover their needs and expectations.

References

[1] World Health Organization, International Classification of Functioning, Disability and Health (IFC). Geneva; 2001.

[2] McFarland DJ, Daly D, Boulay C, Parvaz M. Therapeutic applications of BCI technologies. Brain Comput Interf (Abingdon) 2017;47(1–2):37–52.

[3] McFarland DJ, Vaughan TM. BCI in practice. Prog Brain Res 2016;228: 389–404.

[4] Huggins J, Guger C, Ziat M, *et al.* Workshops of the sixth international brain–computer interface meeting: brain–computer interfaces past, present, and future. Brain Comput Interf (Abingdon) 2017;4(1–2):3–36.

[5] Riccio A, Pichiorri F, Schettini F, *et al.* Interfacing brain with computer to improve communication and rehabilitation after brain damage. Prog Brain Res 2016;228:358–387.

[6] Chaudhary U, Birbaumer N, Ramos-Murguialday A. Brain–computer interfaces for communication and rehabilitation. Nat Rev Neurol 2016;12(9): 513–525.

[7] Lazarou I, Nikolopoulos S, Petrantonakis P, Kompatsiaris I, Tsolaki M. EEG-based brain–computer interfaces for communication and rehabilitation of people with motor impairment: a novel approach of the 21st century. Front Hum Neurosci 2018;31:12–14.

[8] Wagner N, Hassanein K, Head M. Computer use by older adults: a multi-disciplinary review. Comput Human Behav 2010;26:870–882.

[9] van de Watering M. The impact of computer technology on the elderly. http://www.marekvandewatering.com/texts/HCI-Essay Marek van de Watering.pdf.

[10] Morris A, Goodman J, Brading H. Internet use and non-use: views of older users. Univ Access Inform Soc 2007;6:43–57.

[11] Tysnes O, Storstein A. Epidemiology of Parkinson's disease. J Neural Transm (Vienna) 2017;124(8):901–905.

[12] Pringsheim T, Jette N, Froklis A, Steeves T. The prevalence of Parkinson's disease: a systematic review and meta-analysis. Mov Disord 2014;29(13): 1583–1590.

[13] Jankovic J. Parkinson's disease: clinical features and diagnosis. J Neurol Neurosurg Psychiatr 2008;79:368–376.

[14] Hess C, Hallett M. The phenomenology of Parkinson's disease. Semin Neurol 2017;37(2):109–117.

[15] Schapira A, Chaudhuri R, Jenner P Non-motor features of Parkinson disease. Nat Rev Neurosci 2017;18(7):435–450.

[16] Marras C, Chaudhuri R. Nonmotor features of Parkinson' disease subtypes. Mov Disord 2016;31(8):1095–1102.

[17] Kulisevsky J, Oliveira L, Fox S. Update in therapeutic strategies for Parkinson's disease. Curr Opin Neurol 2018;31(4):439–447.
[18] Chaudhuri R, Poewe W, Brooks D. Motor and nonmotor complications of levodopa: phenomenology, risk factors, and imaging features. Mov Disord 2018;33(6):909–919.
[19] Martinez-Martin P. What is quality of life and how do we measure it? Relevance to Parkinson's disease and movement disorders. Mov Disord 2017;32(3): 382–392.
[20] Moreira R, Zonta M, Araujo A, Israel V, Teive H. Quality of life in Parkinson's disease patients: progression markers of mild to moderate stages. Arq Neuropsiquiatr 2017;75(8):497–502.
[21] Cunningham L, Nugent C, Finlay D, Moore G, Craig D. A review of assistive technologies for people with Parkinson's disease. Technol Health Care 2009:269–279.
[22] Begnum M. Challenges for Norwegian PC-users with Parkinson's disease—a survey. In Miesenberger K., Klaus J., Zagler W., Karshmer A. (eds). Computer Helping People with Special Needs. Berlin-Heidelberg: Springer; 2010. p. 292–299.
[23] Begnum M, Begnum K. On the usefulness of off-the-shelf computer peripherals for people with Parkinson's disease. Univ Access Inform Soc 2012;11(4):347–357.
[24] van de Berg M, Castellote J, Mahillo-Fernandez I, de Pedro-Cuesta J. Incidence of spinal cord injury worldwide: a systematic review. Neuroepidemiology 2010;34:184–192.
[25] Hadley M, Walters B, Aarabi B, *et al.* Clinical assessment following acute cervical spinal cord injury. Neurosurgery 2013;72(Suppl 3): 40–53.
[26] Shank CD, Walters BC, Hadley MN. Management of acute traumatic spinal cord injuries. Handb Clin Neurol 2017;140:275–298.
[27] Furlan J, Noonan V, Singh A, Fehlings M. Assessment of disability in patients with acute traumatic spinal cord injury: a systematic review of the literature. J Neurotrauma 2011;28:1413–1430.
[28] Kim D, Lee B, Lim S, Kim D, Hwang S, Yim V, Park J. The selection of the appropriate computer interface device for patients with high cervical cord injury. Ann Rehabil Med 2013;37(3):443–448.
[29] Goodman N, Jette A, Houlihan B, Williams S. Computer and Internet use by persons after traumatic spinal cord injury. Arch Phys Med Rehabil 2008;89:1492–1498.
[30] Badassin V, Shimizu H, Fachin-Martins E. Computer assistive technology and associations with quality of life for individuals with spinal cord injury: a systematic review. Qual Life Res 2018;27(3):597–607.
[31] Parlak Demir Y. Neuromuscular diseases and rehabilitation. 2017. http://dx.doi.org/10.5772/67722.
[32] Morrison B. Neuromuscular diseases. Semin Neurol 2016;36:409–418.
[33] Vilozni D, Bar-Yishay E, Gur I, Shapira Y, Meyer S, Godfrey S. Computerized respiratory muscle training in children with Duchenne muscular atrophy. Neuromuscul Disord 1994;4(3):249–255.

[34] Malheiros R, da Silva T, Favero F, *et al.* Computer task performance by subjects with Duchenne muscular atrophy. Neuropsychiatr Dis Treat 2016;12:41–48.
[35] Climans S, Piechowicz C, Koopman W, Venance S. Survey of Canadian myotonic dystrophy patients' access to computer technology. Can J Neurol Sci 2017;1–5.
[36] Linse K, Aust E, Joos M, Hermann A. Communication matters-pitfalls and promise of high-tech communication devices in palliative care of severely physically disabled patients with amyotrophic lateral sclerosis. Front Neurol 2018;9:603.
[37] Trevizan I, Silva T, Dawes H, *et al.* Efficacy of different interaction devices using non-immersive virtual tasks in individuals with amyotrophic lateral sclerosis: a cross-sectional randomized trial. BMC Neurol 2018;18:209.
[38] Wolpaw J, Bedlack R, Reda D. Independent home use of a brain–computer interface by people with amyotrophic lateral sclerosis. Neurology 2018;91:e258–e267.
[39] Nikolopoulos S, Georgiadis K, Kalaganis K, *et al.* A multimodal dataset for authoring and editing multimedia content: the MAMEM project. Data Brief 2017;15:1048–1056.
[40] Katsarou Z, Plotnik M, Zeilig G, Gottlieb A, Kizony R, Bostantjopoulou S. Computer uses and difficulties in Parkinson's disease. Mov Disord 2016;31(issue S2).
[41] Bostantjopoulou S, Plotnik M, Zeilig G, Gottlieb A, Chlomissiou S, Nichogiannopoulou A, Katsarou Z. Computer use aspects in patients with motor disabilities. Eur J Neurol 2016;23(Suppl):722.
[42] Katsarou Z, Zeilig G, Plotnik M, Gottlieb A, Kizony R, Bostantjopoulou S. Parkinson's disease impact on computer use. A patients' and caregivers perspective. Neurology 2017;88:16(Suppl P6.009).
[43] Hoehn M, Yahr M. Parkinsonism: onset, progression and mortality. Neurology 1967;17(5):427–442.
[44] http://www.mygaze.com/.
[45] https://www.teamviewer.com/en/.
[46] https://www.neuroelectrics.com/products/enobio/enobio-8/.
[47] http://www.shimmersensing.com/products/shimmer3-wireless-gsr-sensor.
[48] Demers L, Weiss-Lambrou R, Ska B. The Quebec User Evaluation of Satisfaction with Assistive Technology (QUEST 2.0): an overview and recent progress. Technol Disabil 2002;14(3):101–105.
[49] Brooke J. SUS: a quick and dirty usability scale. In: P. W. Jordan, B. Thomas, B. A. Weerdmeester, A. L. McClelland. Usability Evaluation in Industry. London: Taylor and Francis; 1996; p. 189–194.
[50] Brooke J. SUS: a retrospective. J Usabil Stud 2013;8(2):29–40.
[51] Bostantjopoulou S, Katsarou Z, Plotnik M, *et al.* Parkinsonian patients experiences operating the computer with their eyes: the MAMEM project. Neurology 2019;92(15):5.8–043.

Chapter 4

Persuasive design principles and user models for people with motor disabilities

Sofia Fountoukidou[1], Jaap Ham[1], Uwe Matzat[1], and Cees Midden[1]

When developing effective assistive technology, it is crucial to focus on how acceptance and continued use of the technology can be optimized considering the (complexity of the) user and his or her situation. Therefore, this chapter describes methods for creating user models and shows how these were applied to user groups (patients with spinal cord injury, Parkinson's disorder and neuromuscular disorders) of a newly developed assistive technology (AT). The user models include user characteristics such as demographics, relevant medical information, computer interaction behaviour and attitudes towards novel assistive devices. Next, this chapter describes persuasive strategies to improve user acceptance and continued use of AT, specifically aimed at motivating individuals with disabilities to learn to operate the AT and to use it, in order to increase their social participation. Also, this chapter shows how empirical research has tested the effectiveness of the proposed persuasive and personalization (i.e., incorporating user model knowledge) design elements. Finally, this chapter shows how the implications of these findings were used to improve the persuasive design requirements of the AT. In sum, this chapter shows how persuasive personalized design principles (implemented into the AT) improve user acceptance (evaluations) and continued use (performance).

4.1 Methods for creating user models for the assistive technology

This section describes methods for creating user models and shows how these were applied to the user groups of a research project to develop AT (i.e., The Multimedia MAnagement/sharing and authoring using your Eyes and Mind (MAMEM) project, see [1]). This AT was developed for patients with spinal cord injury (SCI), Parkinson's disease (PD) and neuromuscular disorders (NMDs). The AT comprised a system that would allow these patients to use a personal computer through eye tracking and electroencephalogram (EEG) computer interface with the ultimate goal to increase the user's social participation. The user models include user characteristics such as

[1]Department of Industrial Engineering and Innovation Sciences, Eindhoven University of Technology, Eindhoven, The Netherlands

demographics, relevant medical information, computer interaction behaviour and attitudes towards novel assistive devices.

The main method used in this project to create user models was user profiling. The user profile is a method of presenting data from studies of user characteristics. In other words, a user profile of the target group contains collective information about mental, physical and demographic data for the user population as well as other characteristics. The goal of using user profiles is to help the team members (i.e., designers) to recognize or learn about the real user by presenting them with a description of his/her attributes. User profile does not necessarily mirror or present a complete collection of a whole user population's attributes. The essence of user profiles is an accurate and simple collection of end-users' characteristics [2].

The user profile is almost the same as a persona, i.e., some kind of fictitious person as a collection of attributes (e.g., age, skills, attitudes, motivation level). However, while a user profile covers a range of characteristics of the target population, a persona uses specific characteristics (e.g., gender, experience and skill level), which are usually derived from the user profile.

In more detail, personas have been defined as hypothetical archetypes of actual users. They are not real people but they represent real people during the design process. Personas have proper names and are represented with pictures. Although they are imaginary, they are defined with significant rigor and precision. The purpose of a persona is to make the users seem more real and help designers to keep realistic ideas of users throughout the design process. Designers and evaluators can refer to the personas when considering design specifics. They put a name, face, and characteristics on users to keep the users in the forefront of design decisions [3].

4.1.1 User profiles

The user profiles for the project were generated from an extensive literature review of our target groups, as well as patients' questionnaire analysis and focus groups with professionals (derived from [4,5]). In these user profiles, each of the three target groups is described covering areas such as disease range and characteristics, level of injury, physical symptoms, emotional functioning, cognitive functioning, motivation, computer operation and assistive devices. Details of these user profiles and sets of personas for the three target groups (SCI, PD and NMDs) can be found in [6].

4.1.2 Personas

Based on these user profiles, personas were developed to help system developers focus on the user of the system. To create personas for each of the three target groups, a list of variables that seem to be the most relevant for the project's scope was made. These variables (i.e., demographic, behavioural) were derived from the user profiles. Next, for each group, we mapped participants' responses to the questionnaires (for all details, see [5]), and summarized in the user profiles, against the selected set of variables. The overall goal of the variable mapping was to find a major pattern for each variable that will form the basis for the personas. The sets of variables have been grouped to four groups for all three groups: demographic, medical, computer use and

Table 4.1 Selected relevant variables to describe users. Example of spinal cord injury user

Demographic	
Name	Arie Cohen
Gender	Male
Age	46
Country of origin	Israel
Marital status	Married
Children	One
Occupation	Unemployed
Education	14 years
Main hobby	Reading
Medical information	
Medical condition	Spinal cord injury
Level of injury	Complete injury in the C3 vertebra and loss of mobility from the neck down
Secondary conditions	Respiratory and breathing issues
Computer use information	
Computer skills	Intermediate
Computer use frequency	Medium
Assistive device use	Typing stick
Frequent computer activities	Recreation, communication
Main computer difficulties	Using two keys at the same time, zooming
Computer skills	Intermediate
Goals and attitudes	
Technology attitude	Technophobe
Main computer contribution	Employment potential, communication

goals and attitudes. Lastly, in order to complete the personas, we added extra details based on the user data (i.e., occupation, marital status).

Table 4.1 provides detailed information about the chosen variables for SCI and how patients with SCI were mapped against them. It is worth mentioning that inclusion and exclusion criteria for the project's pilot trials were taken into account for the variable mapping and consequently for the personas' creation. Hence, according to the inclusion criteria, the SCI audience will be included by individuals suffering from complete or incomplete SCI (levels A, B and C), from C5 level and above. Moreover, according to the exclusion criteria, participants with any psychiatric (e.g., major depression) or cognitive conditions that might interfere with understanding the instruction or with cooperation will be removed. Thus, the variables of cognitive and emotion functioning were not taken into primary consideration for the personas creation.

Next, we presented two sets of personas for each group to represent a wide range of end-users. Below, we present one example of such persona. More specifically, we show a persona description for a male patient with SCI.

<u>Persona 1: Arie</u>

Arie is 46 years old and he lives with his wife in Tel Aviv, Israel. His life changed after a military training accident that caused him an SCI. Now he has a complete tetraplegia at C3. After completing his rehabilitation, he finally managed to sit and to move using a motorized wheelchair. This was one of the biggest accomplishments for Arie, since he struggled a lot to make it happen. He is also mechanically ventilated, 24/7.

Arie requires complete assistance with almost all daily activities. Due to the fact that his wife works full-time, he has a professional caregiver to take care of him most of the day. He is unemployed at the moment, but he receives financial support from the Ministry of Defense. Although he has several friends that meet frequently, he feels most of the time bored and he really misses doing something productive.

His feelings of monotony exacerbated when his son, Timothy, left the house recently. Timothy is 23 years old and he has recently moved to the U.S. to study; thus, they do not see each other very often anymore. Arie is very unhappy because of this and he misses him a lot.

Arie discussed about his feelings with his wife, who insisted that it is time to find an assistive device to help him operate the computer better, in order to be better able to search for a job and to communicate with their son more frequently. Arie does not like changes and he is very cautious when trying new things; that is why he was reluctant to use an assistive device all this time. However, after he started using a typing stick, his computer use has increased. He uses image applications to edit photos and send them to his son through social media networks. He also uses an internet browser to search for job vacancies and sends job applications through email.

Although the typing stick is useful, Arie still faces many difficulties with his computer operation. These difficulties are related to both the type of the assistive device and his medical condition. Specifically, he has trouble using two keys at the same time and zooming. In addition, his medical condition has an impact on the endurance and effectiveness of computer use. As a consequence, he does not use his computer as often as he would desire. Although he would like an assistive device that would provide him comfort and ease of use, he lacks the motivation to search for a more suitable one and to learn how to use it and his proficiency.

4.2 Persuasive strategies to improve user acceptance and use of an assistive device

This section describes persuasive strategies to improve user acceptance and use of AT, specifically aimed at motivating individuals with disabilities to learn to operate the AT and to use it, in order to increase their social participation.

4.2.1 Selection of persuasive strategies

In view of developing a prototype for the current AT, a selection should be made of persuasive design strategies and techniques. The main criterion for this selection is to choose those techniques that are more likely to persuade the target groups to use the AT for managing, authoring and sharing multimedia context.

The core goal of this task is to influence user motivation and make patients willing to use the technology and continue using it in the future. The pilot trials are divided into two phases: Phase I refers to patients testing the platform in a controlled environment (patient centres) to address its feasibility and usability; and Phase II, during which, participants are encouraged to use the AT in their home environments for a fixed period, in order to assess the impact of their multimedia authoring in a less control setting. Hence, the overall desired outcome for the three target groups (SCI, PD and NMD) is 2-fold:

- User *acceptance* and training *engagement* of the AT;
- Keep using the AT to *increase social participation* in (a) Social network activities and (b) Digital productivity.

Section 4.1.1.1 describes the development of the persuasive strategies to realize the objective of increased acceptance and engagement and Section 4.1.1.2 describes the selection of persuasive strategies to increase social participation.

4.2.2 Developing persuasive strategies for Phase I: user acceptance and training

Phase I of the clinical trials included training to use the EEG element, training to use the gaze element, and training for using both. It was divided in two parts: the first part includes the introduction to the platform, setting up the EEG cap on the participants head, and training them with basic tasks, such as mouse pointer, basic windows (operating system) functionalities and basic keyboard operations. The second part of the training taught users how to perform multimedia managing, authoring and sharing using dictated tasks like typing an email and editing a photo.

To come to a selection of the techniques to be incorporated in the system during the first phase, we applied intervention mapping (IM), a framework for developing and implementing health interventions [7]. In the next subsections, we discuss the IM framework in more detail.

Intervention mapping framework

IM is a framework for the development of theory and evidence-based health promotion programmes. It provides guidelines and tools for the selection of theoretical foundations and underpinnings of health promoting programmes, for the application of theory and for the translation of theory in actual programme materials and activities. Past projects showed that IM allows intervention developers to successfully identify (1) behavioural and environmental determinants affecting target health problems and (2) the most appropriate methods and strategies to address the identified determinants.

The IM framework guides the developer through iterative steps, as shown in Figure 4.1.

In the IM framework, steps 1 to 3 provide systematic guidance for the selection of persuasive techniques for the AT. The first step of the IM pertains to the needs analysis, which we performed with the use of focus groups with professionals (for details, see [4]) as well as in the form of questionnaires for the patients and their caregivers (for details, see [5]). The information derived has been also incorporated into the user profiles described above.

Therefore, in the following subsections we focus on the second and third steps of the IM. We define performance objectives, determinants of behaviour and change objectives accordingly (step 2). Based on this, we subsequently map relevant theory- and evidence-based change methods (step 3). At the end of this section, we describe the final selection of persuasive strategies for Phase I.

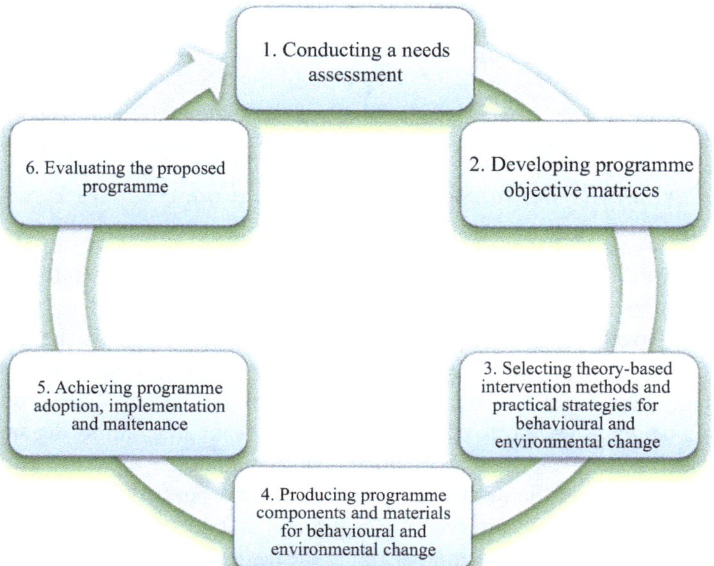

Figure 4.1 Iterative steps of intervention mapping approach (based on Bartholomew et al. 2001)

Performance objectives

Performance objectives have been defined as the detailed breakdown of what the participants must do to express a behavioural outcome [8]. In other words, they are specific sub-behaviours that are necessary to accomplish the desired behaviour or environmental outcome. Performance objectives clarify the exact performance of someone affected by the intervention. **The central question is: What do participants in this programme need to do to accomplish a behavioural outcome?**

Below we define performance objectives related to the first objective, namely, the acceptance and training engagement of the users:

1. Understanding the concept of the current AT project: multimedia management/sharing and authoring using your eyes and mind;
2. Experiencing the benefits of the current project's AT (multimedia management/sharing and authoring using your eyes and mind);
3. Performing the training and dictated task according to the objectives;
4. Creating realistic operation goals and setting personal targets related to these goals;
5. Applying solutions for (un)satisfactory multimedia management, authoring and sharing of the platform;
6. Evaluating the effect of solutions on multimedia operation and achievement of goals and personal targets (i.e., compared to what was planned);
7. Comparing own behaviours with previous own performance or those of other users;
8. Identifying and overcoming barriers in multimedia operation;
9. Maintaining progress in multimedia management/authoring and sharing.

Behavioural and psychological determinants

Determinants are those factors that have been found associated with the performance of the behaviour of the target population or agents that have control or influence over environmental outcomes. Determinants as such form the key to achieving the performance objectives. Personal determinants usually include cognitive factors and capabilities such as skills. Environmental conditions rest outside the individual. Determinants can either create a barrier for or stimulate certain behaviour.

Research in the area of user acceptance of new technology has resulted in several theoretical models, with roots to information systems, psychology and technology. The Unified Theory of Acceptance and Use of Technology (UTAUT) has been formulated to present a unified view of user acceptance, which is based upon conceptual and empirical similarities across the following eight models: (i) theory of reasoned action, (ii) the technology acceptance model, (iii) the motivational model, (iv) the theory of planned behaviour, (v) a model combining the technology acceptance model and theory of planned behaviour, (vi) the model of PC utilization, (vii) the innovation diffusion theory and (viii) the social and cognitive theory [9], and see also [23]. The UTAUT model employs intention and/or usage as the key dependent variable, since the role of intention as a direct predictor of behaviour (e.g., usage) has been well established across different disciplines. Figure 4.2 shows the basic conceptual

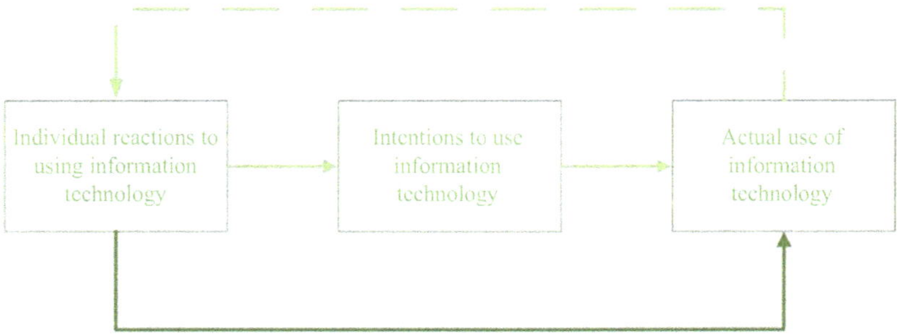

Figure 4.2 Basic concept underlying user acceptance models (adapted from [9])

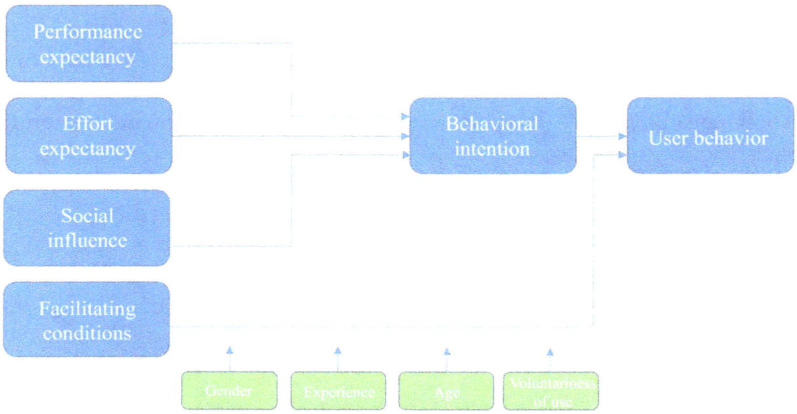

Figure 4.3 A unified model that integrates elements across the eight models (based on [9])

framework underlying the class of models explaining individual acceptance of the technology.

UTAUT provides a useful tool to assess the likelihood for new technology introductions and helps in understanding the drivers of acceptance in order to proactively design intervention (including training, marketing, etc.) targeted at populations of users that might be less inclined to adopt and use new systems.

Therefore, the determinants that will be used within the IM framework stem from the UTAUT model (Figure 4.3). According to this model, there are four core determinants of intention and usage of new technology, each including several constructs (based on the eight integrated theoretical models): performance expectancy, effort, expectancy, social influence and facilitating conditions. Table 4.2 provides the definition of each of these determinants, according to [9], and see also [24].

Table 4.2 Determinants of technology acceptance and use (see [9])

Determinant	Definition
Performance expectancy	The degree to which an individual believes that using the system will help him to attain gains in work.
Effort expectancy	The degree of ease associated with the use of a system.
Social influence	The degree to which an individual perceives that important others believe that he or she should use the new system.
Facilitating conditions	The degree to which an individual believes that the organizational and technical infrastructure exists to support use of the system.

Table 4.3 Performance expectancy: constructs and definitions (adapted from [9])

Construct	Definition
Perceived usefulness	The degree to which a person believes that using a particular system would enhance his/her job performance.
Extrinsic motivation	The perception that users will want to perform an activity because it is perceived to be instrumental in achieving valued outcomes that are distinct from the activity itself, such as improved job performance, pay, promotions.
Job-fit	How the capabilities of a system enhance an individual's job performance.
Relative advantage	The degree to which using an innovation is perceived as being better than using its precursor.
Outcome expectations	They relate to the consequences of the behaviour. They were separated into performance expectations (job-related) and personal expectation (individual goals).

In more detail, performance expectancy includes constructs of perceived usefulness, extrinsic motivation, job fit and relative advantage and outcome expectations (Table 4.3). Effort expectancy includes the constructs of perceived ease of use, complexity and ease of use (Table 4.4). Social influence includes the constructs of subjective norm, social factors and image (Table 4.5). Facilitating conditioning includes constructs such as perceived behavioural control, facilitating conditioning and compatibility (Table 4.6).

The following constructs of each determinant were selected and used for the current project's purposes: perceived usefulness, relative advantage and the outcome expectations (performance expectancy); perceived ease of use and complexity (effort expectancy); subjective norm and Image (social influence); perceived behavioural control, facilitating conditions and compatibility (facilitating conditions).

Table 4.4 Effort expectancy: constructs and definitions (adapted from [9])

Construct	Definition
Perceived ease of use	The degree to which a person believes that using a system would be effortless.
Complexity	The degree to which a system is perceived as relatively difficult to understand and use.
Ease of use	The degree to which using an innovation is perceived as being difficult to use.

Table 4.5 Social influence: constructs and definitions adapted from [9]

Construct	Definition
Subjective norm	The person's perception that most people who are important to him think he should or should not perform the behaviour in question.
Social factors	The individual's internalization of the reference group's subjective culture, and specific interpersonal agreements that the individual has made with others, in specific social situations.
Image	The degree to which using an innovation is perceived to enhance one's image or status in one's social system.

Table 4.6 Facilitating conditions: constructs and definitions

Construct	Definition
Perceived behavioural control	Reflects perceptions of internal and external constraints on behaviour and encompasses self-efficacy, resource facilitating conditions and technology facilitating conditions.
Facilitating conditions	Objective factors in the environment that observers agree make an act easy to do, including the provision and computer support.
Compatibility	The degree to which an innovation is perceived as being consistent with existing values, needs and experience of the potential adopters.

Change objectives

Change objectives are specific goals of the (health) intervention, stating what should change at the individual lever or among environmental agents. Change objectives can be formulated by crossing performance objectives with the determinants. However, crossing performance objectives with determinants is only relevant when determinants affect the performance objective. Based on the performance objectives and determinants described above, we constructed a list of specific behaviour change objectives (for details, see [6]).

Selected persuasion strategies

After the creation of the behaviour change matrix, the next step was the mapping of strategies for each change objective/determinant. The mapping involves selecting strategies that have been proven effective to realize the defined change objectives and describing how they are practically applied. Based on this, a plethora of persuasive strategies was selected for the AT.

In total, seven change objectives were selected as candidates for the AT prototype:

1. Feel that it is easy to become skillful at using the system
2. Feel that the training is fun
3. Know how to state clear goals and tasks
4. Compare performance level with what other users do
5. Show ability to monitor own operation activity (i.e., training tasks)
6. Ask for positive reinforcement on system operation successes
7. Be able to identify barriers in own performance and show how to overcome them

Since each change objective is related to one or more behaviour change theories, multiple persuasive technologies might be appropriate (see [24]). For each change objective selected, several persuasive strategies were chosen (for details, see Table 4.7 of [6]). Tailoring and tunnelling are the most common persuasive strategies and tools selected. By combining different persuasive technologies (e.g., monitoring progress and rewarding user with positive feedback) we believe that we would enlarge the persuasive impact of the system and successfully motivate our target groups to operate multimedia management, authoring and sharing using the AT.

4.2.3 Developing persuasive strategies for Phase II: Social inclusion

In the second phase of the clinical trials of the project, the participants went over the same protocol as in the first phase, but this time in their home environments. The platform was given to them for a fixed period in which they will be encouraged to use it. The core objective is to assess the impact of the AT on multimedia management, authoring and sharing in less controlled settings. In this phase, social network activities (i.e., social media activities) and digital productivity (i.e., online courses taken) are of primary importance.

Since the overall aim of the project is to increase users' potential in social inclusion, strategies are needed, as motivators, for the users to continue using the system and stimulate their online social participation. The desired outcome of the second phase is different from that of the first phase and therefore, the IM framework is not applicable; here, the overall goal is broader, and it is pointless to predefine objectives and determinants of behaviour of potential online activities. All in all, a different persuasive approach needs to be designed.

In the following subsections, we use theories from the domain of social psychology as the foundations of the strategies to motivate user's online social participation and contributions. Next, we describe the selected persuasive strategy of hierarchical

Table 4.7 Updated requirements for persuasive design

No.	Original requirement	Updated requirement
1	System shows how a user is doing on a number of clear and quantifiable criteria	System shows how a user is doing on a number of clear criteria, *and this information is presented in an evaluated way*
2	System shows users' status, progress and achievements	System shows users' status, progress and achievements on a special overview page. *To provide continuous, repeated progress feedback, the system shows a combined, evaluated variable*
3	System encourages or discourages user's behaviour with the use of praises or rewards and punishments (absence of rewards)	System encourages or discourages user's behaviour with the use of praises or rewards
4	System provides positive, evaluative feedback of user's performance	(no change) System provides positive, evaluative feedback of user's performance
5	System provides means for comparing performance with the that of other users	System provides *ambient* means for comparing performance with the that of other users
6	System provides a clear structure among the various levels and tasks	System provides a clear structure among the various levels and tasks, and presents this in a very simple way
7	System provides challenging (though attainable) assignments with clear short-term and long-term goals	System provides challenging (though attainable) assignments with clear short-term and long-term goals; *these assignments only need little amounts of conscious attention*
8	System provides assignments and levels which increase gradually in difficulty, following the training tasks	System provides assignments and levels which increase gradually in difficulty, following the training tasks, *without bothering the user with keeping track of sequences or progress*
9	System provides task instructions in a clear manner	No change
10	System provides opportunities for the user to learn functionalities of the system and develops competences and skills	No change
11	System provides suggestion for carrying out tasks during the system use process	System provides, *in an ambient way*, suggestion for carrying out tasks during the system use process
12	System provides opportunities for the training tasks to be fun	No change

memberships and lastly, present our proposed motivation strategy to accomplish the goal of social inclusion of the current project.

Reciprocation theory

Reciprocation is a basic norm of human society. It states that appropriate rewards are necessary when individuals are invited to do something for someone else. Since the project's goal is individuals to keep using the system, users should benefit from their participation. One overly used technique is to match rewards with the participation level. Simply put, the amount of reward a user gets should be contingent on his/her activity level. At this point, two crucial questions are important: how to precisely measure users' participation and activity and what the reward should be. If the rewards are perceived as unimportant, it can affect users' motivation to continue using the system.

Consistency

According to the consistency theory, making initial public commitments increase the probability that subsequent actions will be consistent with such commitments [10]. Forming this theory for the project, users could be encouraged to make an initial commitment of their online social participation. Thereinafter, they will be reminded of the commitment whenever they do not act in accordance with this commitment. Based on the theory, the user, in an effort to reduce his/her cognitive dissonance, will adjust his/her behaviour accordingly. The central issue here is how to persuade the user to make a public commitment.

Social validation

According to social validation theory, people often choose what to do in a situation by observing the actions of others. If a large percentage of people are in favour of a certain idea, many others would tend to follow their way. Moreover, people who share some short of similarity, can influence the behaviour of one another. Therefore, it is likely to persuade users to increase their online participation, by making known that many others, just like them, performed the same activity, and were rewarded for this.

Theories of discrete emotions

Discrete emotions are universal emotions such as fear, anger, sadness and joy. They have been defined as 'those emotions that have unique appraisals patterns, motivational functions and behavioural associations' [11]. For this project, the theory of fear will be primarily used. According to the theory of fear, a person can feel fear when he/she finds himself/herself in a situation of perceived threat, directed either to the self or to his/her properties. This fear of loss increases the appeal of incoming messages, particularly those enclosing reassuring information [11]. An interpretation of the theory to our motivation strategy could involve provoking fear to the users; for example, by threating that they may lose some of their privileges. Afterwards, they are presented with the information about how to prevail over this issue. As a consequence, the persuasiveness of the information is enhanced and a user becomes more active.

Selected persuasive strategy: hierarchical memberships

The core motivation strategy here is to introduce a set of hierarchical memberships into the system. Users can be given different memberships based on their levels of online activity. The more active a user is, the higher his/her membership level is. The three issues that are explored in the subsections below pertain to the selection of measuring activity indicators, the membership decision and the selected rewards for each group. Lastly, we describe below a detailed example of how the selected persuasive strategy could be applied to the AT.

Selection of measuring activity indicators

In accordance with the questionnaire analysis (see [5]), we expect the users to engage in the following four social activities that have been rated as the most important aspects of computer use categories for all patient groups:

1. Social participation
2. Educational attainment
3. Work/employment potential
4. Recreation and information

Furthermore, each of these four categories contains various social indicators. For example, the interpersonal relationship category includes, among others, increase of online friends and the number of messages sent. A very important aspect here is that the number of indicators (shown to the users) should not be large and should also be concrete and easy for the users to understand (so as to know how to match their activity in accordance with them). Since these will be the target behaviours for each of the four categories, careful consideration is needed in their selection.

Another essential issue is that users of each of the three target groups have different needs and preferences for social inclusion that have to be taken into account. The feedback from questionnaires has pointed out that the three most important computer categories among the group of patients were as follows:

(a) Patients with SCI: productive activities and recreation, social participation, study/recreation and information
(b) Patients with PD: communication, information, social participation
(c) Patients with NMD: social participation, communication and recreation

Although some of the indicators can be chosen to apply to all three groups (such as 'log on the system frequently' and 'stay online') other indicators should be tailored to each of the three groups. This means that the indicators should reflect the most significant social indicators related to computer activities, which are important to the patient groups.

As a result, the core idea is that each group will be presented with different indicators to accomplish, according to groups' preferences for online activities. For example, frequency of SNS use (e.g., Facebook, Twitter) could be a more relevant social indicator for patients with NMD, while for patients with PD, frequency of mail and Skype use would have a higher importance. Based on this, our approach is the following: first, maintain separate numeric values for each user to represent his/her performance on each of the selected indicators per category. For example, a value 'V1' may account for users' frequency of social media use per time unit (e.g., 1 week), in the first category.

Moreover, these indicators can have the same or different weights. Since the importance of the selected indicators for social inclusion can be different, different weights (i.e., W1, W2 and W3) could be introduced per indicator. The weights could be tailored per group; the activities that are more relevant to each group could be assigned with a higher value.

Although in most cases users' activity can be measured accurately, there is a detail that deserves mentioning. We hope that users engage in all categories. In other words, they should not keep performing one of them and skip the others. Therefore, we can put a ceiling value (Ci) for each criterion. If a user's performance value of a certain activity is greater than the ceiling value of that activity, the weight for the excess part (Wi_excess) would be much less than the original one (Wi). The intent of introducing the ceiling values is to stimulate users to be active in four categories, with the same effort. It ensures that the users who always perform one task and ignore the others would not get a high-level membership.

We constructed a variable integrating the four categories using for each an assigned (tailored) indicator that could measure the users' social inclusion (on that category), each with their assigned weights and ceiling value. This integrated variable was used to determine a user's 'membership level' (see the next section).

Users' membership decision

First of all, we should decide how many membership levels we should introduce to the system. If the number of the membership levels is too small the users' activity would not be differentiated well. On the contrary, too many membership levels could be confusing for the users. Generally the number of memberships should not be greater than six and less than three. For the current AT, we selected to rank the users in the system into three levels depending on the overall evaluation of their online activity: gold (top level), silver (middle level) and bronze (bottom level).

In general, becoming a gold member should be relatively more difficult, because gold users, representing the highest level of participation, are not easy to be stimulated further. According to the theory of discrete emotions, their only motivation is trying to maintain their memberships. It is intended to be relatively easier for users to be classified in the silver member category, because these users have the chance to upgrade their membership and at the same time be afraid that they might be demoted. Both possibilities could become their motivation to increase their participation. In addition, according to the social validation theory, the fact that all users can see the participants that managed to get in the top two levels could pressure or stimulate the ones that are on the bronze level, to perform better, thus getting them a promotion.

Users' rewards decision

When users have managed to upgrade their membership, it is really important to have them realize that the system offers some kind of reward for their active participation. Otherwise users would feel there is no meaningful reason to achieve a high membership level and their participation might decline.

What should be the reward for active users? To start with, the membership itself is a kind of reward if it becomes public. Since there are a limited number of this project's platforms to be distributed, they were given to the users in a rotational process. Thus, the simultaneous and 'live' comparison and visualization of the participants was impossible. However, users' membership served as a kind of recognition when it becomes public and can be seen by all users. Gold members gained a high status and silver members gained some social credits as well. Therefore, it was highlighted to the users that the levels of their progress as well as the membership they reached would be visible on the interface to subsequent users.

However, offering only this reward was not enough since not all users are motivated by status and social comparison. A basic rule of thumb is that the reward should deserve or outweigh the users' effort to upgrade their memberships. Consequently, a more 'materialistic' reward could be introduced, such as providing better services for active users. The definition of better services should be what users really need in the system. Previous research rewarded users by altering media items such as sounds, background skin, or user avatar according to user's performance.

However, users of the three groups have different needs, difficulties and requirements; that is why rewards have to be tailored to the group. Based on their status, we rewarded the groups with some extra functionality which is attractive enough to them.

The motivation interface

The AT will introduce three memberships into the system: gold, silver and bronze. A user will be grouped into these three levels according to his/her activity, related to the selected social indicator. A graphical user interface shows the user's membership. On the default scree, right when the user runs the software, a symbolic membership card was displayed, which clearly shows the user's current membership level. If the user clicks on the card, a new window would pop up and show the user's participation during the previous period of time (e.g., days, weeks). The window will describe the proportion of the user's activity in each of the selected indicators (of the four broad categories) instead of the absolute value. This information explains visually why the user is in the current membership class.

In the AT, the users' membership is public among the participants. There will be a visualization panel that shows the hierarchical representation of all (previous) users' nicknames together with their memberships. This representation was included to trigger social comparison and thus stimulate the user to increase participation.

Based on their status we rewarded the active users with some useful extra functionality, tailored to the group. The rewards increased as the user moves to a higher level.

Lastly, users received messages which evaluate their progress, warning them when they are about to downgrade or upgrade their membership.

4.2.4 Conclusions

The section above presented the design of persuasive principles to motivate users towards the use of the AT. This design followed the pilot trials protocol which is divided into two phases; therefore, the persuasive strategies were designed for each phase. For the first phase, which is the user acceptance and training engagement, various persuasive theories were selected and proposals on how they could be applied were made. For the design of the persuasive strategies, the needs and requirements of each of the target groups were taken into consideration. Thus, tailoring was the persuasive technique most frequently used. With regards to the second phase, which refers to the increase of users' social inclusion through online social activities, a persuasive strategy called hierarchic memberships, was designed, considering the particularities of the groups throughout the design process.

The next step is to validate the selected strategies through experiments with users with similar characteristics with our target groups, to incorporate the updated strategies into the prototype.

4.3 Effectiveness of the proposed persuasive and personalization design elements

Also, this chapter shows how empirical research has tested the effectiveness of the proposed persuasive and personalization (i.e., incorporating user model knowledge) design elements. For this, this chapter will describe two user evaluations of the system. That is, firstly, at the three clinical sites a feasibility study has been performed: Phase I trials of the AT (for a detailed evaluation, see [12]). The core goal of this feasibility study was to test whether users from the three patient groups could use the AT, identify usability issues, technology issues, and assess whether they can be effectively trained to use the AT and after such training perform a series of web tasks. Below, we present this evaluation and discuss the implications is has for the persuasive design and personalization of the AT.

Secondly, we studied in a lab study specifically the influence of the persuasive and personalization design. That is, the persuasive and personalization strategies were selected based on a close review of the scientific literature showing what the more effective (and most fitting and relevant) persuasive strategies are. Also, based on the IM approach, we developed a selection system for selecting those persuasive and personalization strategies that best fit within a larger intervention aimed at influencing the current target behaviour (acceptance and use of the AT, and after that, social inclusion behaviour).

To test the actual effectiveness of these strategies, we performed a lab study. In this lab study, we compared the effects of being trained with a version of the AT training software that include all of these persuasive strategies and personalization to the effects of being trained with a version of the AT training software from which most of those elements were removed.

4.3.1 The evaluation of Phase I field trials

As described above, the project conducted Phase I field trials. These field trials were designed to evaluate the first version of the AT with actual users in a controlled environment. The specific purpose was to investigate the feasibility and usability of the AT and the propensity of the participants exposed to it, for adoption purposes. A sample of 34 participants (18 able-bodied participants and 16 patients) was trained to use the AT in a half-day training session supervised by experimenters. The patient sample included six patients with PD, four participants with SCI and six participants suffering from NMD. All had physical disabilities limiting the use of digital devices. In this field trial, participants were trained in using the AT (using the training software), and then performed four dictated tasks that aimed at social inclusion (e.g., writing an email, posting a picture).

Results of this first feasibility study represent the first evidence that the AT can be used effectively by patients, that patients can use it as effectively as healthy users, and that the AT will allow patients to successfully perform social inclusion aimed tasks. More specifically, during these clinical trials, most patients (and also healthy users) expressed strong interest in trying this innovative technology using their mind and eyes. All participants were (with the exception of two participants who dropped out for medical reasons) able to learn to use the device in the basic, intermediate and advanced training tasks, while also showing improvement in the use of the device after practicing in more tasks. Results show that patients learned to use the AT similarly to able-bodied participants. All patients were able to successfully carry out dictated tasks (composing and sending e-mail, posting on social media, watching a video and uploading a photo) defined as important for social inclusion. Their performance on these tasks (with respect to time and accuracy) was not different from able-bodied participants. Importantly, the current findings point out that with the AT, their physical disability tends to not be a hindrance in the use of a computer for social inclusion tasks. Finally, the patients in the sample tended to express satisfaction and interest in using the device, despite some technical difficulties (e.g., repeated necessity of eye-tracker recalibration).

Important for the current report, Phase I field trials also investigated (qualitatively) the feasibility of the persuasive and personalized design of the AT training software. That is, the persuasive design and personalization elements of the training software were included in these field trials: Half of the participants were exposed to the persuasive design elements by training them on how to use the AT with the (original) version of training software that included all persuasive and personalization elements, whereas other participants were trained with a version of the training software from which most of these elements had been removed (but that still retained a potential influence on behaviour as it contained the same structure and training cycles).

Results of Phase I field trials showed that participants were very positive about the persuasive and personalization elements. For example, participants' self-reports showed that these elements added to the fun and enjoyment of usage of the AT. However, these persuasive and personalization design elements did not seem to make a difference in participants' acceptance and use (performance) of the AT. That is, both with respect to acceptance variables (ease of use, perceived usefulness, etc.) and performance variables (speed and accuracy, learning speed), Phase I trials showed comparable findings for participants trained with the two versions of the training software.

However, there are several very clear reasons for this absence of differences in acceptance and use between users trained with the persuasive and personalized training software and the other users. First of all, all participants in Phase I field trials showed to have very high motivation for accepting and using the AT. Also, these participants only used the AT for a very limited amount of time (only 3–4 h), which apparently was brief enough to not cause deteriorations of motivation. This indicates that the persuasive and personalized design could not increase motivation even further. Relatedly, in Phase I field trials, an experimenter is needed to be present in the room to assist the participant. Still, the presence of the experimenter might have had a stimulating effect on the motivation of the participant. However, the persuasion design elements will be especially relevant in Phase II part of the trials which will last for a month in participants' homes, and in which users have to use the system without the presence of an experimenter.

Another potential reason for the absence of clear differences between user acceptance and performance caused by the persuasive and personalized design of the training software is the lack of statistical power of Phase I field trials. That is, these field trials were set up as a feasibility study and had inherent limitations in using larger numbers of patients as participants.

Therefore, next, we conducted a lab study we performed to gather quantitative and more focused evidence for the (potential) effectiveness of the selected persuasive design strategies and personalization.

4.3.2 The evaluation of the assistive technology in a lab study

In addition to studying the feasibility of the AT in Phase I fields trials, we also performed a lab study to specifically investigate the effects of the persuasive design and personalization elements in the AT training software on two crucial outcomes: system evaluation and task performance (for details, see [6]).

To study the effects on system evaluation and task performance, this study investigated the effects of training participants with two different versions of the training software: a version with all persuasive design elements included, and a version of the training software from which most persuasive design and personalization was removed. So, for example, participants who used the version with all persuasive design elements included were instructed on how to use the system through gamified, personalized instructions, while for the participants who used the other version, the same instructions were giving without personalization and gamification elements. After being trained on how to use the AT with one of these two versions of the training software, participants in this study performed two dictated tasks: a Google search task and a YouTube task. Overall, the results of this lab study confirm and extend the results of Phase I trials (as described in [12]).

First of all, the lab study investigated participants' evaluations of the AT. When assessing ease of use evaluations right after completion of the training tasks, results show that the Full Persuasive training software leads to more negative evaluations of the AT. That is, results showed that in their evaluations *right after completing the training tasks*, participants who were trained using the Full Persuasive training software evaluated the perceived ease of use of the AT *more negatively* than participants who were trained using the Limited Persuasive training software. Explanations for this difference may be related to the core consequence of including persuasive design elements: playing the games in the version of the software needed more time (as results also showed), more game elements needed to be understood, and, basically, the tasks within this version of the software were more elaborate (including the persuasive design elements, like feedback, personalization, gaming elements) than the tasks in the Limited Persuasive training software. Thereby, we argue that causing a somewhat lower evaluation on ease of use judgments is not easily avoided when including persuasive design elements (compared to software in which those elements are not present).

Importantly, results could not provide evidence for a difference in ease of use judgments for the two versions of the software, right after the dictated tasks had been completed. This suggests that the detrimental effects on ease of use caused by the extra task elements in the persuasive design version had disappeared, and any disadvantage for ease of use judgments caused by the persuasive design elements had dissipated.

Finally, results could also not provide evidence that participants trained with the persuasive design version of the training software evaluated the AT as more useful than participants trained with the other version. Importantly, results show that the explanation for this is a 'floor effect': all (healthy) participants of our study seem to have evaluated the potential usefulness for themselves of this technology to be very low (in both training software conditions), probably simply because they have much more useful alternatives available (i.e., the mouse and keyboard).

Thereby, these results help understand the lack of (qualitative) differences found in Phase I trials on evaluations of the Full Persuasive training software as compared to the Limited Persuasive training software. The current results suggested that there may not be an advantage (for ease of use judgments) of including persuasive design elements but rather that ease of judgments are negatively influenced, although such more negative ease of use judgments also easily diminish when using the system.

More importantly, the current results help understand how the persuasive design elements help increase user performance. That is, first of all, our analyses show that users trained with the Full Persuasive training software needed more time for completing the training tasks. This finding is closely related to the lower ease of use scores found for this version of the AT: doing more (as more was included) costs more time, and thereby the whole set of tasks was less easy to complete.

Still, using it for a longer time period also seems to have led to more training and better performance.

Indeed, crucially, the current results also showed the advantages of including the persuasive design elements in the AT for a very important performance outcome: accuracy. That is results showed both that participants trained with Full Persuasive training software performed the training tasks better (more accurate) and, perhaps even more importantly, performed the two dictated tasks better (more accurate). As the core goal of the AT is increasing social inclusion activities users perform, this finding provides evidence for the effectiveness of the persuasive design included in the AT training software for increasing such outcome behaviour.

In sum, the lab study gave rise to the following conclusions about first of all AT evaluation, and, secondly, task performance within the AT.

First of all, different from what we expected, including the persuasive and personalization design principles into the AT training software did not lead to more positive AT evaluations, as compared to not including these elements. The Full Persuasive version of the training software was perceived to be less easy to use, and, relatedly, participants needed more time to complete these training tasks. Crucially, we argue that this finding does not mean that these design features should be removed from the AT. Rather, we argue that the comparison made in the current lab study (between the two versions of the training software) was rather specific.

That is, importantly, also the version of the AT training software from which the persuasive and personalization design principles were removed, still comprised skills training included in appropriate and effective training cycles. So, the comparison made in the current lab studies (in hindsight) can be regarded as one between a rather elaborate (with persuasive and personalization included) version of the same skills training, versus a shorter version of the same skills training (with persuasive and personalization elements removed). Many reasons for these lower perceived ease of use judgments can be identified: the increased extensiveness of the tasks (with the persuasive and personalization elements included), the longer time that was needed to complete them (e.g., for reading the feedback messages, or the social comparison tables), and also the potential additional usability issues in these additional task elements (e.g., difficulty of reading personalized feedback message). These characteristics of the persuasive and personalization design may have given rise to cognitive overload (and consequently lower perceived ease of use).

Finally, we conclude that the perceived ease of use of the training software could (although not necessarily) be improved. For this, the included persuasive and personalization design principles could be screened for elements that take unnecessary time or might otherwise lower perceptions of ease of use. Still, we argue that not too much should be changed in the persuasive and personalization design, because of the more important advantages these design elements show to have for the performance of users of the AT.

Second, and more importantly, we conclude that the persuasive and personalized design of the AT is effective in influencing user performance. The accuracy of users trained with the persuasive personalized version of the training software improved (as compared to users trained with the Limited Persuasive version) both on the training tasks themselves (as found in the lab study) as well as on the dictated tasks (as found in the lab study and suggested by the results of Phase I).

4.4 Implications for persuasive design requirements

Based on the outcomes of Phase I field trials and the lab study reported (on the persuasive and personalization design strategies specifically), we were able to presents updates and extensions earlier insights. That is, the user profiles and personas (Section 4.1), and also the persuasive strategies selected for improving user acceptance and continued use (Section 4.2) could be ameliorated based on these findings. Overall, the main conclusion of Phase I field trials was confirmed by the main conclusion of the lab study: the persuasive and personalization design elements of the AT can be effective for improving acceptance (motivation, etc.) and use (performance, especially in accuracy) and thereby improve users effectiveness in using the computer system for social inclusion tasks. Thereby, the findings of these two user studies had great value to improve the persuasive and personalization design elements of the AT.

4.4.1 Implication for user profiles and personas

In both Phase I field trials and also the lab study reported in this document, indications were found that the complexity of the AT's training task (especially the version including the persuasive and personalization design principles) contained a lot of cognitive tasks to be performed by the user (e.g., process feedback, play and understand task and game elements, etc.). Even though evidence was found that performance improved (i.e., better accuracy), the abundance of additional tasks may have negative effects. For example, the user evaluations of the AT may be negatively influenced by this cognitive overload.

The user profiles and personas for the three cohorts (patients with SCI, PD and NMD) covered the following six areas: disease range and demographic characteristics; physical functioning; emotional functioning; motivational aspects; cognitive functioning; and computer and AT operation.

The results of the field trials clearly show that the created user profiles cover most of the essential user attributes. For example, regarding the physical symptoms of PD, the user profile contained information about posture/loss of postural reflexes. As it was observed in the field trials, indeed, a PD user had difficulties in holding his body posture, leading to usability problems with the eye-tracking device. Another example that provides evidence for the rigor with which the profiles were created comes from the NMD patient groups. The NMD profile discusses the gradual mobility reduction of the patients with NMD as a physical symptom and the consequences for psychosocial functioning. Specifically, in computer use part of the profile it is mentioned that the use of computer validates their ability to think, respond and function well. Indeed, findings from field trials provided similar evidence: NMD users prefer to make use of any ability they have in using their hand as long as they have it. In conclusion, the derived requirements based on the findings of the trials were all included in the profiles and personas.

However, one limitation is that although the cognitive functioning is included in the profiles, it was excluded from the personas, giving rise to more the psycho-social aspects rather than the cognitive aspects of an ATAT use, like the AT. The reason for this was that the limited information in the literature regarding cognition (of these patient groups) and AT use, but also this was due to the fact that cognitive functioning was not included in the project's first patient questionnaires. This lack of consideration of cognitive functioning might have been the reason for the decreased ease of use found in our lab study, which compared the Full Persuasive to the Limited Persuasive training software.

Based on this argumentation, we provided an update of profiles per patient group, adding a focus on the aspect of cognitive functioning. This was mainly useful for updating the persuasive design training, as well as the persuasive strategies included in Phase II trials.

One important note to be made is that the cognitive aspects to be taken into consideration in the following subsection pertain to both patients with muscular disorders as well as to healthy individuals, since one of the exclusion criteria of the project's trials was that cognitive function of users has to be intact in order to be able to operate the AT successfully. In fact, Phase I field trials presented no evidence for statistical differences in responses between able-bodied and patients in variables in terms of system perceptions, beliefs, satisfaction and ease of use. So, although the personas we created could remain as they are, based on Phase I fields trials and the lab study, we propose (below) an updated general cognitive user profile. These additions are relevant for all three patient groups (as well as healthy users), because all three patient groups have comparable mental characteristics in the sense that they might suffer from cognitive overload when tasks are too demanding (and the exclusion criterion of no cognitive deficits was used for all three patient groups).

4.4.2 Updated cognitive user profile

Based on the two user evaluation studies (Phase I field trial and the lab study), we present below an addition and update of cognitive user profiles (for all three patient groups). This update does not apply to patients with any cognitive deficits (i.e., memory or attention impairments).

It became evident in Phase I that participants need clear instructions about how to use AT. This made clear that the AT is not readily intuitively figured out. Proper training and learning is very important in leading the user to a successful learning experience. Below we describe an extension of variables to be described in the user profiles emphasizing cognitive aspects that will pave the way for a meaningful update of persuasive design training method, to be used in the home trials too (such a cognitive user profile is specifically tailored to the project's patient groups and lies on the cognitive theory of multimedia learning) (see also [13]):

- Cognitive overload: The AT is mainly used with the eyes. At the first learning steps, users learn both the functionalities of the system (i.e., different icons) as well as the interaction between them and the system. A potential problem to be considered from this learning situation is that the processing demands evoked by the learning task may exceed the processing capacity of the cognitive system, called cognitive overload. The split-attention effect is a learning effect with detrimental learning effects when cognitive load is high. It is apparent when the same modality (e.g., visual) is used for various types of information within the same display. To learn from these materials, learners must split their attention between these materials to understand and use the materials provided.

- Limited capacity: users are limited in the amount of information that can be processed in each channel at one time.
- Active processing: Users are engaged in active learning by attending to relevant incoming information, organizing selected information into coherent mental representations and integrating mental representation with other knowledge.

These cognitive user variables extend the user profiles. That is, the cognitive characteristics of each of the three patient groups can now be evaluated and taken into account using also the new insights generated by these two evaluation studies.

In general, we saw that especially patients with muscular disorders might face difficulties with memory, attention and processing speed (both for patients with PD and NMD). So, one general implication (at least for PD and NMD, but probably also for patients with SCI) of the current findings is that any persuasive and personalization design elements should refrain from taxing memory, attention and/or processing speed.

Below, we present for each of the three patient groups the cognitive functioning element of their user profiles, and the implications the current findings have for these user profiles.

4.4.2.1 SCI Cognitive functioning

Cognitive function can be normal, but a substantial number of patients with SCI have significant deficits in one or more cognitive domains: moderate attention and processing speed deficits, mild deficits in processing speed, executive processing difficulties, or moderate memory impairments.

Implications based on current evaluation studies: As described above, especially for these patients, the persuasive and personalization design should take into account cognitive limitations, and use more influencing strategies that demand less cognitive resources (both in memory as in processing speed).

4.4.2.2 PD Cognitive functioning

Mild cognitive impairment (MCI) in PD individuals is associated with increasing age, disease duration and disease severity. The frequency of cognitive dysfunction is from 36% at the time of diagnosis to as high as 93% in more advanced stages of the disease. The most frequently encountered domains of cognitive dysfunction involve executive functions, memory, visuospatial skills, attention, and mental processing speed. Preserved functions include basic attentional processes and many language abilities (particularly comprehension).

Implications based on current evaluation studies: Next to the importance of exclusion criteria (no cognitive impairments), the current evaluation studies also stress that for PD individuals cognitive overload caused by the abundance of persuasive and personalization strategies included may certainly occur. Therefore, also for these patients, limitations to necessity of memory and cognitive processing in the persuasive games are needed. Also, the games might also be limited in the extent to which they need visuospatial skills.

4.4.2.3 NMD cognitive, learning and neurobehavioral functioning

A substantial number of patients from the NMD population have a cognitive impairment. Cognitive skills do not deteriorate over time. Cognitive deficits documented in older children and adults mainly pertain to verbal skills. Vision-spatial skills, long-term memory and abstract reasoning skills are not affected. Patients with NMD have been characterized as being easily frustrated, easily distracted, and have poor attention span.

Implications based on current evaluation studies: Next to limiting the strain on memory, processing speed, and visuospatial capabilities (needed for patients with SCI and PD), for NMD individuals the persuasive games need to be optimized for causing no or very limited frustration, and needing only a limited attention span.

4.4.3 Updated requirements for personalization

In general, the two evaluation studies showed favourable results as for the personalization strategies included in the design of the AT training software. We propose two sets of updates to the requirements for the personalized persuasive design elements in the next version of the training software.

First, the current two evaluation studies show that the personalization included is effective, but at the same time also suggest that further limitations of the necessity of using memory, cognitive processing, etc., might be helpful for optimizing the effectiveness of all persuasive strategies (also the personalization persuasive strategies). Therefore, we propose to check in this perspective all personalization now included in the training software. Using the participant's first name, for example, seems to be effective, but must be done only for a limited number of times. Likewise, the personalized feedback messages might be effective but should be limited in length and complexity.

Second, as mentioned by various participants in Phase I field trials, the current AT's interfaces contained possibilities for customization. That is, also the AT training software contained personalized persuasive strategies (e.g., using the participant's first name, and adapting feedback messages to the participant's age and gender), this personalization was done by the AT itself. An additional personalization strategy is to allow users of a system to customize it to their own preferences. That is participants might be allowed to set the background colour of the interface, choose certain graphics, set other issues like response speed, etc. Indeed, earlier research presented evidence for the effectiveness of customization as a persuasive strategy (see [14]).

4.4.4 Updated requirements for persuasive design

Finally, we will discuss the implications for the requirements for the persuasive design of the AT training software of the two user evaluations (Phase I field trials and the lab study reported in this document). Most importantly, as argued above, in both Phase I field trials and also the lab study reported in this document, indications were found that the complexity of the AT training task (that included the persuasive and personalization design principles) gave rise to cognitive overload. Indeed, we presented earlier scientific research [6] that investigated the effectiveness of most of the persuasive and personalization principles *separately*. For example, based on the research of social norm activation and related interventions (see [15]), we [6] proposed to include various kinds of social norm activation interventions. And although some earlier studies have investigated combinations of a few persuasive strategies (see, e.g., [16]), scientific research has not investigated the effects of combining larger numbers of persuasive strategies with one another. Indeed, dual process theories of persuasion (e.g., [17]) disentangle more elaborate, conscious and controlled mental processes (central processing) from less elaborate, more unconscious and less controlled (peripheral processing). Such theories (e.g., the Elaboration Likelihood Models [17]) would argue that presenting too many persuasive strategies that need to be processed through central processing will lead to overload, may lead to interferences between these persuasive strategies and diminish their effectiveness.

Now, although the current user evaluation study results suggested limited (Phase I trials) to negative (lab study) effects of the persuasive and personalized design principles on user evaluations of the AT, results also showed that the persuasive and personalized design was effective in stimulating performance improvement (on accuracy). Therefore, our main conclusion for updating the requirements for the persuasive design entails that only limited changes should be made, as the current design seemed to be effective on the most important variable: behaviour change.

Still, changes in the persuasive and personalized design principles that improve user evaluations and leave unchanged (or even improve) the effectiveness of the AT training software for influencing behaviour (performance accuracy) can provide a positive contribution. Therefore, we propose two kinds of improvements for the persuasive design principle requirements: Simplification (more peripheral cognitive influencing strategies), and more positive elements. That is, based on dual process theories of persuasion, as argued above, we propose to adapt some of the selected influencing strategies towards more peripheral processing: less elaborate, more unconscious and less controlled influencing strategies may lead to less cognitive load. Thereby, a set of influencing strategies that includes (next to already incorporated, more central influencing strategies) also more peripheral influencing strategies may have a more positive influence on user evaluations of the AT and lead to more positive perceptions of ease of use. In other words, we propose to include in the AT persuasive personalized design more ambient persuasive technology strategies (see [18]), that influence user behaviour from the 'ambient' environment without the necessity of the conscious attention of the user. For example, we propose to replace factual feedback (e.g., presenting a score) with evaluated feedback (e.g., a colour between red and green, or a flower in a particular state of opening up). That is, factual feedback needs more elaborate cognitive processing, whereas evaluated feedback already has been processed (evaluated) and causes less cognitive load for the user.

Also, we propose that negative evaluations within the persuasive principles may have influenced the perceived ease of use of the AT. Psychological research showed that negative associations may spread through what is known as the 'halo'-effect [19] to related judgments. Research findings by [20] (see also [25]) showed that hedonistic elements of user interfaces (e.g., negative feedback) can influence (e.g., lower) ease of use perceptions. Still research [21] on the effectiveness of evaluative feedback showed that negative feedback can be more effective than positive feedback for changing user behaviour. Therefore, we also propose to restrict the number of negative evaluations in the persuasive and personalized, but not to abandon negative feedback.

Based on these analyses, we propose the following update of the requirements for persuasive design of the training tasks. In Table 4.7, we present the original requirement and the updated requirement.

4.4.5 Implications for Phase II persuasive design strategies

Earlier, we proposed a separate set of persuasive strategies to be included in the extended use situation of Phase II trials. That is, in the second phase of the clinical trials of the project, the participants will go over the same protocol as in the first phase, but this time in their home environments. The platform will be given to them for a fixed period in which they will be encouraged to use it. The core objective of Phase II trials will be to assess the impact of the AT on multimedia management, authoring and sharing in less-controlled settings. In this phase, the research focused on the user's social network activities (i.e., social media activities) and digital productivity (i.e., online courses taken).

Since the overall aim of the project is to increase users' potential in social inclusion, strategies are needed, as motivators, for the users to continue using the system and stimulate their online social participation. The specific target of the persuasive and personalization strategies included in Phase I training software was to stimulate acceptance and use (although use within the limited time frame of 3–4 h). After comparable initial training (that may also be repeated over the weeks to improve performance), in Phase II trials, participants will also (mainly) use the AT for actual and extensive web browsing (over a period of 4 weeks). In these web browsing activities, social inclusion related activities (e.g., using social media) will be stimulated through the persuasive elements of the AT. Thereby, the desired outcome of the second phase is different from that of the first phase and therefore, for Phase II trials, we need additional persuasive design elements.

We used the following list of persuasive strategies that can be incorporated into the AT's web-browsing interface to stimulate social inclusion behaviour (for details, see [6]): Reciprocation (responding likewise when receiving something), Consistency (showing consistency in attitudes and behaviours), Social validation (doing what others do), Theories of discrete emotions (people are sensitive to specific emotional appeals).

These persuasive strategies were included by incorporating in Phase II trial software the concept of 'hierarchical memberships' because in this persuasive design element the above-mentioned persuasive strategies are combined. The core motivation strategy here is to introduce a set of hierarchical memberships into the system. Users can be given different memberships based on their levels of online activity. The more active a user, the higher their membership level.

Finally, we argue that indeed social activity indicators should be measured in Phase II to steer this persuasive design element of hierarchical group membership and give user rewards. See [6] for a detailed example of how the selected persuasive strategy could be applied to the AT.

Based on the two evaluation studies the following implications can be identified for the persuasive design elements to be added to Phase II trials: In general, the two evaluation studies present evidence supporting that also these additional persuasive strategies will be effective. Even though these strategies were a part of Phase I trials, or of the lab study, the proven effectiveness of the persuasive and personalization elements of the training software makes it probable that also these additional strategies will add to the overall effectiveness of the AT for stimulating social inclusion behaviour.

Also, it is important to limit in the implementation of these additional persuasive strategies (e.g., the hierarchical group membership) the extent to which they rely on and need user memory, processing capacity, visuo-spatial capacity and the extent to which they might cause frustration. The considerations presented above for the already included persuasive and personalization design elements should be used also to optimize the to-be-included Phase II persuasive design elements.

4.4.6 Conclusions

Based on the outcomes of Phase I field trials and the lab study (on the persuasive and personalization design strategies specifically), we present an update of the proposed persuasive and personalization strategies as they were included in the training software for Phase I. Overall, the main conclusion of Phase I field trials and the lab study was that the included persuasive and personalization design elements are effective in improving user performance (that is, task accuracy). For implementation of these requirements into the final version of the training software to be used in Phase II field trials, the updated requirements presented in Table 4.7 can be used.

The IM approach seems optimal for distilling what the behavioural steps should be for users of the AT. That is, participants were able to successfully complete the training cycle (as was shown in Phase I field trials and also in the lab study), and after performing the training tasks, all participants could successfully complete the dictated tasks (again, in both studies).

Importantly, because Phase II trials will allow the participant to use the AT for an extended period of time (4 weeks), the effectiveness of the motivators included in the persuasive and personalized design elements (the ones included in Phase I, and also the additional ones presented for Phase II) will potentially strongly increase. That is, both in Phase I field study but also in the lab study, the effectiveness of the persuasive and personalized design elements for increasing user acceptance and use was limited. Indeed, there are several very clear reasons for this limitation of differences in acceptance and use between users trained with the persuasive and personalized training software and the other users. First of all, all participants (especially in Phase I field trials, but also in the lab study) showed to have very high motivation to accept and use the AT. Also, these participants only used the AT for a very limited amount of time (only 3–4 h in Phase I field study, and 30 min in the lab study) that may have been very short for that high motivation to diminish. This makes clear that (as results show) the persuasive, personalized design could (but to a limited extent) increase motivation even further. However, the persuasion design elements will be especially relevant in Phase II part of the trials which will last for a month in participants' homes. Internal motivation of the user may be lost or diminish within that time frame, and the persuasive and personalization design elements will be much more important and have the possibility of increasing or retaining motivation for accepting the system and to keep on using it for social inclusion activity.

Based on the current report, the persuasive and personalized design principles can be further improved, and the persuasive and personalization design elements for Phase II field trials can be developed.

4.5 Summary

In sum, this chapter shows how persuasive personalized design principles (implemented into the AT) can be used to improve user acceptance (evaluation) and use (performance). When developing AT like the current, taking into account the user is crucial, and this research shows how this can be done. Building on user profiles and personas, very effective personalized persuasive technology can be developed. Using persuasive design principles and user models helps create technology that is inherently more easily adopted, and that motivates the user to continue using it.

References

[1] The Multimedia MAnagement/sharing and authoring using your Eyes and Mind (MAMEM) research project. url: http://www.mamem.eu.

[2] Liu, Y., Osvalder, A. L., & Karlsson, M. (2010). Considering the importance of user profiles in interface design (pp. 61–80). INTECH Open Access Publisher.

[3] Henry, S. L. (2007). Just ask: integrating accessibility throughout design. Lulu.com.

[4] Clinical requirements for the MAMEM platform for each of the patient cohort, MAMEM Consortium, August 2015. url: http://www.mamem.eu/wp-content/uploads/2015/09/D6. 1_Clinical_requirements_Final.pdf

[5] Definition of pilot trials with the participation of patients, MAMEM Consortium, November 2015. url: http://www.mamem.eu/wp-content/uploads/2015/11/D6.2_ClinicalRequirements_PilotTrialsDefinition_Final.pdf

[6] Report on persuasive design principles, user models and profiles. MAMEM Consortium, 2016. url: http://www.mamem.eu/wp-content/uploads/2016/05/D5.1_PersuasiveDesignPrinciples_final.pdf
[7] Bartholomew, L. K., Parcel, G. S., Kok, G., Gottlieb, N. H., & Fernandez, M. E. (2011). Planning health promotion programs: an intervention mapping approach. San Francisco: Jossey-Bass.
[8] Bartholomew, L. K., Parcel, G. S., Kok, G., & Gottlieb, N. H. (2001). Intervention Mapping: Designing theory and evidence-based health promotion programs. Mountain View, CA: Mayfield Publishing.
[9] Venkatesh, V., Morris, M. G., Davis, G. B., & Davis, F. D. (2003). User acceptance of information technology: toward a unified view. MIS quarterly, 425–478.
[10] Dillard, J., & Harmon-Jones, C. (2002). A Cognitive Dissonance Theory Perspective on Persuasion. In J. P. Dillard & M. Pfau (Eds). The Persuasion Handbook: Developments in Theory and Practice, 99–111.
[11] Nabi, R. L. (2002). Discrete emotions and persuasion. In J. P. Dillard & M. Pfau (Eds). The persuasion handbook: Developments in theory and practice, 289–308.
[12] Interim report on pilot experiments based on half the subjects, including updated clinical requirements and definitions of clinical trials. MAMEM Consortium, July 2017. http://www.mamem.eu/wp-content/uploads/2017/08/D6.4_Interim-report-on-pilot-experiments_Final.pdf
[13] Mayer, R. E. (2002). Teaching for meaningful learning. Upper Saddle River, NJ: Prentice-Hall.
[14] Orji, R. Vassileva, J., & Mandryk, R. L. (2014). Modeling the efficacy of persuasive strategies for different gamer types in serious games for health. User Modeling and User-Adapted Interaction, 24, 453-498.
[15] Cialdini, R. B., & Trost, M. R. (1998). Social influence: social norms, conformity, and compliance. In D. Gilbert, S. Fiske, & G. Lindzey (Eds.) The handbook of social psychology, (4th edition) vol. 2, pp. 151–192. New York: McGraw-Hill.
[16] Ham, J. R. C., Cuijpers, R. H. & Cabibihan, J. J. (2015). Combining robotic persuasive strategies: the persuasive power of a storytelling robot that uses gazing and gestures. International Journal of Social Robotics, 7(4), 479–487.
[17] Petty, R. E. & Cacioppo, J. T. (1986). From communication and persuasion: central and peripheral routes to attitude change. New York: Springer.
[18] Lu, S., Ham, J., & Midden, C. (2014). Using ambient lighting in persuasive communication: The role of pre-existing color associations. Conference proceedings of Persuasive 2014, Padova, Italy.
[19] Thorndike, E.L. (1920). A constant error in psychological ratings. Journal of Applied Psychology 4, 25–29.
[20] Kakar, A. K. (2014). When form and function combine: hedonizing business information systems for enhanced ease of use. 47th Hawaii International Conference on System Sciences, vol. 00, no. , pp. 432–441, 2014.

[21] Midden, C. & Ham J. (2014). The Power of Negative Feedback from an Artificial Agent to Promote Energy Saving Behavior. Conference proceedings of the International Conference on Human–Computer Interaction, Crete, Greece.

[22] Fogg, B.J. (2003). Persuasive technology: using computers to change what we think and do. San Francisco, CA: Morgan Kaufmann Publishers.

[23] Bandura, A. (1986). Social foundations of thought and action: A social cognitive theory. Prentice-Hall, Inc.

[24] Venkatesh, V. & Bala, H. (2008). Technology acceptance model 3 and a research agenda on interventions. Decision Sciences, 39, 273–315.

[25] Hirschman, E. C., & Holbrook, M. B. (1982). Hedonic consumption: emerging concepts, methods and propositions. Journal of Marketing, 46, 92–101.

Part II

Algorithms and interfaces for interaction control through eyes and mind

Chapter 5
Eye tracking for interaction: adapting multimedia interfaces

Raphael Menges[1], Chandan Kumar[1], and Steffen Staab[1,2]

This chapter describes how eye tracking can be used for interaction. The term eye tracking refers to the process of tracking the movement of eyes in relation to the head, to estimate the direction of eye gaze. The eye gaze direction can be related to the absolute head position and the geometry of the scene, such that a point-of-regard (POR) may be estimated. We call the sequential estimations of the POR *gaze signals* in the following, and a single estimation *gaze sample*. In Section 5.1, we provide basic description of the eye anatomy, which is required to understand the technologies behind eye tracking and the limitations of the same. Moreover, we discuss popular technologies to perform eye tracking and explain how to process the gaze signals for real-time interaction. In Section 5.2, we describe the unique challenges of eye tracking for interaction, as we use the eyes primarily for perception and potentially overload them with interaction. In Section 5.3, we survey graphical interfaces for multimedia access that have been adapted to work effectively with eye-controlled interaction. After discussing the state-of-the-art in eye-controlled multimedia interfaces, we outline in Section 5.4 how the contextualized integration of gaze signals might proceed in order to provide richer interaction with eye tracking.

5.1 Tracking of eye movements

In this section, we provide a basic understanding of the eye anatomy, the technologies to track its movement as eye gaze, and the processing of raw gaze signals.

5.1.1 Anatomy of the eye

The eyeball is covered mostly by a white protective tissue, called sclera. The colored part in the center of the eye is the iris, which encloses a hole, called pupil. The iris can shrink or grow the pupil, thus less or more light enters the eye. Iris and pupil are covered by the transparent cornea. A flexible lens is placed directly behind the iris

[1]Institute for Web Science and Technologies, University of Koblenz-Landau, Koblenz, Germany
[2]Web and Internet Science Research Group, University of Southampton, Southampton, United Kingdom

which refracts the light, such that it falls focused on the rear interior surface of the eye, the retina. The retina is covered by two types of photosensitive cells. The rods are sensitive to brightness, and the cones are sensitive to chromatic light. There is a spot of high density in cones with a diameter of 1°–2° on the retina, which is called fovea. The eye moves so that the light from the focused object falls into the fovea. Thus, our perceived focus of sight is sensed in the fovea region. Figure 5.1 shows a schematic illustration of the eye.

Each eye moves with the help of three antagonistic pairs of muscles that provide the eye with three degrees of freedom. One muscle pair performs horizontal movements, one muscle pair performs vertical movements, and the third muscle pair performs rotational movements around the direction of the view. The eye movements can be categorized into saccades or fixations [1]. Saccades are fast movements under 100 ms of both eyes in the same direction. These movements can be jump-like, gradual and smooth, or random. Fixations are times of about 100–600 ms for which an eye rests at a particular POR.

The diameter of the fovea defines the portion in visual angle of sharp sight, which is what we are interested in for eye tracking. We can estimate the visual angle of an object [2] with

$$A = 2 \arctan \frac{S}{2D}. \tag{5.1}$$

S is the size of the object and D is the distance to the object.

The anatomy indicates two limitations of eye tracking which cannot be simply overcome with better technology. First, we need calibration of an eye-tracking system for each individual. The eyeballs of different people have a variance in size, but more importantly, the fovea is not located directly opposite of the pupil in the optical axis but 4°–8° higher. Thus, the visual axis must be determined by a calibration procedure in order to find the POR. Second, the eye does not necessarily move such that the light of the object-of-interest falls in the perfect center of the fovea. Thus, the fovea diameter of about 1° limits the accuracy of an eye-tracking system to $\pm 0.5°$ [3]. In fact, this is most often the best-case accuracy that manufacturers provide in their

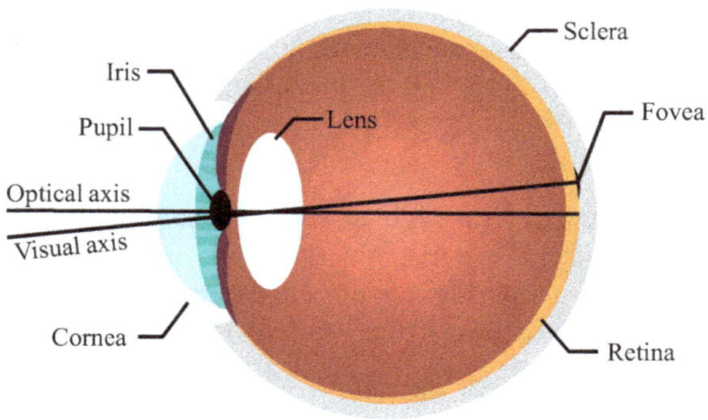

Figure 5.1 Schematic illustration of the anatomy of a human eye

Table 5.1 Specifications of commercial video-based eye-tracking devices for interaction purposes

Model	myGaze n	4C	Skyle
Manufacturer	Visual Interaction	Tobii AB	eyeV
Sampling rate (Hz)	30	90	20–40
Operating distance (cm)	40–100	50–95	45–65
Head box	50 × 30 cm at 65 cm	40 × 30 cm at 75 cm	30 × 12 cm
Accuracy	0.4°	n/a	1–2°
Precision (RMS)	0.05°	n/a	n/a
Connection	USB 3.0	USB 2.0	USB 3.0
Screensize (inches)	10–27	Max. 27	Max. 24

specifications, see Table 5.1. In comparison, the index fingernail also covers about 1° of the view at arm's length distance. This means, at a distance of 70 cm, and a visual angle of 1°, an uncertainty of 1.2 cm. An uncertainty of 1.2 cm translates to about 44 pixel on a 24 in. monitor at a resolution of 1,920×1,080 pixel. Additionally, photosensitive cells in the eyes only react to changes in the amount of light they receive, and in lack of changes quickly adapt and stop responding, the eyes need to move during long fixations, introducing micro-saccades [4].

5.1.2 Techniques to track eye movements

Three techniques have become popular to measure eye movements [2]. Each technique has its own motivations but also limitations with regard to eye-controlled interaction.

5.1.2.1 Electro oculography

In electro oculography (EOG), electrodes are attached to the skin around the eyes to measure an electric field that exists when the eyes rotate. The movement of an eye can be estimated by recording small differences in the electric potential of the skin around the eye. Specific electrode designs can record horizontal and vertical movements and, thus, can be used to estimate eye gaze [5]. However, the signal can change even when there is no eye movement, especially through external factors like nearby electric devices. The EOG offers a cheap, easy, but invasive method to record large eye movements. The EOG technique is not well suited for the use-case of interaction, yet, it is frequently used by clinicians. The advantage of the EOG technique is its ability to detect eye movements even when an eye is closed, e.g., while a person sleeps.

5.1.2.2 Scleral search coils

In scleral search coils (SSC) [6], a coil of wire is attached to the eye. Voltage is induced in the coil and movements within a magnetic field can be measured. Small coils of wire are embedded in a modified contact lens to measure human eye movements. The advantage of SSC is the high accuracy and the nearly unlimited resolution in time. The disadvantage of SSC is that it is a highly invasive method, as it requires a participant to put contact lenses into her eyes. Thus, this method is mostly used in medical and psychological research. Similarly to EOG, the SSC technique tracks eye movements only in relation to the head and provides no indication about the spatial POR.

5.1.2.3 Video oculography

In video oculography, a single or multiple camera setups with no or multiple light sources are used to determine the movement of the eyes from captured images [7]. Mostly, infrared light is used to cause reflections on the cornea, which are visible as glints in the captured images. As the cornea has a spherical shape, the glints stay in the same position independent of the eye gaze direction. A vector between pupil and glints can be calculated, which is used to estimate the direction of the eye gaze. This technique of video-based eye tracking is the most widely used method in commercial eye-tracking systems (see Table 5.1).

The video-based eye gaze estimation can be problematic for dark brown eyes where the contrast between the brown iris and the black pupil is very low and it makes the pupil difficult to detect in the captured image. An additional infrared light source can be placed close to a camera, such that the emitted light reflects from the retina straight back toward the camera. The reflected light makes the pupil to appear white in the captured image and provides more contrast to the iris, which helps in detecting the pupil on the captured image. The effect is well known as "red eye" when photographing faces with a flash. Modern systems combine a single or two cameras with an array of carefully placed light sources. The light sources can flicker periodically to illuminate the eyes only for certain image captures. For example, two light sources might be used to produce glints in one image while capturing a dark pupil, and a third light source, which is placed close to the camera, produces an image with a bright pupil effect. The images are recorded consecutively and can be processed directly on the image-processing chip of the eye-tracking device. Modern eye-tracking systems vary in the frequency to report eye gaze coordinates from 30 up to 1,000 Hz. Interaction requires a frequency that is comparable with the refresh rate of the screen; thus, we recommend to use systems that sample eye gaze at 60 Hz. Eye detection and tracking remain a very challenging task due to several unique issues, including illumination, viewing angle, occlusion of the eye, and head pose.

Eye-tracking systems using the video-oculography technique are available as table- and head-mounted devices. In the context of our research, we primarily use table-mounted eye-tracking devices (Figure 5.2), because these are calibrated to provide POR estimations in the coordinate system of the computer screen.

5.1.3 Gaze signal processing

Eye-tracking systems produce gaze signals, which require processing to be used for interaction purposes.

5.1.3.1 Calibration of the eye-tracking system

A direct calculation of the POR from the captured image would not only need the spatial geometry of the eye-tracking system, the relative display placement, and the eye position but also the radius of the eyeball and offset of the fovea, which is specific to

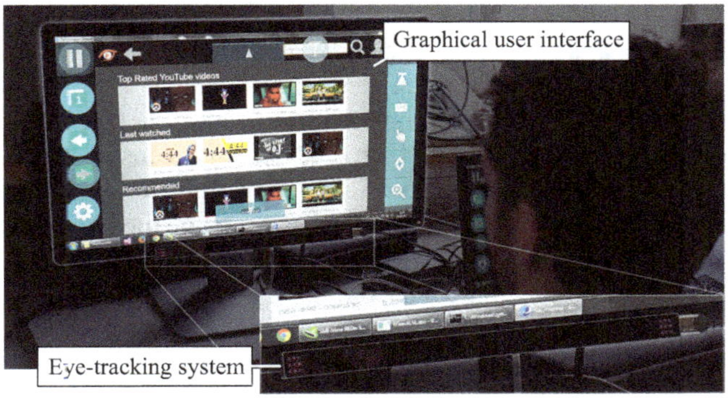

Figure 5.2 Photo of an eye-tracking system in use

an individual. Commercial systems provide a dedicated mode for calibration, during which the screen presents a stimulus that is rendered by the eye-tracking system. The user is asked to follow a symbol on the screen using the eyes, often in the form of a moving dot. The dot stops for a short time at the coordinates of predefined calibration points. This calibration procedure provides reference data for known fixated calibration points on the screen. The reference data is used to calculate the parameters for the mapping of the glint-pupil vector to positions on the screen. Calibrations usually make use of five to nine calibration points, depending on the desired accuracy. Quality of the model can be tested with reference data by comparing the mapped gaze signals to the true position of the calibration points.

5.1.3.2 Error modeling of gaze signals

Besides lighting, visual angle, and iris color, also calibration drift, tremor, and microsaccades of the eye movements, which are in a magnitude of 0.1°, contribute to the inaccuracy of gaze signal estimation. Error in gaze signal estimation can be modeled using the two values of precision and accuracy. Precision is defined as the ability of the eye-tracking system to reliably reproduce the POR at a fixation. The precision error is perceived as high-frequency noise in the gaze estimation. Accuracy is defined as the average difference between the object coordinate and the measured gaze coordinate. The accuracy error is perceived as bias in the gaze signal estimation.

5.1.3.3 Filtering of gaze signals

Filtering of gaze signals has the purpose to improve the precision and to delineate between fixations and saccades. Some eye-tracking systems already incorporate filtering, yet, the manufacturers often do not exactly specify their implementation. There are four popular approaches to filter gaze signals and identify fixations [8], as described next.

- **Velocity-threshold identification.** A point-to-point velocity is calculated between the coordinates of sequential gaze samples. Saccadic eye movements are considered to have a velocity of more than 300°/s, whereas fixations are considered to have a velocity of less than 100°/s. A static threshold criterion can be applied to separate fixations and in-between saccades. For each fixation, a centroid of the gaze samples is calculated. The time of the first gaze sample is the start of the fixation, whereas the time from the first gaze sample in the fixation until the time of the last gaze sample in the fixation is considered the duration of the fixation. However, the static threshold is prone to over-segmentation of gaze signals into fixations. Yet, velocity-based filters are the most popular, as they are effective and simple to implement.

- **Hidden-Markov-model identification.** Saccades and fixations can be modeled as probabilistic state machines using a hidden-Markov-model. Two states are implemented. One represents the velocity distributions of gaze samples for saccades and the other the velocity distributions for gaze samples of fixations. The hidden-Markov-model filter allows a more robust identification than fixed-threshold methods, but the filter is more difficult to implement.
- **Dispersion-threshold identification.** Gaze points that belong to a fixation tend to cluster closely together. One can go through the gaze signal over time and calculate the spatial spread of the coordinates of gaze samples in a time window. When the spatial spread is below a preset threshold, this time window is considered a fixation. A window whose content is considered a fixation can be expanded until the spatial dispersion of the covered gaze samples is too high. For each time window its spatial centroid represents the fixation coordinate.
- **Area-of-interest identification.** Previous methods to identify fixations can identify them at any location on the screen. In contrast, area-of-interest identification considers only gaze signals that occur within specified target areas. The method uses a duration threshold to distinguish fixation in target area from passing saccades in those areas. The target areas have to be defined manually or by another algorithm, which is the reason why the filter is not generally applicable. However, the filter might provide a higher level insight into the gaze signals in regard to the screen content.

5.1.3.4 Online filtering of gaze signals for eye-controlled interaction

Filtering of gaze signals for interaction must be fast, as only a small-time overhead to the estimation of the POR may be added. A fixation must be recognized in the gaze signal, while new gaze samples are coming in. The incoming gaze samples might extend the fixation or be part of a saccade toward another fixation. Furthermore, usually only the current fixation is of interest for interaction purposes. A basic attempt to smooth the gaze signal is to collect samples over time and to average their coordinates [9], toward a reliable POR for every update of the interface. This basic smoothing idea can be performed for a sliding time window, to cover, e.g., the current user fixation but not the whole gaze signal history. By declaring a gaze sample at the current time t as X_t, N as window size, and w as applied weight per sample, we can define the following formula to calculate a weighted average of the gaze signal:

$$\hat{X}_t = \sum_{i=0}^{N-1} \frac{w_i}{\sum_j w_j} X_{t-i}. \tag{5.2}$$

The weight can be defined by an additional kernel function, which takes the age of each sample into account. Literature [9] names three common kernels to calculate a weight per sample in the window:

- Linear kernel, $w_i = 1$. Every sample is weighted equally.
- Triangular kernel, $w_i = N - i + 1$. Latest sample is weighted the highest, oldest with one.
- Gaussian kernel, $w_i = e^{-((i-1)^2/2\sigma^2)}$, where $\sigma = \sqrt{(-(N-1)^2/2\ln(0.05))}$ is defined to assign the oldest sample in the window a weight of 0.05.

The Gaussian kernel is reported [9] to deliver the best results.

Saccade detection. The presented filtering assumes that all gaze samples inside the sliding time window belong to a single fixation. However, the window might contain samples from two or more fixations when the user performs a saccade. Averaging these samples would produce a fixation somewhere in the middle between the contained fixations, possibly on a region which the user never had fixated. Therefore, the sliding window for filtering is limited to the current fixation by using a spatial threshold [10]. The spatial threshold works similarly to the velocity-threshold identification and separates fixations by the spatial distance of sequential gaze samples. The current fixation is then defined as gaze samples which distance is successive below spatial threshold, starting from the latest gaze sample.

Outlier correction. Eye-tracking systems may produce single outliers, e.g., when a reflection is shed on the camera or the data transfer suffers from an error. This may produce a single outlying gaze sample, which would prohibit a proper filtering of a fixation. Therefore, when going from the latest to the oldest sample within the sliding window, for each gaze sample that is classified to belong to another fixation than the current one, the previous gaze sample is also checked if it belongs to the current fixation. If the previous gaze sample belongs to the current fixation, the looked-at gaze sample is discarded as an outlier and the filtering is continued at the previous gaze sample [11]. Otherwise, the gaze sample collection is stopped and the weighted average of the collected gaze samples is calculated as the current fixation. The described filtering can improve the precision of the gaze signal and should be applied for any interaction purpose.

5.2 Eye-controlled interaction

Eye gaze is a natural means of communication between humans [12]. Bolt has already proposed in 1981 to extrapolate the communication toward human–computer interaction [13]:

> At the user/observer interface level, interactivity is extended to incorporate where the user is looking, making the eye an output device.

However, in traditional human–computer interaction, the process of selection and perception works in parallel. Physical inputs like mouse and keyboard are used to select interactive elements in an interface and the eyes observe the selection process on the interface. In a naive implementation of eye-controlled interaction, the eye gaze triggers an interactive element that it dwells upon. In this scenario, a user often would select elements by accident while browsing through the available options. The overload of inspection and selection is referred to as Midas touch [14]. There are two fundamental approaches to deal with this overload. Either, the conventionally parallel process of selection and perception is sequentialized allowing unimodal input with eye gaze only. Or, another input device is used to perform selections, allowing multimodal input that is potentially even more efficient and enjoyable than using an interface with traditional input devices.

Moreover, an eye-tracking system provides gaze signals with limited precision and accuracy. As shown in the previous section, precision can be improved with filtering. However, there is yet no universal approach to bring the accuracy of gaze signals on par with mouse pointing. Thus, interfaces must consider the limited accuracy of at most $\pm 0.5°$. The limited accuracy renders it infeasible to simply replace mouse pointing with eye-controlled pointing or a keyboard with eye-controlled typing.

Both, Midas touch and limited accuracy make it inevitable to develop distinct interaction means to interpret gaze signals as input signals. In this section, we outline selection methods with eye tracking and explain how these selection methods are combined to interactions that emulate mouse and keyboard events.

5.2.1 Selection methods

The most intuitive method to select an area using eye gaze might be the blinking of the eyes. A user could look at the option to select, closing one or both eyes for a short period of time, so as to select the looked-at element. However, there are various flaws with blinking as a selection method. First, it is not easy to distinguish between a natural blink and an intentional blink solely from the gaze signal. The user might blink with only one eye, which is not convenient. An intentional blink with both eyes would have to last longer than a natural blink, rendering it still as tedious process. Second, even if blinking with one eye would be sufficiently convenient, the gaze signals are disturbed by blinking. Eye tracking does not work when the eyes are closed, that is why blinking has been considered infeasible for selection purposes [14].

Another simple but much more popular selection method is the so-called dwell time, i.e., if the user's fixation exceeds a predefined threshold, a selection is triggered [14]. The dwell time selection method can be improved with visual feedback about the dwelling progress, a preview of the effect of the selection in the area of dwelling, and audio output. Moreover, dwell time can be adjusted to fit the user's expertise in the interface and may even be automatically deduced for distinct areas of an interface [15].

Lately, eye gestures have become popular. Eye gestures are eye movements that perform a predefined pattern, similar to the pattern lock on phones. The benefit of eye gestures is that they are not relying on accurate gaze signals, because only relative distances of consecutive gaze samples are considered. However, eye gestures might be perceived as not natural when a user is asked to make eye movements without a corresponding stimulus. The eye is not used to move without a point to focus on. Thus, early attempts to use eye gestures for selection [16] or typing [17] required a lot of training. More recent approaches, however, make use of dynamic elements in interfaces as stimuli which a user may follow for selection [18]. Eye gestures bear a huge potential for all kinds of eye-controlled interaction, as they allow gaze-based selection with something as small as a wristwatch [19].

A lot of research has considered additional input devices for selection, leaving only the pointing control to eye tracking. Obviously, eye tracking has been combined with mouse [20,21] and keyboard [22]. Recently, also touch [23,24], and voice [25] modalities have been integrated with eye gaze.

5.2.2 Unimodal interaction

The most challenging use of eye tracking as an input method is using eye gaze alone for unimodal interaction. Using eye gaze alone allows for contact-free interaction, even in noisy environments. It may be used for public ticket machines, information monitors, clinical environments, and to improve accessibility in general. Interaction with eye gaze alone has to cope with both the Midas touch and the limitations in accuracy.

5.2.2.1 Eye pointing

A simple but effective approach for pointing with eye gaze is to employ a multistep approach of dwell time and zooming. Lankford [26] has proposed a four-step approach to emulate mouse events on a classical desktop interface. The user first looks at the area of interest, which contains the target. After a dwell time, a window pops up near the center of the screen. The region around which the user was fixating on appears magnified in this window. The user then fixates on the point in the window where they wished to have performed a mouse action. After another dwell time, a menu pops up, offering six different mouse actions as buttons. Again, after a dwell time on the button corresponding to the desired mouse action, the pointing process is completed. Further work has proposed different approaches for the zooming, e.g., in the appearance of a fisheye lens [27], yet, the principle remains the same.

Besides zooming into the whole interface, there are approaches to zoom or scale only some parts of the interface. The eye movements as caused by the zoom or scale functionality are then used to improve the accuracy. Špakov and Miniotas [28] have proposed a system that selects items in a drop-down menu more accurately. The item on which the POR is sensed by the eye-tracking system does vertically expand, while the text within the item stays on the same screen position. When there is an inaccuracy in the estimation of the eye gaze, the actual POR might be on an item that is below or above the expanded item. The expansion of the item would move the desired item on the screen and a user's eye gaze would follow that movement. This can be recognized as a relative shift in the gaze signals. Then the gaze signals are adapted accordingly and the desired item can be selected.

Another approach is to separate pointing and selection not only time-wise but also spatially, which allows for reducing dwell times drastically. Lutteroth *et al.* [29] mark interactive elements, e.g., hyperlinks on webpages, with unique colors. On the side of the screen, each color occupies a distinct area. A user can first look at the desired hyperlink, and then next the area which is at the side of the screen with the color corresponding to the highlight color of the hyperlink.

Eye gestures may eliminate the need for dwell time, why they are potentially more efficient to use. GazeEverywhere [18] by Schenk *et al.* is an approach using eye gestures to perform selections and improve accuracy. After the fixation at a POR on the screen, two moving dots appear above and below the fixated point. Movement of dots and relative eye gaze can be compared to know that the user focuses on one of the dots and mouse click on the targeted coordinates is performed. Furthermore, the offset of POR and dot, orthogonal to the movement of the dot, can be entered into an offset grid that improves the accuracy of gaze signals. Because only the offset in one direction can be detected at once, the moving direction of the dots has to alternate between a horizontal and vertical moving direction.

5.2.2.2 Eye typing

Dwell time as a selection method can be also used to implement an eye typing keyboard. Lankford [26] has introduced an eye-controlled keyboard that displays a button for each key, which can be selected via dwell time. Additionally, a dictionary provides choices that match the characters typed by the user. The choices are displayed on further buttons on the screen, which can be selected using dwell time.

Over the last years, researchers put much effort to accelerate eye typing with better integrated suggestions in order to save required dwell times. Diaz-Tula *et al.* [31] improve the throughput by augmenting keys with a prefix, to allow continuous text inspection, and suffixes to speed up typing with word prediction. The system is designed to limit the visual information to the foveal region to minimize eye movements. They both integrate word suggestions in a frame on the right side of the keyboard as well as include prefixes around the key to exploit the foveal region of visual perception. Current works even suggest integrating suggestions into the keys [32]. A pilot study reveals that 54.4% of all the word suggestions were selected via gazing on the keys, one participant achieved a maximum of 92.6% (Figure 5.3).

There are also systems to integrate eye gestures instead of dwell time into eye typing. Kurauchi *et al.* [33] have designed a keyboard that is inspired by swiping on touch screen keyboards. An eye gesture indicates the first and last characters of a word, while the in-between characters are entered by glancing over the keys in the keyboard. An eye-controlled version of the well-known Dasher keyboard [34] has been also successfully evaluated.

5.2.3 Multimodal interaction

Multimodal interaction usually uses gaze signals to pick one from the available targets, while the selection is performed with another modality. Multimodal interaction has to cope with the limitations of accuracy in eye tracking, but not Midas touch.

(a) (b)

Figure 5.3 Eye typing interface of GazeTheKey [32]: (a) suggestions are embedded to keys and (b) multi-step dwell time is used for selections of firstly the letter and secondly the suggestion

Gaze signals can be combined with physical pointing devices. Zhai *et al.* [20] developed a system which wraps a cursor toward the POR and allows for fine adjustments and selections with manual input devices, e.g., relative mouse movement and mouse clicks. They deduced that the input speed is similar to using the mouse only, however, the users subjectively tended to feel faster using the multimodal pointing techniques. Further developments have extended the approach with a touch-sensitive mouse [21].

Gaze signals can be also combined with a physical keyboard. Kumar *et al.* [22] proposed a look-press-look-release method. A user looks at the desired target. A key press on the keyboard lets us the system magnify the area around the desired target. A user then can refine her selection in the magnified area and release the key, which confirms the selection. This approach is very fast and reliable.

Gaze signals have been also combined with touch input. Pfeuffer and Gellersen [23] have presented various combinations of gaze signals and touch input. They redirect touches to the gaze target and, thus, provide whole-screen reach while only using a single hand for both holding the touch screen device and touch input selections.

5.2.4 Emulation software

Eye pointing and eye typing may be combined into an emulation software that can emit mouse and keyboard events. Emulation software can provide people with motor impairment opportunity to control a broad set of computer applications. Various commercial emulation softwares with eye tracking are available, e.g., Tobii Windows Control, myGazePower, or Microsoft Windows Eye Control. In addition, both researchers [26] and open-source developers [30] have created similar projects (Figure 5.4). However, emulation is slower than native control with physical devices, more error-prone, and adds more mental load to a user [35]. The emulation software

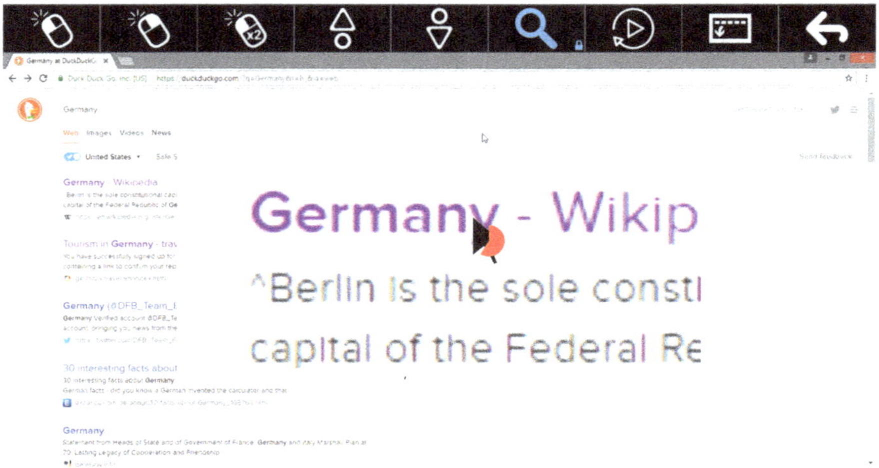

Figure 5.4 Pointing in the OptiKey software [30]. The pointing is implemented as a two-step dwell-time process. The first dwell time triggers a magnification, the second dwell time confirms the selection of the fixated element

acts as a command-translation layer between the eye-tracking environment and interfaces that expect traditional input device events. A user must first imagine which event from the physical input device the interface expects and emulates the corresponding input device accordingly. There are various programming interfaces to improve the workflow with an emulation, however, these programming interfaces often lack a complete cover of the available options. Hurst *et al.* [36] have studied the coverage of accessibility APIs, finding approximately 25% of the widgets missing from the API's view of many common interfaces. Thus, we argue that only interfaces which are designed for eye-controlled interaction can yield the full potential of gaze signals as an input channel.

5.3 Adapted multimedia interfaces

Eye-controlled interactions can be composed toward interfaces to access and manipulate multimedia content. In this section, we discuss adaptation of interfaces to be operated with gaze signals, which we call *gaze-adapted interfaces*. First, we take a look at applications with a single-purpose whose interfaces have been adapted to be operated with gaze signals. Then, we present a framework for eye-controlled interactions, which allows for designing custom eye-controlled interfaces. Last, we show how interactive elements in the Web can be mapped to eye-controlled interfaces. The mapping of interaction elements from the Web to eye-controlled interfaces allows for accessing websites through eye movements in a unified way.

5.3.1 Adapted single-purpose interfaces

Researchers have created and evaluated various multimedia interfaces that were designed for eye-controlled interaction. These interfaces are mostly prototypes, which are limited in their functionality for a single purpose and often not maintained after the experiments. Nevertheless, their designs provide insights into the potential of interfaces that are adapted to work effectively with gaze signals as input.

5.3.1.1 Drawing with eye movements

Drawing allows people to express their creativity and feelings. Especially, people with motor impairment, and, even more specifically, children with motor impairment are considered a target group for eye-controlled drawing applications. Hornof and Cavender [37] developed EyeDraw, an application that enables children with severe motor impairment to draw with their eyes. The authors propose an interface with two states. One state allows for free looking and the other state for drawing. A user defines the start and end of a line through fixations instead of drawing pixel-by-pixel like a brush using eye gaze. Transition between the states is performed via dwell time on the canvas. The application allows for drawing lines, circles, rectangles, polygons, and stamps. The tools are available as dwell time buttons that surround the canvas, see Figure 5.5. A grid of dots can be displayed on the canvas to help a user fixating a POR. Furthermore, a dedicated mode for looking has been implemented. Any dwell time interaction on the canvas can be paused through the selection of a button in the interface.

The idea of a drawing application with gaze signals has been enhanced in multiple works. Heikkilä [38] has developed EyeSketch. The application allows the drawing of shapes that can be moved and resized, and their color attributes can be changed. The tools for moving and resizing are controlled with gaze gestures and blinking.

Figure 5.5 Interface of EyeDraw 2.0 [37]

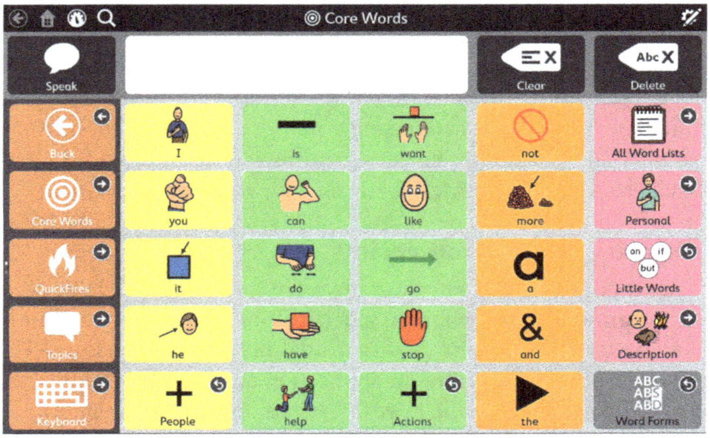

*Figure 5.6 Interface of Tobii Snap + Core First for Windows**

5.3.1.2 Writing with eye movements

Inserting text for data processing, documentation, and communication is the traditional purpose of desktop computers. Communication has been in the focus of eye-controlled applications for people with motor impairment. There are various commercial grid-based applications available. They offer buttons in a grid-like interface, which can be selected through dwell time. The buttons represent common phrases, which allow to formulate queries of everyday communication. There are systems with a deep hierarchy of buttons, customizable grid layouts, eye typing, and combinations with speech and touch input. See Figure 5.6 for a representative example of an application with grid-like interface.

Recent research in writing with gaze signals has rather focused on multimodal interaction as it bears huge potential for a broad user audience. Beelders and Blignaut [39] have enhanced Microsoft Word with speech and eye gaze as input. They have combined speech commands for formatting, cursor movement, text selection with speech dictation, and eye gaze for pointing, e.g., cursor placement in the document and eye typing. The system is highly configurable, as different selection methods for eye pointing are available, like dwell time, blinking, or via key press. However, it was found that the eye gaze and speech-interaction technique caused a significantly higher error rate than the traditional keyboard.

Sindhwani *et al.* [40] have developed ReType, a system for quick text editing with keyboard and eye gaze. The system is gaze-assisted and attempts to erase the need for mouse pointing in certain scenarios. It allows common editing operations, while the hands can remain on the keyboard. A text editor is enhanced with a specific mode to retype. The mode to retype can be entered via a hotkey or via heuristics on

*Image source of Tobii Snap + Core First for Windows: https://www.tobiidynavox.com/globalassets/pictures/software/snap/product-listing-images/scf-2.jpg.

the gaze signals. A user can start to type an edit string, which is matched with the complete text. Phrases with low string distance are highlighted as candidates to be replaced with the edit string. The user looks at the desired candidate and presses a key to apply the edit.

5.3.1.3 Gaming with eye movements

Games have the potential to introduce many people in a playful way to interaction with eye tracking. Furthermore, the gaze signals of players can be used to retrieve a deeper understanding of human perception. However, it is often difficult to adapt games for eye-controlled interaction due to their complexity and frequency in input commands.

Games for entertainment. Vickers *et al.* [41] have evaluated locomotion tasks in games, in which an avatar can be moved by players over a terrain. They have divided the screen into multiple rectangular zones which control the movement of the avatar, e.g., moving it forward or changing its direction. The zones are triggered on gaze signals that fall into the extents of a zone. This interface design allows to steer an avatar in a conventional computer game. However, the interaction with eye tracking is slower and more error-prone than with input by mouse and keyboard. In addition, games usually require broader set of interactions, which is not yet covered by this approach.

Chicken Shoot is a classic two-dimensional shooter game, in which the player must hit flying chickens with a shotgun. Usually, it is played with a mouse. Aiming with the cross-hair is performed through the movement of the mouse, and shooting and reloading with the mouse buttons. Isokoski *et al.* have translated gaze signals into mouse and keyboard events and then inject these into the game. The cross-hair is changed to follow the eye gaze. The shooting and reloading via mouse buttons have been replaced with a switchable automatic machine gun. An off-screen target above the screen region can be fixated to switch the gun on and off. After four to five trials, most participants outperformed the mouse and keyboard control condition in the final score. Especially, the faster positioning of the cross-hair by eye gaze compared to the manual positioning improved the score.

Isokoski *et al.* [4] also have discussed EyeChess, developed by Špakov. Turn-based games and other games that do not require high-frequency control can be easily adapted for interaction via gaze signals. However, the user experience can be significantly improved by adapting the interface itself for eye-controlled interaction. The elements in the interface of EyeChess have been made large enough to be selectable through dwell time without additional magnification. Furthermore, the elements provide feedback about selection and feature points to focus on the center. The changes in comparison to a mouse-operated chess application are not large, yet, they are evaluated as critical to allow a satisfactory control with eye movements. See Figure 5.7 for a screenshot of the interface of EyeChess.

Games with a purpose. In the context of research, games with a purpose are popular to offer knowledge to players and to retrieve feedback from them in a playful way. In combination with eye tracking, saliency information from eye gaze on images can be

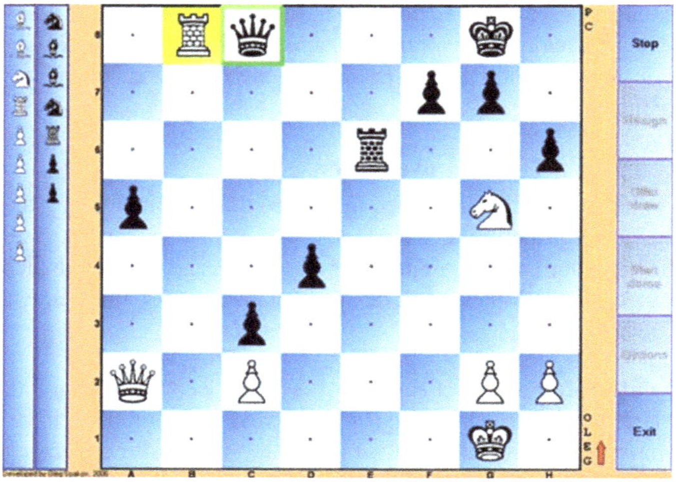

Figure 5.7 Interface of EyeChess [4]

gathered by users looking at presented images. The saliency data may be useful to tag images and improve image search in the future.

Walber *et al.* [42] have developed a two-dimensional eye-controlled game, eyeGrab, for image classification. Players classify a number of images according to their relevance for a given tag. While the game is entertaining for players, the aim is to enrich the image context information. Players are asked to look at images falling down the screen. The task is to classify the images as relevant or irrelevant to a given tag. A player can select an image by fixation and classify the image by dwelling on a corresponding button. A player receives points for each correctly categorized image, negative points for each false categorized image, and no points for images that fell off the screen without classification by the player. The speed of the images that fall is increased over time, in order to increase the difficulty of the game.

We have developed a three-dimensional successor to the game, Schau genau! [43]. In the game, the gaze signals are used to control an avatar in appearance of a butterfly, see Figure 5.8 for the setup and a screenshot of the game. The task of a player is to collect flowers with the avatar in order to gain points. The flowers are spawned in the distance and the avatar flies constantly toward the flowers over a meadow. The user interface of the game consists only of two panels on the bottom of the screen. The collected points are displayed in the green panel on the right. The current multiplier is displayed in the purple panel on the left. The game terminates when the avatar is caught by spider web spawned analogously to the flowers. During the game, both game speed and spider-web density linearly increase to make the game more difficult. The research purposes of the game are threefold: we have integrated control styles with different levels of intelligence, we educate the players about flower species as the serious game part, and we collect gaze signals on photos of flowers.

(a) (b)

Figure 5.8 The Schau genau! [43] arcade-box was placed at a state horticulture show in Germany for half a year: (a) arcade-game like box and (b) screenshot from the gameplay

The controls of the avatar with gaze signals are of special interest in this game. The avatar moves accordingly to the active control style on a two-dimensional plane, which is placed parallel to the screen. Three mappings of gaze-on-screen to world position of the avatar on that plane, featuring different levels of intelligence, have been defined and were randomly assigned to players. The first approach is the interpretation of the gaze on screen as avatar position, similar to the mouse emulation. The second approach features a grid-based positioning, similar to the first approach but with very coarse fixation filtering. The third approach supports the player by making use of the knowledge about the virtual world. The eye gaze is checked to lay upon a flower in the distance, and the avatar is moved toward the future collision point with the flower. The visual position of the flower in the distance and the future collision point are not the same, as the perspective projection moves objects in the distance closer to the center of the scene. When the eye gaze is upon the avatar, its opacity is reduced to enable players to see what would have been hidden by the avatar.

The serious part of the game educates the players about flower species, who are rewarded with an increase of the multiplier for knowledge. After a certain time interval, the game state switches from normal game into a picture mode (Figure 5.9(a)), in which one tag and two images of flowers are presented to the player. The player shall select the image that depicts the flower which corresponds to the displayed tag. For each correct selection, the multiplier is increased. During this decision process, gaze signals on the two images are collected. These gaze signals may provide insightful details in future, like which portion of the image led the player to identify the flower's species.

In this immersive eye-controlled game environment, several interaction elements were included with respect to the size, shape, and visual feedback. The photo in Figure 5.9(b) shows the game screen for the player inserting a nickname for the high score table. The alphabet is displayed on the top where the player can scroll horizontally through the letters by fixating on specific letters. The fixated letter moves toward the center of the screen and enlarges until a dwell time is over and the letter

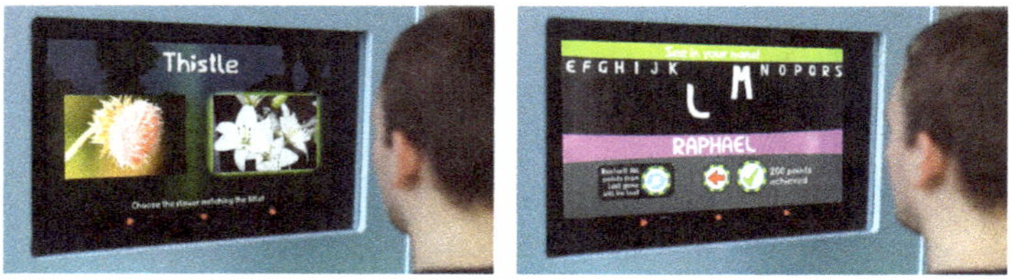

(a) (b)

Figure 5.9 Schau genau! [43] game with eye-tracking system mounted below the monitor (the three red illumination units are visible): (a) picture mode and (b) entry of a nickname

is selected. If the player fixates on another letter in the meantime, the previous letter is scaled down again. On the bottom of the screen, the player can either confirm the input or delete the last written later by individual buttons. All these interface and interaction components of Schau genau! were very well received by the participants yielding to high usage and explicit positive feedback.

The game has been presented in an arcade box made of wood. The box includes a height-adjustable chair in front of a screen, which is placed behind glass. A Tobii EyeX eye-tracking device is attached to the lower part of the screen's frame. For the sole purpose of starting and aborting the game, a single red buzzer has been placed on a tray between the chair and the monitor. Nearly 3,000 completed sessions were recorded on a state horticulture show in Germany, which demonstrates the impact and acceptability of eye-controlled interaction among lay users as implemented in our interface. The control style had a statistically significant, yet very small effect on high scores: games with direct positioning led to the lowest, games with indirect positioning to the highest scores. This is another hint about the importance to consider gaze signals not as replacement of mouse pointing but to treat it specifically in each context.

5.3.1.4 Social media with eye movements

As of today, social media allows to share multimedia at a big scale. One of the biggest social networks is Twitter, a platform that allows users to share their opinion, to follow the posts of other users, and to post media on a personal wall. Other users may respond to entries on another user's personal wall and forward entries on the same of their own. Social media platforms may act as an open window to the world for people with motor impairments—if they are able to use such services. Usually, Twitter is accessed via a Web interface or smartphone applications. We have developed a stand-alone Twitter application with an eye-controlled interface [35].

Design and positioning of interactive elements are a significant aspect of gaze-adapted interfaces. Unintended activation of interaction confuses users and has negative impact on the user experience.

Figure 5.10 Interface of the eye-controlled interface in our Twitter application [35]

The most important part of Twitter† is the personal wall. The personal wall contains posts from users that the user follows. See Figure 5.10 for our implementation of a gaze-controlled personal wall. The posts are presented in the center of the screen, called *content area*. A user can select a post in the content area by just focusing it, e.g., while reading the content. When a post at the bottom or top is selected, the system scrolls the post toward the center of the view. This behavior allows intuitively to browse through the available posts. The currently selected post is colored in dark gray, which informs the user about the selection and connects the post to the available actions in the *action bar*. The action bar is placed on the right-hand side for the content area and provides contextual actions for the selected post. This spatial separation of content and actions allows a user to scroll through the content without triggering unintended actions. Most of the functionalities of Twitter have been implemented. Each selected post can be forwarded, responded to, liked, and unliked, and one can visit the profile of the user who created the post. Actions that require textual input automatically present a virtual keyboard for eye typing, which works with dwell time keys.

In addition to the personal wall from Figure 5.10, there are five other screens that can be accessed with the global navigation on the top of the screen. Most screens of the interface share the concept of *content area* and *action bar*:

- send a tweet
- discover trends
- visit own user profile
- private messaging for direct user communication
- search and view other profiles

†https://twitter.com

In order to develop an eye-controlled interface for an existing application, one needs low-level access to the functionalities. Twitter offers a public representational state transfer (REST) API[‡] for developers to access a user's information and to submit tweets and messages. For this, a user has to activate communication over this REST API,[§] which is performed in our application at the first successful login.[‖] The API follows the OAuth[¶] protocol. There are various libraries[**] available to access the REST API in a convenient way in different programming languages. Since our application is written in C++ language, we decided to delegate the twitcurl[††] library for all communication with the social network.

We have performed a comparative evaluation with 13 participants. The experiment was designed to compare objective and subjective user experience between using the mobile webpage with emulation and our eye-controlled interface. The participants were asked to perform specific tasks representing the common social media usage: to write a post and to publish it, to find a particular user and follow another user, to find and like a certain tweet about a specific topic, and to explore the application like one would do for social media browsing (5–10 min). We reported that our eye-controlled interface is regarded as more intuitive and easier interpretable by the participants than the emulation. Despite its novelty, the novel interface performed well on usability analysis and required less mental demand from the participants. The result implies that gaze-adapted interfaces with a tight integration of gaze signals must be considered for enhanced usability and performance of eye-controlled interfaces.

5.3.2 Framework for eye-controlled interaction

Most adapted single-purpose interfaces share similar mechanisms, like dwell time buttons and feedback about selections. Thus, we have developed a framework, called eyeGUI [44], that makes it easy to compose eye-controlled interfaces using interface elements that feature these kinds of mechanisms.

5.3.2.1 Gaze-adapted interface with eyeGUI

To cater with the eye-tracking accuracy limitations and interaction issues like Midas touch problem [14], eye-controlled applications primarily depend on the presentation, manipulation, visual cues, and feedback of elements in the interface. The eyeGUI framework provides a variety of interface elements, like buttons, keyboards, images, text displays, and text editors, which allow to compose interfaces for many different kinds of applications. The interface elements can be grouped into layouts and further ordered with grids, stacks, and scrollable overflows. The elements can be customized in their size, appearance, or behavior, e.g., buttons can be given an arbitrary icon that

[‡]https://dev.twitter.com/rest/public
[§]https://apps.twitter.com
[‖]https://dev.twitter.com/oauth/application-only
[¶]https://dev.twitter.com/oauth
[**]https://dev.twitter.com/overview/api/twitter-libraries
[††]https://github.com/swatkat/twitcurl

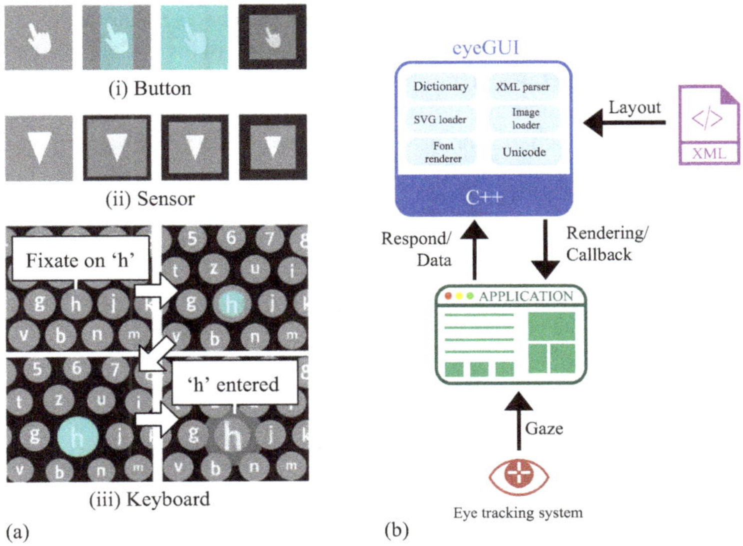

Figure 5.11 eyeGUI! [44] framework for eye-controlled interfaces: (a) interface elements in eyeGUI and (b) architecture of eyeGUI

scales automatically when the overall size of the interface changes, though the ratio of height and width of the image stays the same.

All interface elements in eyeGUI are designed especially for eye tracking in their size, appearance, and interaction, e.g., buttons get activated when the gaze hits them for a set dwell time, and they shrink after triggering to provide a user with feedback about the activation. A colored overlay increasing in size works as a visual representation of the remaining dwell time until the activation. The screenshots in Figure 5.11(a) show the interaction with three different elements from the eyeGUI framework. On the top (i), the states of a button are depicted. After the user fixates on the button, a cyan overlay fills the button. When the dwell time is over, the cyan overlay completely covers the button, vanishes, and the button performs a press animation. In the second row (ii), the interaction with a sensor is shown. A sensor is an interaction element that instantly reacts to gaze and spawns signals to the application. The longer the gaze lays upon the element, the stronger the signals become. This instant interaction is useful, e.g., for smooth scrolling of content. The four images at the bottom (iii) show different stages of eye typing with magnifying effect in the character selection. Besides the interface elements, the framework also offers other eye-tracking-specific features, like displaying the gaze path during operation.

5.3.2.2 Architecture of eyeGUI

The framework is developed in C++ 11 and uses OpenGL for rendering. Each layout and its contained elements are defined in XML files. Furthermore, colors, sounds, dwell times, and animations can be adjusted in a style sheet. Each interface element can be assigned classes from the style sheet and classes in the style sheet can inherit properties from each other in a hierarchical manner. All events that are spawned from interactive interface elements are sent to registered listeners, which can handle incoming events on their own behalf. The listeners can be created in the custom application environment in order to interact with external APIs. Eye-tracking devices (e.g., SMI REDn Scientific, and Tobii EyeX) would send raw gaze signals to the application, which implements a receiver and filter for the gaze signals. A filtered gaze signal is then passed to the eyeGUI framework, which handles interaction and calls events in the interface (e.g., button pressed), and the application would react to those user interactions at calls to the registered listeners. For development purposes, the control paradigm can be switched to mouse-control to emulate gaze signals. Figure 5.11(b) depicts the architecture of the eyeGUI framework.

The integration of eyeGUI is similar to other OpenGL user interface libraries like ImGui[‡‡] and AntTweakBar.[§§] A developer is free to choose how to create a window and in which way to initialize the OpenGL context. Before the render loop is entered, the GUI object for eyeGUI must be instantiated and an arbitrary number of layouts from XML files can be added. During the render loop, for every frame, the most recent gaze sample is used to update eyeGUI, which provides feedback as whether the input has been used by any layout. Based on that feedback, a developer can decide how to update the custom application content. This enables developers to overlay their own rendering with eyeGUI and use the gaze signal not only to allow interaction with eyeGUI but also with their custom interface. All functions are accessible through a single header file with C++ functions and the memory allocation for displayed images and other media content is handled automatically.

The eyeGUI framework has been used by us in two eye-controlled applications. First, the eye-controlled client to access Twitter from the previous section. Second, GazeTheWeb, an eye-controlled Web browser that is discussed in the next section.

5.3.3 Adaptation of interaction with multimedia in the Web

The Web offers a broad range of services and applications, e.g., shopping, communication, learning, and working. Although such tasks appear to be very different, the underlying data structure on the client side is built with HTML, CSS, JavaScript code, and the interactions with the pages are designed to be performed with mouse, keyboard, or solely by touch devices. Therefore, a Web browser is able to cover modern computer tasks in a unified fashion, since Web interfaces are written in standardized languages and anticipate the same set of input devices. Hence, we want to adapt the interfaces in the Web uniformly to be conveniently eye controlled. We have developed GazeTheWeb [45], a framework for full-featured Web browsing with gaze signals as sole input. It integrates the visual appearance and control functionality of webpages in an eye-tracking environment. GazeTheWeb has the potential to make a huge number

[‡‡] https://github.com/ocornut/imgui
[§§] http://anttweakbar.sourceforge.net

of Web-based services and applications feel like being designed for eye-controlled interaction.

The Web application environment is built upon the Chromium Embedded Framework[||] with eye-controlled interface [44] overlays for traditional menu functions like page navigation, URL input, bookmarks, and tab management. For webpage interaction, the system examines the location of selectable objects on webpages, such as text inputs, select fields, videos, hyperlinks, buttons, scrollable sections, and edit boxes. The system represents these objects with explicit or implicit indicators to be accessed by eye gaze input. We achieve the adaptation through knowledge about interface semantics, such as the position, size, properties, state, and interaction means of the elements in an application interface. To retrieve the semantics, introspection of the interface allows us to track object properties such as type, location, and status within a known, yet dynamic, system.

5.3.3.1 Eye-controlled Web browsing

There have been some attempts in research to integrate eye tracking to the Web environment. The Text 2.0 framework [46] had introduced a novel benchmark to mix eye-tracking data with Web technology, where gaze-responsive webpages can be implemented via interpreting a new set of gaze handlers (e.g., *onFixation*, *onGazeOver*, and *onRead*) that can be attached to parts of the Document Object Model (DOM) tree and behave similar to existing HTML and JavaScript mouse and keyboard event facilities. Wassermann *et al.* [47] have built upon this concept to enable eye gaze events in eLearning environments. Although it is a pertinent guideline for the Web developers to include eye gaze interactions in their application, it does not resolve the problem of browsing the current Web with eye-controlled interactions. Hence, there is a need of Web extraction methodology to identify the input and selectable objects in Web so that it can be revised to eye-controlled interactions. There have been some elementary approaches in this direction to identify basic elements such as scrolling and hyperlinks [16,29]; however, Web is much more complicated and requires a comprehensive and scalable approach to identify the interactive elements, and to design suitable interaction for all browsing functionalities. None of the approaches allow for a full browsing experience, e.g., they have not integrated eye typing to allow entering text on webpages.

5.3.3.2 Introspection of dynamic web page interfaces

A core contribution of GazeTheWeb is the novel real-time observation of the DOM tree, for changes in interaction elements [48]. Retrieval, classification, and tracking of elements on the webpages, while the webpage is loaded and updated, are desirable to provide an eye-controlled interface that corresponds to the displayed webpage. Today's Web makes heavy use of dynamically loaded content, thus, simple polling of DOM tree parsing at a specific time of execution (like end of initial page loading) is not sufficient. We propose to use direct callbacks from JavaScript into C++ in

[||] https://bitbucket.org/chromiumembedded/cef

combination with JavaScript-side observation of the DOM tree for dynamic changes. This mechanism is combined with a Mutation Observer, which efficiently observes changes in the DOM tree.

5.3.3.3 Gaze-adapted interaction with webpages

Clicking on hyperlinks to navigate to different pages is an essential component of user's Web browsing behavior. For such navigational task, basic dwell time selection strategy does not work, because hyperlinks might be very close to each other. This makes it impossible to precisely choose intended links with fixations and dwell time. In GazeTheWeb, the accuracy problem is tackled using traditional multistep dwell time with magnification. A click mode can be activated with a button in the interface. All hyperlinks on the webpage are highlighted, in order to support the user in the decision process of which hyperlink to select. The magnification after the first dwell time is centered at the fixation. Another fixation on the magnified web page that exceeds the dwell time triggers a click event on the webpage. The system provides visual feedback with a shrinking circle around the click. If there is no hyperlink element at the POR, but there exists a hyperlink in nearby region, the click coordinates are moved to the center of the nearest hyperlink to perform the click.

Typically, webpages exceed the available vertical screen space and users need to scroll the webpage within their viewport to reveal lower parts of the page. Various approaches have been published, like a two-step approach where a user first selects to scroll and is then presented with an element to control speed [16], or an automatic scrolling feature with respect to the POR on the screen [49]. We provide the user with a choice between manual approach with sensor elements that instantly react to fixations, and an automatic approach, where the webpage content beneath the POR is constantly centered on the screen.

Whereas hyperlinks and scrolling are also reasonably supported in the existing eye-controlled Web browsers [16], we have especially adapted the interaction with more complex interaction elements, like text inputs, select fields, and videos. Each of these elements is overlaid with a gaze-sensitive icon on the webpage and can be selected through a fixation. The system knows the type of the selected element and switches to a dedicated interface mode that allows a user to spawn the events as desired by the element through eye movements. See Figure 5.12 for a webpage displayed in GazeTheWeb, augmented with gaze-sensitive icons, and the corresponding interface modes in Figure 5.13.

(a) A text input is a webpage interaction element that allows a user to insert text for a search query, information transactions like filling a form or sending a chat message. In conventional interaction, text insertion is directly performed via keyboard. However, in gaze-based interaction, sophisticated eye-typing techniques are required to translate gaze signals into keystrokes. Most eye-typing interfaces provide one dwell-time sensitive key per available letter, why eye typing can be considered as a complex interaction. We provide the user with an interface mode featuring a virtual keyboard for eye typing, including the option to directly submit a query without any additional selection on the webpage. Additionally,

Figure 5.12 *Complex interaction elements on a webpage are augmented with gaze-sensitive icons to handle (a) text input, (b) select fields, and (c) videos*

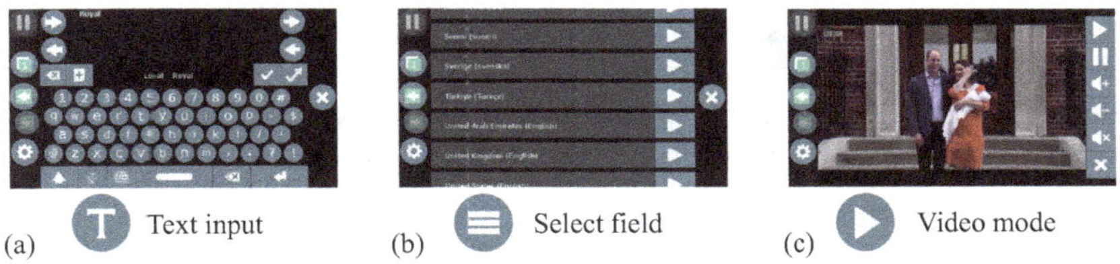

Figure 5.13 *Gaze-sensitive icons on a webpage are associated with gaze-controlled interface modes for sophisticated gaze-based interaction: (a) virtual keyboard for text input, (b) list of options from select field, and (c) mode with video controls*

other semantics of the text input can be queried for further optimizations of the virtual keyboard, e.g., to automatically provide a discrete and secure way to insert passwords or to determine the expected language to support the user in the typing task.

(b) A select field is a webpage interaction element that allows the user to select among predefined options, like date of birth, search result filters, or localization. The select field offers multiple options for the user to choose from, and hence we categorize it as a complex interaction element. We present the available options in an interface mode with a button associated with each option, to allow for a convenient gaze-based interaction. If the number of options extends beyond the vertical screen space, the currently inspected option is vertically moved to the center of the screen and the interface mode automatically reveals further options in the direction of movement through scrolling.

(c) A video is a webpage interaction element that embeds video content into a webpage. Videos offer multiple options for control, e.g., play, pause, or skip, which makes the video element a complex interaction element according to our definition. Videos are especially cumbersome to interact with in an emulation, as the controls of video on a webpage usually disappear when no mouse movement is detected over the video by the Web browser. However, mouse movement in emulation is only inherently performed through positioning the mouse cursor to emulate a mouse click event. Instead, we suggest an eye-controlled interface mode that shows the video instead of the webpage viewport and features associated with buttons to control the video.

The interaction with the previously listed complex interaction elements has been gaze-adapted for the scope of our experiments, however, adaptation is not limited to text inputs, select fields, and videos. In a similar fashion, interface semantics about less common but standardized interaction elements like sliders or audio players might be defined and their interaction adapted with dedicated eye-controlled interface modes.

The frequent usage of GazeTheWeb during the MAMEM field trials, including the visits of various Web domains, pages, and the activities (e.g., clicks and text entry), indicates that users were able to interact with the variety of pages and perform desired browsing operations [50]. Furthermore, the longer stay duration and revisits of popular platforms like Facebook, Gmail, YouTube (incorporating complex interaction elements in Facebook posting, mail body, video control) imply that GazeTheWeb could effectively adapt the associated elements and webpages for gaze interaction.

5.3.3.4 Challenges and limitations

We have developed an effective architecture to include gaze signals for interaction in modern webpages. However, Web is complex and it brings several challenges due to the non-standardization or unintended use of Web technologies. Here we describe some of these limitations and future challenges to improve the interaction further.

Semantic information. For a convenient experience with eye input, we aim to provide different options to end users, for example, the text input action can apply the typed text to the input field or it can be directly submitted for processing (e.g., search query). However, in several instances, input submission depends on the input of other fields, for example, a password cannot be submitted without providing a username first, or a registration form requires multiple entries. In such scenarios, deactivating the submit button or automatic transformation to a "next-text-input-field" function would provide a more user-friendly interaction. To accomplish this, we need precise semantic information about the input fields, which is hard to extract from the DOM tree. We could employ trivial rules, for example, in the scenario of username and password combination, search for DOM nodes that have the tag "password" and other input fields within the same form. However, this is just a heuristic and does not work out for all cases. A more robust approach is required to gauge the semantic information.

Text input field density. Specific forms for registration or communication tools feature a high number of text input fields, which are represented by gaze-sensitive icons in GazeTheWeb. These gaze-sensitive icons have to be rendered at a minimum size in order to cope with the eye-tracker accuracy. Hence, in the scenario of complex forms, the icons might overlap with each other and interaction becomes difficult and error-prone.

Extensive use of CSS and JavaScript. The abuse of modern Web technologies limits the impact of DOM tree parsing. A well-known example is the search input field on the Google page.¶¶ At the time of our assessment, there are at least two text input fields stacked onto each other for the search box. It appears that one field is responsible for displaying suggestions in gray color and the other one is used for the actual user input. The webpage uses the CSS mechanism of *z*-index to advise the browser to render the actual text input field in front of the other, which is not detectable in simple DOM tree parsing. One has to do additional CSS property look-ups in order to find the text input field to be filled by the keyboard. It is not just sufficient to check the *z*-index, since there may be multiple accessible text input fields and some hidden ones, which are not accessible to a user but share the same *z*-index. At present, we do an in-depth investigation to resolve such issues for popular websites, however, a universal solution is required to deal with these scenarios.

For highly customized interaction contexts (e.g., map navigation, document editing and game controls), we imagine a future research approach would employ automatic interaction-template matching, i.e., to cater to different Web services with similar interaction behavior. This would optimize the engineering effort required to suffice GazeTheWeb adaptation for highly customized pages. For example, there are common patterns of interactions for map navigation (panning and zooming), regardless of the website that is offering the corresponding service like Google Maps, Bing Maps, or Open Street Maps. An intelligent interaction-context recognition would detect the context of map navigation on these sites and offer a unified, gaze-adapted interface mode for map navigation. The interface mode would be the same for all map services, yet spawn interaction events as expected by the different services.

5.4 Contextualized integration of gaze signals

Gaze signals can be used not only to mimic traditional input devices and interactions but to allow for novel interactions with multimedia, which are specific to eye gaze [1].

5.4.1 Multimedia browsing

The act of scrolling is strongly coupled with the user's ability to engage with information via the visual channel. Therefore, the use of implicit eye gaze information is a natural choice for enhancing multimedia content browsing techniques. For scrolling and reading, we propose using natural eye movements to control the motion of the interface for the user. In an automatic scrolling mode, the user has a smoother and more

¶¶https://www.google.com

natural reading experience, because the scrolling is supported via implicit observation of a user's POR. The scroll direction is determined by noting the quadrant where the user is currently looking. The scroll speed is proportional to the distance from the center of the screen.

5.4.2 Multimedia search

Another focus of our work is to employ implicit gaze patterns for an improved search experience. Essentially, a user implicitly requests better visual media, like images and videos. This feedback is inferred from gaze signals, while the user looks at the media. This could significantly reduce a user's effort for explicit refinement of search results. There have been some preliminary studies related to the use of implicit gaze information in image retrieval [51]. However, an eye-controlled interface provides a more natural scenario of gathering implicit feedback to enhance the results, while users are actively engaged in gaze-driven searching and browsing.

5.4.3 Multimedia editing

Gaze signals allow to recognize a user's region of interest. We plan to use such fixation data to enhance multimedia interaction, for example, to identify important content and use it for multimedia editing, such as image cropping. The goal is to create appealing crops without explicit interaction. Furthermore, precise identification of relevant image content without explicit interaction is a vital feature. It lets us analyze and quantify the viewing behavior on images and how users select the region of interest for image editing. It also lets us analyze other useful functionalities, such as the automatic creation of snapshots or thumbnails for adaptive Web documents. In this context, we have already conducted experiments and employed human fixation patterns to identify the most salient region of images, because defining a good crop requires a model that explicitly represents important image content. Our analysis of the Schau genau! game data implies that human fixations are very particular in identifying important image content. Implicit gaze feedback is also relevant in personalizing the user experience—especially given that the sensory input performance largely depends on the individual's ability to process the cognitive information. For example, some users might be able to easily select the target with smaller dwell times compared to the time needed by others.

5.5 Summary

Eye-tracking hardware becomes more and more available. Table-mounted eye trackers that are designed for gaming purposes are already fairly cheap, with prices of about 150 dollars. Headsets for augmented and virtual reality like the Microsoft HoloLens 2 also include eye tracking and might have a big impact on the availability of gaze signals in the near future. In this chapter, we have provided a comprehensive overview about eye-controlled interactions for multimedia interfaces. We have started with the anatomy of the eye and technologies to track eye movements, have discussed selection methods, and have described the design of eye-controlled interfaces. We conclude that eye-controlled interaction requires dedicated interfaces for good user experience, because of the overload in perception and control, and gaze signal estimation errors. We argue that the Web has many standardized elements for which we can make dedicated eye-controlled interface modes to support a broad range of applications. Upcoming interface toolkits must consider gaze signals for a deeper integration. The Windows API*** already includes handling of gaze signals; thus, eye tracking will be even more available in the future. Especially, the interaction with multimedia can benefit from gaze signals. They potentially allow for magic-like interaction, for example, with automatic scrolling or automatic cropping of content-of-interest. Gaze signals are already used as unimodal interaction means to improve accessibility and bear a huge potential for everybody's life with multimodal interaction and well-designed interfaces.

References

[1] Kumar C, Menges R, Staab S. Eye-Controlled Interfaces for Multimedia Interaction. IEEE MultiMedia. 2016;23(4):6–13.

[2] Duchowski AT. Eye Tracking Methodology: Theory and Practice. Berlin: Springer-Verlag; 2007.

[3] Drewes H. Eye Gaze Tracking for Human–Computer Interaction; 2010. Available from: http://nbn-resolving.de/urn:nbn:de:bvb:19-115914.

[4] Isokoski P, Joos M, Špakov O, *et al.* Gaze Controlled Games. Universal Access in the Information Society. 2009;8(4):323–337. Available from: http://dx.doi.org/10.1007/s10209-009-0146-3.

[5] Gips J, Olivieri P. EagleEyes: An Eye Control System for Persons with Disabilities. In: Proc. 11th Int. Conf on Technology and Persons with Disabilities; 1996. p. 13.

[6] Van Der Geest JN, Frens M. Recording Eye Movements with Video-Oculography and Scleral Search Coils: A Direct Comparison of Two Methods. Journal of Neuroscience Methods. 2002;114(2):185–195. Available from: http://www.sciencedirect.com/science/article/pii/S0165027001005271.

[7] Ohno T, Mukawa N, Yoshikawa A. FreeGaze: A Gaze Tracking System for Everyday Gaze Interaction. In: Proceedings of the 2002 Symposium on Eye Tracking Research & Applications. ETRA'02. New York, NY: ACM; 2002. p. 125–132. Available from: http://doi.acm.org/10.1145/507072.507098.

[8] Salvucci DD, Goldberg JH. Identifying Fixations and Saccades in Eye-Tracking Protocols. In: Proceedings of the 2000 Symposium on Eye Tracking Research & Applications. ETRA'00. New York, NY: ACM; 2000. p. 71–78. Available from: http://doi.acm.org/10.1145/355017.355028.

[9] Feit AM, Williams S, Toledo A, *et al.* Toward Everyday Gaze Input: Accuracy and Precision of Eye Tracking and Implications for Design. In: Proceedings

***https://docs.microsoft.com/en-us/windows/uwp/design/input/gaze-interactions

of the 2017 CHI Conference on Human Factors in Computing Systems. CHI'17. New York, NY: ACM; 2017. p. 1118–1130. Available from: http://doi.acm.org/10.1145/3025453.3025599.

[10] Schwarz J, Hudson S, Mankoff J, *et al.* A Framework for Robust and Flexible Handling of Inputs With Uncertainty. In: Proceedings of the 23rd Annual ACM Symposium on User Interface Software and Technology. UIST'10. New York, NY: ACM; 2010. p. 47–56. Available from: http://doi.acm.org/10.1145/1866029.1866039.

[11] Kumar M, Klingner J, Puranik R, *et al.* Improving the Accuracy of Gaze Input for Interaction. In: Proceedings of the 2008 Symposium on Eye Tracking Research & Applications. ETRA'08. New York, NY: ACM; 2008. p. 65–68. Available from: http://doi.acm.org/10.1145/1344471.1344488.

[12] Kobayashi H, Kohshima S. Unique Morphology of the Human Eye and Its Adaptive Meaning: Comparative Studies on External Morphology of the Primate Eye. Journal of Human Evolution. 2001;40:419–435.

[13] Bolt RA. Gaze-Orchestrated Dynamic Windows. In: Proceedings of the 8th Annual Conference on Computer Graphics and Interactive Techniques. SIGGRAPH'81. New York, NY: ACM; 1981. p. 109–119. Available from: http://doi.acm.org/10.1145/800224.806796.

[14] Jacob RJK. What You Look At Is What You Get: Eye Movement-Based Interaction Techniques. In: Proceedings of the SIGCHI Conference on Human Factors in Computing Systems. CHI'90. New York, NY: ACM; 1990. p. 11–18. Available from: http://doi.acm.org/10.1145/97243.97246.

[15] Chen Z, Shi BE. Using Variable Dwell Time to Accelerate Gaze-Based Web Browsing with Two-Step Selection. International Journal of Human–Computer Interaction. 2019;35(3):240–255. Available from: https://doi.org/10.1080/10447318.2018.1452351.

[16] Porta M, Ravelli A. WeyeB, an Eye-Controlled Web Browser for Hands-free Navigation. In: Proceedings of the 2nd Conference on Human System Interactions. HSI'09. Piscataway, NJ: IEEE Press; 2009. p. 207–212. Available from: http://dl.acm.org/citation.cfm?id=1689359.1689396.

[17] De Luca A, Denzel M, Hussmann H. Look into My Eyes!: Can You Guess My Password? In: Proceedings of the 5th Symposium on Usable Privacy and Security. SOUPS'09. New York, NY: ACM; 2009. p. 7:1–7:12. Available from: http://doi.acm.org/10.1145/1572532.1572542.

[18] Schenk S, Dreiser M, Rigoll G, *et al.* GazeEverywhere: Enabling Gaze-Only User Interaction on an Unmodified Desktop PC in Everyday Scenarios. In: Proceedings of the 2017 CHI Conference on Human Factors in Computing Systems. CHI'17. New York, NY: ACM; 2017. p. 3034–3044. Available from: http://doi.acm.org/10.1145/3025453.3025455.

[19] Esteves A, Velloso E, Bulling A, *et al.* Orbits: Gaze Interaction for Smart Watches Using Smooth Pursuit Eye Movements. In: Proceedings of the 28th Annual ACM Symposium on User Interface Software and Technology. UIST'15. New York, NY: ACM; 2015. p. 457–466. Available from: http://doi.acm.org/10.1145/2807442.2807499.

[20] Zhai S, Morimoto C, Ihde S. Manual and Gaze Input Cascaded (MAGIC) Pointing. In: Proceedings of the SIGCHI Conference on Human Factors in Computing Systems. CHI'99. New York, NY: ACM; 1999. p. 246–253. Available from: http://doi.acm.org/10.1145/302979.303053.

[21] Drewes H, Schmidt A. The MAGIC Touch: Combining MAGIC-Pointing with a Touch-Sensitive Mouse. In: Proceedings of the 12th IFIP TC 13 International Conference on Human–Computer Interaction: Part II. INTERACT'09. Berlin: Springer-Verlag; 2009. p. 415–428. Available from: https://doi.org/10.1007/978-3-642-03658-3_46.

[22] Kumar M, Paepcke A, Winograd T. EyePoint: Practical Pointing and Selection Using Gaze and Keyboard. In: Proceedings of the SIGCHI Conference on Human Factors in Computing Systems. CHI'07. New York, NY: ACM; 2007. p. 421–430. Available from: http://doi.acm.org/10.1145/1240624.1240692.

[23] Pfeuffer K, Gellersen H. Gaze and Touch Interaction on Tablets. In: Proceedings of the 29th Annual Symposium on User Interface Software and Technology. UIST'16. New York, NY: ACM; 2016. p. 301–311. Available from: http://doi.acm.org/10.1145/2984511.2984514.

[24] Kumar C, Akbari D, Menges R, MacKenzie S and Staab S. TouchGazePath: Multimodal Interaction with Touch and Gaze Path for Secure Yet Efficient PIN Entry. In 2019 International Conference on Multimodal Interaction (ICMI '19), Wen Gao, Helen Mei Ling Meng, Matthew Turk, Susan R. Fussell, Björn Schuller, Yale Song, and Kai Yu (Eds.). ACM, New York, NY, USA, 329–338. DOI: https://doi.org/10.1145/3340555.3353734

[25] Sengupta K, Ke M, Menges R, *et al.* Hands-Free Web Browsing: Enriching the User Experience with Gaze and Voice Modality. In: Proceedings of the 2018 ACM Symposium on Eye Tracking Research & Applications. ETRA'18. New York, NY: ACM; 2018. p. 88:1–88:3. Available from: http://doi.acm.org/10.1145/3204493.3208338.

[26] Lankford C. Effective Eye-Gaze Input into Windows. In: Proceedings of the 2000 Symposium on Eye Tracking Research & Applications. ETRA'00. New York, NY: ACM; 2000. p. 23–27. Available from: http://doi.acm.org/10.1145/355017.355021.

[27] Ashmore M, Duchowski AT, Shoemaker G. Efficient Eye Pointing with a Fisheye Lens. In: Proceedings of Graphics Interface 2005. GI'05. Waterloo, Ontario, Canada: School of Computer Science, University of Waterloo; Canadian Human–Computer Communications Society; 2005. p. 203–210. Available from: http://dl.acm.org/citation.cfm?id=1089508.1089542.

[28] Špakov O, Miniotas D. Gaze-Based Selection of Standard-Size Menu Items. In: Proceedings of the 7th International Conference on Multimodal Interfaces. ICMI'05. New York, NY: ACM; 2005. p. 124–128. Available from: http://doi.acm.org/10.1145/1088463.1088486.

[29] Lutteroth C, Penkar M, Weber G. Gaze vs. Mouse: A Fast and Accurate Gaze-Only Click Alternative. In: Proceedings of the 28th Annual ACM Symposium on User Interface Software and Technology. UIST'15. New York,

NY: ACM; 2015. p. 385–394. Available from: http://doi.acm.org/10.1145/2807442.2807461.

[30] Sweetland J. Optikey: Type, Click, Speak; 2016. Available from: https://github.com/OptiKey/OptiKey.

[31] Diaz-Tula A, Morimoto CH. AugKey: Increasing Foveal Throughput in Eye Typing with Augmented Keys. In: Proceedings of the 2016 CHI Conference on Human Factors in Computing Systems. CHI'16. New York, NY: ACM; 2016. p. 3533–3544. Available from: http://doi.acm.org/10.1145/2858036.2858517.

[32] Sengupta K, Menges R, Kumar C, *et al.* GazeTheKey: Interactive Keys to Integrate Word Predictions for Gaze-Based Text Entry. In: Proceedings of the 22nd International Conference on Intelligent User Interfaces Companion. IUI'17 Companion. New York, NY: ACM; 2017. p. 121–124. Available from: http://doi.acm.org/10.1145/3030024.3038259.

[33] Kurauchi A, Feng W, Joshi A, *et al.* EyeSwipe: Dwell-Free Text Entry Using Gaze Paths. In: Proceedings of the 2016 CHI Conference on Human Factors in Computing Systems. CHI'16. New York, NY: ACM; 2016. p. 1952–1956. Available from: http://doi.acm.org/10.1145/2858036.2858335.

[34] Tuisku O, Majaranta P, Isokoski P, *et al.* Now Dasher! Dash Away!: Longitudinal Study of Fast Text Entry by Eye Gaze. In: Proceedings of the 2008 Symposium on Eye Tracking Research and Applications. ETRA'08. New York, NY: ACM; 2008. p. 19–26. Available from: http://doi.acm.org/10.1145/1344471.1344476.

[35] Kumar C, Menges R, Staab S. Assessing the Usability of Gaze-Adapted Interface against Conventional Eye-Based Input Emulation. In: 2017 IEEE 30th International Symposium on Computer-Based Medical Systems (CBMS); 2017. p. 793–798.

[36] Hurst A, Hudson SE, Mankoff J. Automatically Identifying Targets Users Interact with During Real World Tasks. In: Proceedings of the 15th International Conference on Intelligent User Interfaces. IUI'10. New York, NY: ACM; 2010. p. 11–20. Available from: http://doi.acm.org/10.1145/1719970.1719973.

[37] Hornof AJ, Cavender A. EyeDraw: Enabling Children with Severe Motor Impairments to Draw with Their Eyes. In: Proceedings of the SIGCHI Conference on Human Factors in Computing Systems. CHI'05. New York, NY: ACM; 2005. p. 161–170. Available from: http://doi.acm.org/10.1145/1054972.1054995.

[38] Heikkilä H. EyeSketch: A Drawing Application for Gaze Control. In: Proceedings of the 2013 Conference on Eye Tracking South Africa. ETSA'13. New York, NY: ACM; 2013. p. 71–74. Available from: http://doi.acm.org/10.1145/2509315.2509332.

[39] Beelders T, Blignaut P. The Usability of Speech and Eye Gaze as a Multimodal Interface for a Word Processor. In: Ipsic I, editor. Speech Technologies. Rijeka: IntechOpen; 2011. Available from: https://doi.org/10.5772/16604.

[40] Sindhwani S, Lutteroth C, Weber G. ReType: Quick Text Editing with Keyboard and Gaze. In: Proceedings of the 2019 CHI Conference on Human Factors in Computing Systems. CHI'19. New York, NY: ACM; 2019. p. 203:1–203:13. Available from: http://doi.acm.org/10.1145/3290605.3300433.

[41] Vickers S, Istance H, Hyrskykari A. Performing Locomotion Tasks in Immersive Computer Games with an Adapted Eye-Tracking Interface. ACM Transactions on Accessible Computing. 2013;5(1):2:1–2:33. Available from: http://doi.acm.org/10.1145/2514856.

[42] Walber T, Neuhaus C, Scherp A. EyeGrab: A Gaze-Based Game with a Purpose to Enrich Image Context Information. In: Proceedings of the 2nd European Workshop on Human–Computer Interaction and Information Retrieval, Nijmegen, The Netherlands, August 25, 2012; 2012. p. 63–66. Available from: http://ceur-ws.org/Vol-909/poster8.pdf.

[43] Menges R, Kumar C, Wechselberger U, *et al.* Schau Genau! A Gaze-Controlled 3D Game for Entertainment and Education. Journal of Eye Movement Research. 2017;10:220. Available from: https://bop.unibe.ch/JEMR/article/view/4182.

[44] Menges R, Kumar C, Sengupta K, *et al.* eyeGUI: A Novel Framework for Eye-Controlled User Interfaces. In: Proceedings of the 9th Nordic Conference on Human–Computer Interaction. NordiCHI'16. New York, NY: ACM; 2016. p. 121:1–121:6. Available from: http://doi.acm.org/10.1145/2971485.2996756.

[45] Menges R, Kumar C, Müller D, *et al.* GazeTheWeb: A Gaze-Controlled Web Browser. In: Proceedings of the 14th Web for All Conference. W4A'17. ACM; 2017. Available from: http://dx.doi.org/10.1145/3058555.3058582.

[46] Biedert R, Buscher G, Schwarz S, *et al.* The Text 2.0 Framework: Writing Web-Based Gaze-Controlled Realtime Applications Quickly and Easily. In: Proceedings of the 2010 Workshop on Eye Gaze in Intelligent Human Machine Interaction. EGIHMI'10. New York, NY: ACM; 2010. p. 114–117. Available from: http://doi.acm.org/10.1145/2002333.2002351.

[47] Wassermann, B, Hardt, A and Zimmermann, G. Generic gaze interaction events for web browsers. In: Proceedings of the 21st international conference on World Wide Web, Vol. 9; 2012.

[48] Kumar C, Menges R, Müller D, *et al.* Chromium Based Framework to Include Gaze Interaction in Web Browser. In: Proceedings of the 26th International Conference on World Wide Web Companion. WWW'17 Companion. Republic and Canton of Geneva, Switzerland: International World Wide Web Conferences Steering Committee; 2017. p. 219–223. Available from: https://doi.org/10.1145/3041021.3054730.

[49] Kumar M, Winograd T. Gaze-Enhanced Scrolling Techniques. In: Proceedings of the 20th Annual ACM Symposium on User Interface Software and Technology. UIST'07. New York, NY: ACM; 2007. p. 213–216. Available from: http://doi.acm.org/10.1145/1294211.1294249.

[50] Menges R, Kumar C and Staab S. Improving User Experience of Eye Tracking-Based Interaction: Introspecting and Adapting Interfaces. ACM Transactions on Computer–Human Interaction. New York, NY: ACM; 2019;26(6): 37:1–37:46. Available from: http://doi.acm.org/10.1145/3338844.
[51] Oyekoya O, Stentiford F. Perceptual Image Retrieval Using Eye Movements. In: Proceedings of the 2006 Advances in Machine Vision, Image Processing, and Pattern Analysis International Conference on Intelligent Computing in Pattern Analysis/Synthesis. IWICPAS'06. Berlin: Springer-Verlag; 2006. p. 281–289. Available from: http://dx.doi.org/10.1007/11821045_30.

Chapter 6
Eye tracking for interaction: evaluation methods

Chandan Kumar[1], *Raphael Menges*[1], *Korok Sengupta*[1], *and Steffen Staab*[1,2]

Eye tracking as a hands-free input method can be a significant addition to the lives of people with a motor disability. With this motivation in mind, so far research in eye-controlled interaction has focused on several aspects of interpreting eye tracking as input for pointing, typing, and interaction methods with interfaces, as presented in Chapter 5. In this regard, the major question is about how well does the eye-controlled interaction work for the proposed methods? How efficiently can pointing and selection be performed? Whether common tasks can be performed quickly and accurately with the novel interface? How different gaze interaction methods can be compared? What is the user experience while using eye-controlled interfaces? These are the sorts of questions that can be answered with an appropriate evaluation methodology. Therefore, in this chapter, we review and elaborate different evaluation methods used in gaze interaction research, so the readers can inform themselves of the procedure and metrics to assess their novel gaze interaction method or interface.

We discuss the common methodological denominators and guidelines for eye tracking experiments in Section 6.1. In Section 6.2, we outline how the evaluation of eye tracking as input for atomic interactions like pointing, selection, and typing is performed. Besides reviewing the general metrics, we also discuss the cognitive load metric that we introduced in our recent work [1]. In Section 6.3, we discuss the methodology to evaluate complete eye-controlled application interfaces. We specifically describe how we have evaluated our gaze-controlled Web browser [2], incorporating comparative and feasibility evaluations.

6.1 Background and terminology

Evaluation in eye-tracking research conforms to the methodology from the interdisciplinary field of human–computer interaction (HCI) [3] and experimental psychology [4]. A user study is an experiment with human participants, and the

[1] Institute for Web Science and Technologies, University of Koblenz-Landau, Koblenz, Germany
[2] Web and Internet Science Research Group, University of Southampton, Southampton, United Kingdom

methodology includes all the choices one makes in undertaking the user study. The choices pertain to the research questions, study design, variables, participants, ethics, data collection, and analysis. In the following, we discuss these terminologies in the context of experiments involving eye tracking for interaction.

6.1.1 Study design

Based on the goal and hypothesis about the interaction method or interface, researchers undertake different types of studies to evaluate the performance and usability of the proposed system.

Lab studies involve the user carrying out a set of constrained tasks in a controlled environment. The key element is to test the differences in the developed system while keeping all other factors constant, i.e., the controlled environment implies that there should be no other differences between the conditions than the introduced differences of the proposed methods. This means that the experimenter must control all other factors, including lighting, temperature, noise, and instructions, given to the participants. To assist with this, lab studies are usually carried out in a dedicated usability laboratory. There are several advantages of lab studies in eye-tracking-related experiments. First of all, lab studies are fairly cheap to perform and provide a concrete comparative analysis of the performance of a gaze interactive system or technique. More importantly, the most common research questions in eye-tracking-related approaches are rather comparative, such as "What value of dwell time yields better performance?", "Is user performance better using mouse click, gaze-based dwell confirmation, or touch selection?", "Does a filtering algorithm improve the user performance compared to no-filtering or compared to previous filtering techniques?", or "Does the gaze-adapted interface help completing the tasks efficiently than the generic interface?". These kinds of questions can be best answered using a controlled experiment. Furthermore, the performance and accuracy of eye-tracking systems are highly subjective to environmental influences such as lighting, display, seating position, and distance. If those factors are not controlled, the difference in results could be due to the external factors, rather than the method or interface to be tested. However, lab studies are not very suitable to provide the understanding of how eye-tracking systems fit into everyday life. Eye tracking can be used for novel interaction concepts, and not many users are equipped with eye-tracking systems in daily life environment; hence, evaluating the end-user acceptability is not trivial.

Field studies allow to assess the difficulties, challenges, and acceptance in real-world environment from end-user's perspective. Field studies are generally carried out at users' environment that may be their workplace, home, or leisure environment, depending on the tasks to be supported by the technology. Field studies can last for a couple of hours to several days, weeks, or months, depending on the resources and the research/commercial goal of the study. It is important to investigate eye-tracking systems acceptability as a future interaction technology; however, there are not many such studies reported in the literature. In Section 6.3, we describe the experimental setup and procedure of how we have conducted a field study at a home environment for a period of one month. Such studies, which last for a longer period, are also categorized as longitudinal study.

Longitudinal studies focus on the learnability aspect, i.e., the role of time and experience in the usability of a system. A longitudinal study may be of duration of anywhere from a few days to several decades and may be conducted by either running repeated measurements in a lab setting or in a field setting.

6.1.1.1 Within-subjects and between-subjects designs

If you are comparing different conditions (interaction method or interfaces), i.e., if the evaluation is comparative, the experiment can be performed with within-subjects or between-subjects conditions. When all participants are tested on all conditions, the design is called a within-subjects design. It is also termed "repeated-measures design," as the measurements gathered under one condition are repeated on the same participants for the other conditions. In a between-subjects design, each participant is tested under one condition only. One group of participants is tested under one condition and a different group under the other condition.

The choice of design highly depends on the goal of the evaluation. For controlled lab studies, a within-subjects design is generally preferred, because effects due to the behavioral differences of participants are minimized as they are likely to be consistent across the conditions. For example, a participant who often moves her head would exhibit such behavior consistently across experimental conditions affecting the accuracy of gaze estimation. For field studies, or longitudinal studies generally, between-subjects design is preferred, because in real-life environment it is not natural to ask participants to use two different methods to perform similar tasks. For example, the Schau genau! [5] game, described in Section 5.3.1 of Chapter 5, was placed in a public area and had three different control types. We could not ask each player to play the game three times to directly compare the different control types; hence, a between-subject comparison was performed.

6.1.1.2 Counterbalancing

For within-subjects designs, an important factor is learning. Because two or more methods are compared and all participants use one method first followed by the others, an improvement with later might be because of the practice on the first method. In this regard, counterbalancing is useful where the grouping is done according to a *Latin Square*. Counterbalancing may be specially critical in an eye-tracking setup, as users are mostly novices of the technology and could have a significant learning in the first trials. Latin Square [6] ordering is used mostly to ensure that there is no bias for within-subjects design. The (balanced) Latin Square orders participants for different experimental scenarios ensuring that a learning effect from one scenario does not impact the overall experiment in totality.

6.1.2 Participants

Persons who participate in an experiment are called *participants*. Most empirical evaluations with eye tracking as input include 5–25 participants [8,9], although the exact number to use is arguable. In general, if the number of participants is low, large performance differences may occur but the differences might not be statistically significant. In contrast, if there are large numbers of participants, small performance differences of no practical relevance might indicate statistical significance. The best practice is to review the literature and use approximately similar number of participants used in the prior research with a similar methodology [4,10,11].

In the ideal scenario, participants are selected randomly from the population of people in the intended community of users. For instance, eye tracking as an assistive technology should ideally involve the *target-group participants* in an evaluation. The target-group participants in such eye-tracking experiments should involve representative groups from patients of amyotrophic lateral sclerosis, aphasia, neuromuscular disease, Parkinson's disease, spinal cord injuries, and cerebral palsy. (More detailed discussions on target-group users are available in Chapters 2–4.) Although there are some notable works that conducted usability studies with target-group participants [8,12], the most common approach in practice is to invite people conveniently available, e.g., students at a local university or colleagues at work. In some cases, prescreening is necessary if, for example, users are required with a specific level of expertise or experience with aspects of the interaction. This includes having no prior exposure of eye tracking, if such is important in the evaluation. In all cases, it is recommended to report on demographic data, i.e., number of participants, their gender, age, and prior experience with eye tracking or the experience with related interfaces to be tested. The use of glasses or corrective lenses is also noted for eye-tracking experiments as it could have an impact on accuracy. Before each condition, a recalibration of the eye-tracking system is recommended and the reached accuracy should be recorded for later reporting.

6.1.2.1 Ethics

One important aspect of conducting evaluation with human participants is ethics. Usually, there are formal regulations on ethics of your institution which you have to follow, but ethics in principle should be taken care by everyone who carries out experiments. A good ethics code includes steps such as treating participants with respect (you are testing the technology and not the participants), pilot testing (to make sure everything works in the actual study as expected with the least overhead to the participants), and informed consent (you provide information about what will happen and get the approval of participants by signing the consent form). More specifically for eye-tracking studies, the recorded data might contain sensitive eye-gaze information of user attention; hence, this needs to be clarified in consent form regarding which data being recorded during the experiment. For eye-tracking studies, if you are inviting participants with motor disability, you should be specially careful of accessibility guidelines and clinical protocol. One example is Helsinki ethical approval [13].

6.1.3 Experimental variables

Extending the discussion and terminology on study design, the two primary variables in HCI and experimental research are independent and dependent variables [3].

For the research to be reproducible, it is highly recommended to clearly define all experimental variables as part of the evaluation methodology in any work.

6.1.3.1 Independent variables

Independent variables are the values that are targeted to be evaluated in the experiment, in order to analyze the conditions. Those variables are called "independent," because they are controlled by the experimenter and are completely independent of participants, i.e., a participant cannot do anything to influence an independent variable being evaluated. In eye-tracking research, "input method" would be an independent variable in studies comparing (gaze vs. mouse), (dwell vs. blink), or (eye movement vs. head movement) as different conditions. Generally speaking, any factor influencing the performance could be hypothetically an independent variable in the experiment. Examples include interface element size (big vs. small), visual effect (feedback vs. no feedback), fixation filtering (on vs. off), gender (male vs. female), age (old vs. young), and so forth. In eye-tracking-related evaluations, various kinds of independent variables are used in literature [8,11,14].

6.1.3.2 Dependent variables

A dependent variable is any measurable estate of the interaction involving an independent variable. It is called dependent because it "depends on" actions by the participant. Any observable, measurable aspect of the interaction is a potential-dependent variable. However, speed and accuracy in executing the tasks are the most general dependent variables used in experiments. For example, throughput (bits/s) is a common measure to evaluate gaze-based pointing and selection, which includes both the speed and accuracy in participants' responses. For gaze-based text entry, most common dependent variables are speed (words per minute (WPM)), error rate, backspace usage, and keystroke savings (we provide more details on pointing and typing measures in Section 6.2). It is advised to separate the dependent variable name from its units, for example, typing speed is a dependent variable with units "WPM"; however, the unit could also be "characters per minute" but the dependent variable of typing speed would be the same. For interface evaluation, task completion time is the most common dependent variable (we provide more details on interface evaluation measures in Section 6.3).

6.1.3.3 Further variables

We discussed earlier in lab study design of eye-tracking experiments that consistent environmental conditions for the experimental setup are particularly important.

Control variables include all circumstances that are kept constant. It is to confirm that there are no variations of independent variables due to external factors, and the outcomes of dependent variable are largely based on independent variable conditions. Example of control variable includes ambient lighting, seating arrangement, display settings, or interface elements, like text fonts.

A confounding variable is a variable that influences both the dependent and independent variables, causing a spurious association. A circumstance that varies systematically with an independent variable confounding is a causal concept, and as such, cannot be described in terms of correlations or associations [15]. In general, the guideline is that any such variable should be controlled or randomized to avoid misleading results. For example, a study comparing the eye-typing performance on a QWERTY layout vs. a novel layout which the participants have never seen before. Hence, a performance difference may emerge, but prior experience is a confounding variable. The experiment needs to state such variables and take precautionary actions such as providing training and randomizing the order.

6.1.4 Measurements

Another common terminologies in reporting evaluation results are objective and subjective measurements.

Objective measurement is something that is measured consistently and yields from quantitative data recorded while participants perform the designated tasks. For example, how effectively someone can perform a set number of tasks in a controlled environment, i.e., it is closely related to the dependent variable.

Subjective measurement is about measuring what people say or they feel about the interaction experience. It is very important to listen to the feedback from participants, because eye tracking is a novel interaction mechanism for most of them, and the usage is very subjective to human behavior and responses. Furthermore, it can be a physically and cognitively demanding task to operate with an eye-tracking system. Therefore, it is very common in eye-tracking experiment to use a survey to answer open-ended questions, ranking an experience (how they judge speed and accuracy) based on feelings, and more.

Both objective and subjective measurements are keys in reaching our goal and evaluate the suggested method or interface. Researchers want to support users with faster and accurate inputs, at the same time they want users to feel good about using the system. For subjective assessment, developing a questionnaire is a key aspect, which means that getting the choice of questions, their format, ordering, etc. are often time consuming but also imperative to report a full picture of the user experience. Besides this, there are some general questionnaires used for subjective assessment discussed in the following:

- The system usability scale (SUS) [16] is a simple ten-point questionnaire that helps in understanding a subjective assessment of usability. The ten levels of SUS cover system usability, need for support, training, and complexity. The scale is generally used after the participants complete the experiments. They are asked to record their immediate responses to each of the points on a Likert scale.
- NASA's Task Load Index (NASA-TLX) [17] has been used in a wide variety of experiments to assess the usability, stress, or effectiveness of performance in regard to perceived mental load. This is a qualitative metric where participants provide a rating to six subjective scales to show their mental load. The six subjective scales look into the following:
  - Mental demand: How mentally demanding was the task?
  - Physical demand: How physically demanding was the task?
  - Temporal demand: How hurried or rushed was the pace of the task

- Overall performance: How successful were you in accomplishing what you were asked to do?
- Effort: How hard did you have to work to achieve your level of performance?
- Frustration level: How insecure, discouraged, irritated, stressed, and annoyed were you?

- Heuristic evaluation is an informal usability testing method for user interfaces. Nielsen *et al.* [18] define heuristic evaluation of an interface as "*simply looking at the interface and passing judgement according to ones own opinion*". Heuristic evaluation is a fast and easy method to test user interfaces in order to recognize failures with respect to intended purposes and can be performed by the designer. Following the guidelines of Nielsen *et al.* [19], we proposed to adapt the questionnaire for eye-tracking environment [20]. In the following subsections, we would provide some examples of how the heuristics were adapted for eye-typing experiment (Section 6.2) and for browser application evaluation (Section 6.3).

6.1.4.1 Statistical analysis

In the analysis, we are interested in knowing whether there is a significant difference in the dependent variable when the participants are confronted with the experimental conditions. Thus in HCI research, usually significance tests are performed on the outcome of the dependent variables, which is sometimes difficult because of the low number of participants. The dependent variables are usually measured as interval or ratio values, e.g., execution times or words entered per minute. Furthermore, we assume that the experimental data is balanced, i.e., the sample sizes are equal across all tested methods. Generally, the p-value should be below or equal to 5% to consider a significant difference between the conditions.

We generally distinguish between parametric and nonparametric tests. Parametric tests are those that make assumptions about the parameters of the population distribution from which the samples are drawn. In most of the cases, the assumption is about the normality of the population distribution. We recommend to first check the histogram plot of the dependent variable as measured across the participants and draw a conclusion from the shape of the histogram. There are also statistical measures like the Shapiro–Wilk test. However, the null hypothesis of the Shapiro–Wilk test is that the samples are taken from a normal distribution. We cannot reject the p-value over the threshold as well as we cannot validate it. A p-value greater than the threshold still may indicate a normal distribution of the values of the dependent variables. In addition, the effect size should be reported alongside each significance result. Statistical significance is the probability that the observed difference between groups is due to chance. A bigger sample size makes it more probable to find a significant difference between the groups. While a p-value can inform the reader whether an effect exists, the p-value will not reveal the size of an effect. Thus, both the substantive significance (effect size) and statistical significance (p-value) are to be reported for a dependent variable.

The t-test is a parametric test to compare the means of two groups. It is assumed that the distribution of the population of the sample data is normal. A paired t-test can be performed when the participants using both conditions are matched; thus, we can look at the differences in each participant's outcome of a dependent variable. An unpaired t-test just considers the means of two independent groups of participants using one condition or the other. As effect size Cohen's d can be reported for both paired and unpaired t-tests.

Analysis of variance (ANOVA) is a common test to estimate the difference in means among more than two conditions. Similar to a t-test, the test is parametric and requires the population from which the samples are drawn to conform to a normal distribution. Furthermore, homogeneity of variances must be checked for between-subjects designs and sphericity for within-subjects designs. The effect size is called eta squared and calculated as the square of the correlation ratio. Refer to your statistics package of choice for further instructions on how to perform those tests and to calculate the effect size.

If normality cannot be assumed, there are nonparametric tests that do not require the samples to be drawn from a normally distributed population. Yet, the nonparametric tests still assume the equality of variances. A Mann–Whitney's U test compares the medians of the two groups, not the means, and works with ranks instead of the actual values of the dependent variable. Analogous to a paired t-test, a Wilcoxon signed-rank test is similar to a Mann–Whitney's U test, but it works on the differences between the ranks of dependent variable outcomes per participant. For three or more conditions without normality assumption about the samples, a Friedman test or Kruskal–Wallis test can be performed. Table 6.1 presents a summary of statistical methods used in eye tracking evaluation.

6.2 Evaluation of atomic interactions

The characteristics of eye gaze as "what you look at is what you get" [21] had initiated the initial comparisons of eye gaze as a substitute to conventional input mechanisms like mouse, pen, touch, or keyboard. Therefore, several methods for better pointing and typing with eye gaze input are an imperative aspect of gaze interaction research, as discussed in Section 5.2 in Chapter 5. In the following, we discuss the most common methodologies used for the evaluation of gaze-based pointing and typing methods.

6.2.1 Evaluation of gaze-based pointing and selection

As a pointing device, an eye tracking system typically emulates a mouse. Similar to point-select operations with a mouse, the eye can spatially orientate user intention and perform select operation using dwell, blink, or gestures (more details in Section 5.2 in Chapter 5). Evaluation of eye trackers for pointing and selecting operation has largely followed the methodology pertaining to the conventional issues for computer input using a mouse and the unique characteristics of eye-tracking system and gaze interaction.

6.2.1.1 Objective measures

The conventional measures such as speed and accuracy are obviously applicable to the eye-tracking scenario for evaluation and comparison. Similar to pointing devices,

Table 6.1 Statistical methods used in eye tracking evaluation

Study design		Normality assumed	Normality not assumed
Within-subjects design	Two conditions	Paired t-test	Wilcoxon test
	Three or more conditions	Repeated-measures ANOVA	Friedman test
Between-subjects design	Two conditions	Unpaired t-test	Mann–Whitney test
	Three or more conditions	ANOVA	Kruskal–Wallis test

"throughput" (TP) has been introduced by MacKenzie [22] as Fitts' law for eye tracking. It is considered a primary performance metric (dependent variable) in various eye tracking evaluations till date [9,23–26]. Throughput conforms to ISO 9241-9 to uniform guidelines and testing procedures for evaluating computer-pointing devices [27,28]. It is measured in bits per second, and it is a composite measure that includes both the speed and accuracy in performance [29]. The equation for throughput is Fitts' "index of performance" except using an effective index of difficulty (ID_e). Specifically,

$$TP = \frac{ID_e}{MT},$$

where MT is the mean movement time, in seconds, for all trials within the same condition and

$$ID_e = \log_2\left(\frac{D}{W_e} + 1\right),$$

ID_e, in bits, is calculated from D, the distance to the target. W_e is the effective target width with

$$W_e = 4.133 \times SD_x,$$

SD_x is the standard deviation in the selection coordinates measured along the axis from the default position to the center of the target.

The standard mouse throughput varies from about 3 up to 5 bits/s [30,31]. The performance of eye gaze input is found to be lower than the mouse and other conventional pointing mechanisms like pen or touch [11]. However, the throughput value is highly subjective to experimental variations (e.g., display size and experience of participants). Figure 6.1 shows an example of throughput comparison where input method is the independent variable (gaze vs. head vs. mouse) [32] and (gaze vs. touch vs. mouse) [31]. Various methods of eye pointing, e.g., multistep magnification [33], fish-eye lens magnification [34], or smooth pursuit of visual targets [35], have been evaluated to improve the pointing performance and accuracy. There are also case studies where eye gaze outperforms mouse in specific scenarios such as a gaming environment [36].

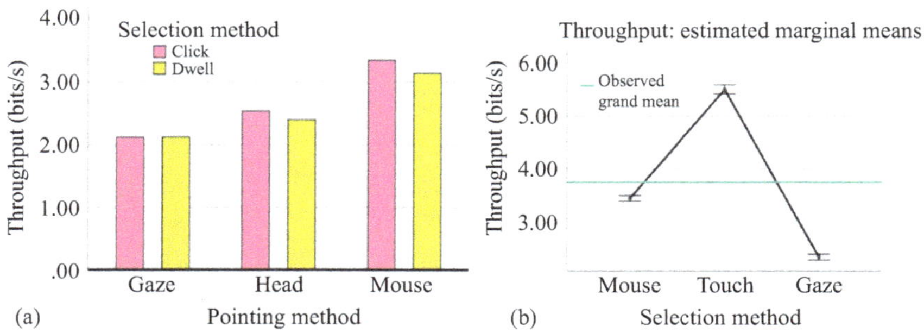

Figure 6.1 Throughput comparison of eye gaze vs. other input modalities for target pointing and selection tasks: (a) Hansen et al. [32] and (b) Rajanna and Hammond [31]

6.2.1.2 Subjective measures

ISO 9241-9 has also laid out the methodology for subjective assessment of user comfort with input devices, such as mice, trackballs, touchpads, joysticks, or pens. The eye tracker as a pointing device also falls within the scope of this standard. However, the generic ISO 9241-9 assesses comfort using a questionnaire soliciting Likert-scale responses to 12 items, and this needs to be customized for the characteristics of a specific device. Zhang and MacKenzie [28] have adapted the questionnaire as per eye-tracking comfort and fatigue (Figure 6.2). Besides this, researchers have used customary questionnaire depending on the nature of tasks; most often, it involves Likert-scale questionnaire on how would you rate speed, accuracy, comfort, and learnability aspects of the presented methods [9,37] to participants.

6.2.2 Evaluation of gaze-based text entry

Similar to its assessment to accomplish pointing and selecting operations, eye tracking has also been considered a substitute for a keyboard, to allow for text entry [8]. The most common design to evaluate gaze-based text entry is to conduct lab studies, in which participants are asked the enter phrases using proposed methods [38,39]. There are also some longitudinal studies analyzing the long-term learning effect [40,41]. However, the evaluation measures are common across all these studies, as discussed in the following.

6.2.2.1 Objective measures

In terms of objective measures, the eye typing methods use the dependent variable of speed and accuracy with the most common units being WPM, keystrokes, and error rate.

- WPM forms one of the most basic metrics for evaluating text entry. WPM is calculated as

$$WPM = \frac{|S - 1|}{T} \times 60 \times \frac{1}{5},$$

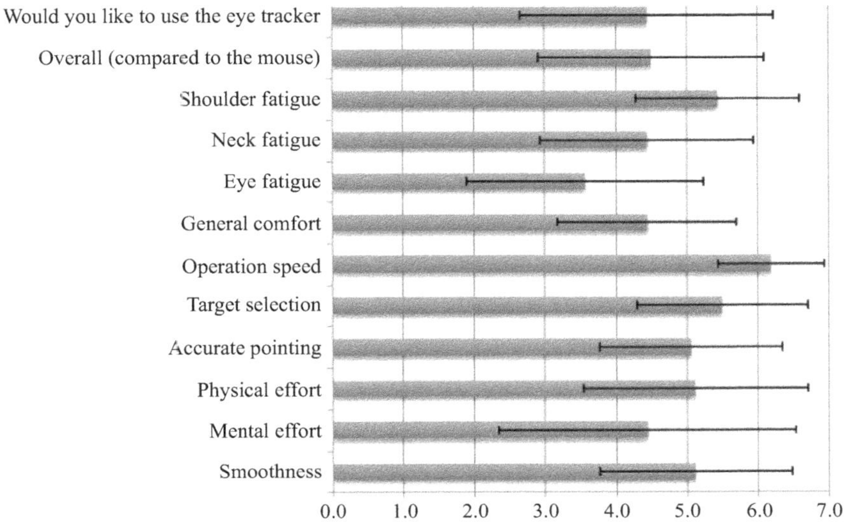

Figure 6.2 Questionnaire for subjective assessment of eye tracking as input modality [28]. Response 7 was the most favorable, response 1 the least favorable

where S is the length of the transcribed string, including spaces, and T is the time in seconds from the entry of the first character to the entry of the last character, and 5 represents the average characters in a word [10]. In Figure 6.3, we showcase how WPM is usually reported taking an example of our eye-tracking experiment [42], where the independent variable was three different keyboard designs having suggestions at different positions.

- Accuracy and error rate are measured as mean square difference (MSD), which is a metric for the number of errors left in the transcribed text.

$$MSD\ error\ rate = \frac{MSD(P,T)}{\overline{S_A}} \times 100\%,$$

where $MSD(P, T)$ is the minimum string distance between the presented and transcribed strings, and $\overline{S_A}$ is the mean length of the alignment strings [43].

- Keystrokes measure is defined as the number of key activations required, on an average, to generate a character for text for a given entry technique in a given language. Measurement of keystrokes gives an indication of the efficiency of text entry system. The efficiency can be attributed to the keyboard design, the language model, or both. In text entry systems, keystrokes are mostly measured as keystrokes per character [44].
- Besides the conventional metrics, we introduced direct estimation of cognitive load as an additional metrics used in our recent research [1]. In comparison

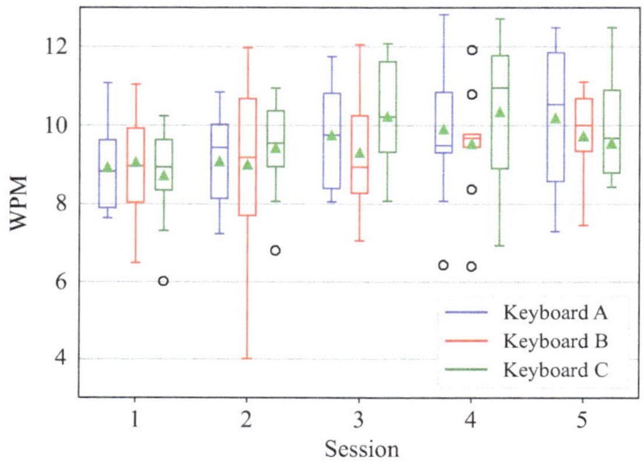

Figure 6.3 Words per minute measured across five sessions for three different keyboard designs. Keyboard A, a traditional keyboard design with predictions placed on top of the keyboard layout area. Keyboard B, an interspaced prediction positioning where predictions appear on top of every row of keys, which is currently being used. Keyboard C, featuring on-key predictions that try to bring the predictions to the visual fovea

to the traditional text entry approaches, gaze-based typing involves natural eye movements that are highly correlated with human brain cognition. Employing eye gaze as an input can lead to excessive MD, and we argue the need to include cognitive load as an eye-typing evaluation measure. Cognitive load is defined as the effort or the load imposed on the memory by the cognitive process involved in learning [45]. We investigated the text entry use case to understand whether cognitive load can be measured by a noninvasive process and would that help our investigation. Electroencephalogram (EEG) signals can be used to investigate cognitive load from a noninvasive direction [46]. For our research purposes, we used an Emotiv EPOC+ wearable device and applied short-time Fourier transform to the EEG signal time series, in order to evaluate the cognitive load of each participant during the experiments [47]. We computed the average value of the spectral power ratio in the Beta band [48] of the EEG signal from all 14 channels within a time window (Figure 6.4). The average value of the spectral ratio served as an indicator of the cognitive load within each particular time window. For a detailed analysis on how we performed cognitive load analysis on keyboard designs and related interaction, we recommend reading [1].

6.2.2.2 Subjective measures

Similar to the pointing method, it is common practice to provide a customary questionnaire for text entry assessment; most often, it involves Likert-scale questionnaire on how a participant rates speed, accuracy, comfort, and learnability aspects of the used text entry methods [8,10,38]. NASA-TLX has also been used by various

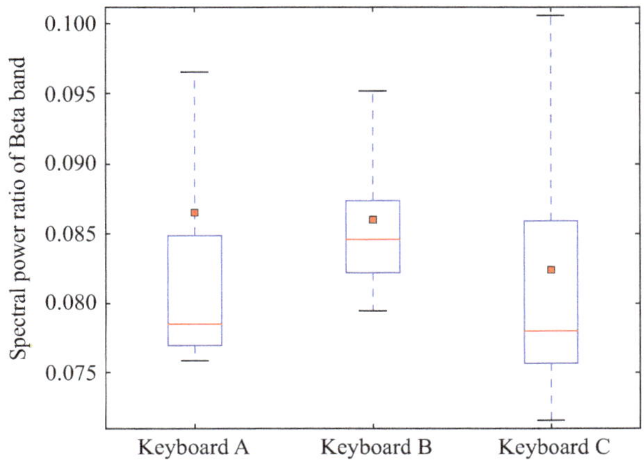

Figure 6.4 Comparison of overall cognitive load of participants using three keyboard designs during the experiment. The x-axis marks the keyboard design, while the y-axis denotes the spectral power ratio of the Beta band of EEG signals, which indicates the level of cognitive load of the participant. Each data entry in the box-plot corresponds to the spectral power ratio value of one-time window. The horizontal bar in the middle of the box shows the median value, while the red square shows the mean value

researchers [49]. Besides this, we formulated a heuristic questionnaire [42] that would give us insights into how participants felt when using the system. Participants were asked to give a grade between zero and ten. The questions are listed next:

- Was the design intuitive? (If there was no guidance, would you have figured it out easily?)
- How comfortable was to use eye tracking on the keyboard designs (A/B/C)?
- How will you rate the design of the keyboard?
- How easy was it recover from the errors made?
- How easy was it to control the keyboard?
- How close were the features of the gaze-based keyboard in comparison to conventional keyboard?
- How good is the visibility of the interaction elements? (Interaction elements include keys, suggestions, and typing area.)

6.3 Evaluation of application interfaces

In the previous section, we discussed the approaches assessing eye tracking to accomplish primary interactions such as target selection or text entry. However, users do not use these methods in isolation, and the acceptance of technology primarily depends on how these atomic interactions let users interact with applications through their interfaces as a whole. Hence, it is rather interesting to investigate how gaze interaction impacts the user experience in terms of both objective and subjective measurements. In this section, we go beyond the atomic interactions and aim to assess the gaze interaction performance for interface control.

So far, the evaluation of eye-controlled interfaces lacks a universal benchmark. Commercial systems like Tobii Dynavox Windows Control* or Visual Interaction myGaze Power† have not published any resources on how or even if they evaluated the usability of their eye-controlled interfaces. Most of the research prototypes aim to validate whether users are merely able to accomplish fundamental tasks or not, as evaluated in various kinds of formative studies. For example, to evaluate the drawing application discussed in Section 5.3.1, authors conducted lab studies to validate if participants are able to draw lines, circles, or basic diagrams, and then ask their opinion on how easy or difficult was the drawing procedure [12,50]. Similarly, the Web browsing applications by Abe *et al.* [51] and the WeyeB [52] prototype had no comparative benchmark. The authors validated if users were able to accomplish basic Web browsing operations like clicking or scrolling on webpages by measuring success and failure rates. We argue that it is rather interesting to investigate how interfaces impact the user experience in terms of objective user performance in task execution, subjective usability, and workload impression. In this regard, we have proposed a comparative evaluation methodology to evaluate eye-controlled interfaces, e.g., applications such as a Twitter client [53] and a Web browser [54], GazeTheWeb.

Moreover, there is a need to move beyond the concept of eye tracking being a useful input method, to being usable in daily use of applications. Hence, it is imperative to conduct field studies on assessing how eye-controlled interfaces would be used by target-group users in a real-world environment. In this regard, we also discuss a field study of GazeTheWeb in subsequent section. This would lead to eye tracking being more acceptable as an interaction technology for end-users.

To showcase the procedure of a comparative and feasibility evaluation methodology of eye-controlled interfaces, we describe the methodology we used to evaluate our eye-controlled Web browser [2], which includes both the comparative evaluation as a lab study and the feasibility evaluation as a field study.

6.3.1 Comparative evaluation

For the evaluation of eye-controlled application interfaces, the choice of a comparative baseline is a crucial aspect. Especially, since the goal is to investigate how proposed enhancements improve the user experience of controlling an application, rather than specific performance measures related to pointing or typing. In our study related to GazeTheWeb, the aim has been to investigate how the gaze-adapted Web browser interface improves the user experience as compared to conventional method of Web browser control using eye tracking. The conventional method to employ eye tracking

*https://www.tobiidynavox.com/software/windows-software/windows-control
†http://www.mygaze.com/products/assistive-products/mygaze-power

for controlling any kind of application interfaces is the *emulation approach*, i.e., to emulate mouse and keyboard devices through an additional command-translation layer to provide indirect control of application interfaces through eye gaze. The emulation approach has existed for several years, e.g., the Eye-gaze Response Interface Computer Aid system [55,56]. The commercial systems like Tobii Dynavox Windows Control or Visual Interaction myGaze Power incorporate similar mechanisms. Microsoft has also integrated eye-controlled mouse and keyboard emulation into the Windows 10 operating system.[‡] There are also open-source alternatives, like OptiKey [57], which work effectively with affordable eye tracking systems available in the market. OptiKey has received high praise in recent years as a tool to assist gaze-based computer access.[§] In our study, we used OptiKey as an instance of the emulation approach to control the popular Google Chrome browser.[||]

6.3.1.1 Methodology

The central aim of the evaluation methodology was to assess if GazeTheWeb allows for effective Web browsing experience compared to controlling a browser with conventional emulation approach, i.e., Google Chrome with OptiKey. For this purpose, we conducted a lab study with 20 participants (9 females and 11 males) in the age range from 23 to 31 years (average = 25.55, sd = 2.16). All those participants were students at our university with no prior experience in operating GazeTheWeb or OptiKey. Seven participants wore corrective lenses (three females and four males). As per the ethics guidelines, we got the approval of each participant on the informed consent form prior to the study. The participants were paid each an amount of ten euro for their effort after the study. The participants were instructed to perform tasks involving common activities to search and browse the Web for finding specific information about German cities and subsequently to bookmark and access these bookmarks later.

The independent variables were the systems itself (GazeTheWeb vs. OptiKey and Google Chrome) (Figure 6.5). The dependent variables were related to task efficiency, i.e., the completion time for the entire task and for the high granular browsing operations such as typing, searching, scrolling, link clicking, back navigation, selecting bookmark, and marking bookmark. As per the eye-tracking control variables, we kept the dwell time constant (1 s), provided similar environment, with artificial illumination and blocking of sunlight and fixed distance of chair from the screen. The study employed a within-subjects design, as all participants used both the systems. Counter-balancing was used to eliminate any bias of one system over the other.

6.3.1.2 Objective measures

As mentioned before, the task completion time was objectively measured as a dependent variable to compare the performance. The participants were on average over 135 s faster with GazeTheWeb than with OptiKey to complete the overall tasks.

[‡] https://support.microsoft.com/en-us/help/4043921/windows-10-get-started-eye-control
[§] http://www.businessinsider.com/an-eye-tracking-interface-helps-als-patients-use-computers-2015-9, https://alsnewstoday.com/2016/03/08/article-for-als
[||] https://www.google.com/chrome

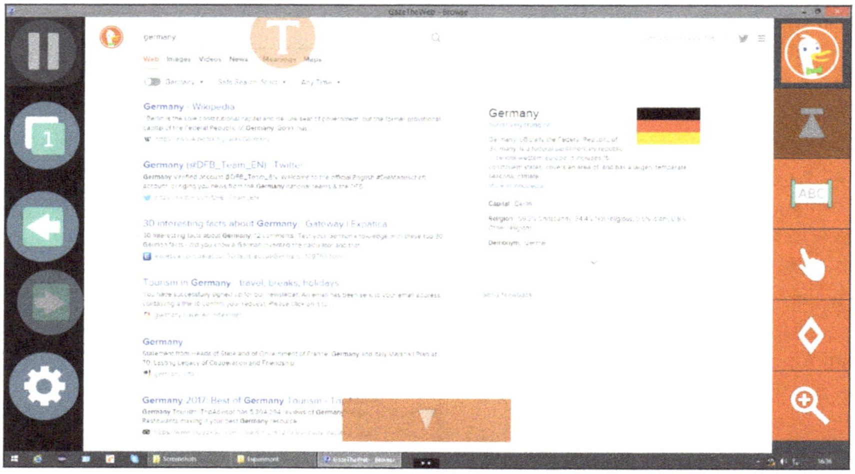

(a)

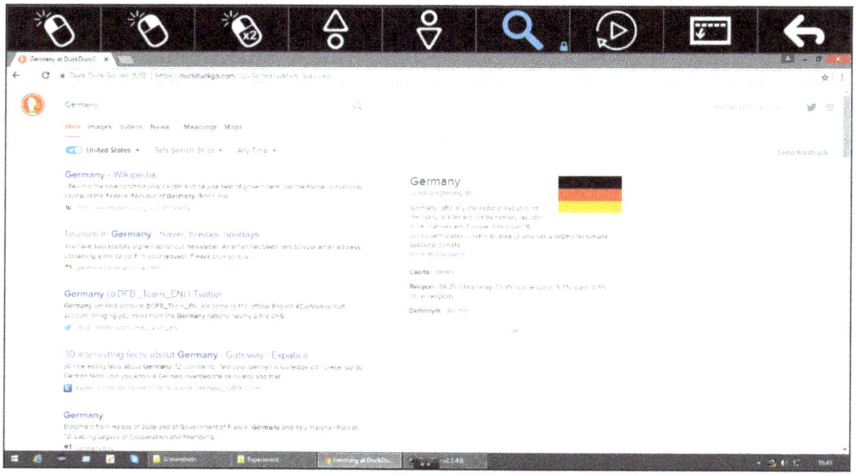

(b)

Figure 6.5 Screenshots of both systems from the comparative evaluation: (a) GazeTheWeb and (b) OptiKey and Google Chrome

Furthermore, the time differences between both systems appear to be normally distributed, according to a Shapiro–Wilk test with $p = 0.05$ threshold. This allowed us to assess the significance of the differences by a paired t-test calculating the two-tailed p-value. We reported a significant difference in the completion time for GazeTheWeb (average = 261.91 s, sd = 49.25) and OptiKey (average = 397.43 s, sd = 130.76), with

$t(19) = -5.23$, $p = 4.78E^{-5}$, and a high effect size of Cohen's $d = 1.17$. See Figure 6.6 for a box plot showing the consistent times the participant required to fulfill all tasks in GazeTheWeb, whereas the times for OptiKey showed much higher variety.

We also compared the timings for different browsing activities in GazeTheWeb with the timings in OptiKey as shown in Figure 6.7. The timings for typing and hyperlink clicking appear to be similar, whereas submission of search, back navigation, and bookmark management have been achieved faster with GazeTheWeb than using OptiKey. We reported a significant improvement in times for GazeTheWeb over

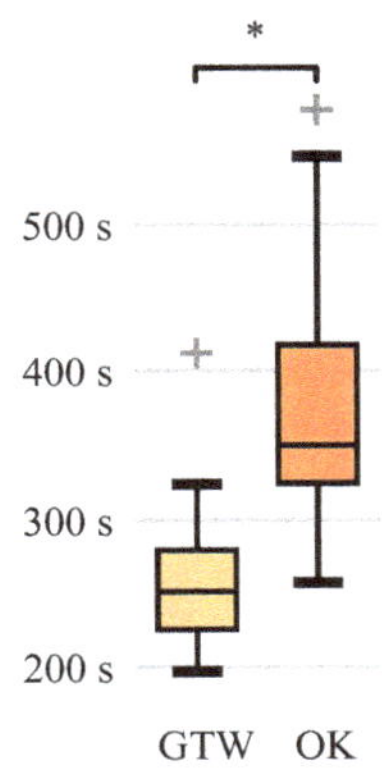

Figure 6.6 Task completion time

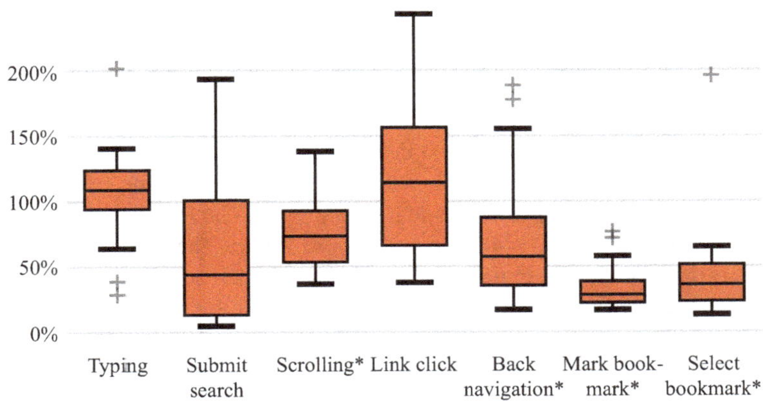

Figure 6.7 Times in GazeTheWeb in comparison to times in OptiKey as baseline. A percentage of 100 means a participant needed the same time in GazeTheWeb as in OptiKey, 50% means a participant needed only half the time with GazeTheWeb than with OptiKey, for the same task. The star symbol marks a significant difference between both systems in the reported dependent variable

OptiKey for scrolling ($W = 20$, $Z = -3.17$, $p < 0.05$, $r = 0.71$), back navigation ($W = 41$, $Z = -2.39$, $p < 0.05$, $r = 0.53$), marking ($W = 0$, $Z = -3.92$, $p < 0.05$, $r = 0.88$), and selection ($W = 14$, $Z = -3.4$, $p < 0.05$, $r = 0.76$) of bookmarks.

6.3.1.3 Subjective measures

The SUS questionnaire was used to measure the overall usability of applications, where the participants answered on a five-point Likert scale from *strongly disagree* to *strongly agree*. The NASA-TLX questionnaire was used to assess subjective workload, which contains six components: Mental demand, physical demand, temporal demand, performance, effort, and frustration. For each component, the participant specified the most applicable scores on a scale from 1 (low) to 7 (high). The custom gaze interaction design heuristics evaluation for eye-controlled interfaces was used [18,20], which could be answered on a scale from 1 *strongly disagree* to 10 *strongly agree*. Participants were also asked if they have any general feedback on what they liked, disliked, and further comments for improvement.

The average SUS usability score for GazeTheWeb was 77.13 in contrast to 55.0 for OptiKey, indicating an over-average acceptability rate of the gaze-adapted system of GazeTheWeb among the participants. The difference between the SUS scores by the participants appears to be normally distributed, according to a Shapiro–Wilk test with $p = 0.05$ threshold, and we have performed a paired t-test to assess the significance of the higher rating for GazeTheWeb. There was a significant difference in the overall SUS scores for GazeTheWeb (average = 77.13, sd = 16.08) and OK (average = 55.0, sd = 19.36), $t(19) = 3.6, p = 0.0019$.

The consistently better ratings of GazeTheWeb over OptiKey in terms of NASA-TLX mental workload, physical demand, level of effort, and sense of stress and irritation are presented in Figure 6.8. The feeling of success in accomplishment of the task was also slightly better for GazeTheWeb in comparison to OptiKey, since a lower value means a higher feeling of success in NASA-TLX questionnaire design.

The results of the gaze interaction design heuristics questionnaire are shown in Table 6.2, where we can see that actions like the ease of recovering from errors (#5) in GazeTheWeb receive better scores than in OptiKey. The intuitive factor (#3) and ease of hyperlink navigation (#4) are rated better for GazeTheWeb.

Furthermore, we received interesting insights from participants' open comments after usage, where participants stated that they liked the intuitive interaction aspect of GazeTheWeb, some example statements are: *"It was easy to navigate and also very easy to get used to start working with it"* emphasizes the good usability of GazeTheWeb, whereas about OptiKey a participant reported *"Too cumbersome and physically exhausting. Requires many clicks which are tiring for the eyes"*. Some participants explicitly reported about OptiKey on their cumbersome interaction experience, e.g., *"Multiple click for every action"* or *"Reduce left click mechanism using any other intuitive means. For example, let's say I am accessing the bookmarked URLs it should give easy access to the links by reducing the number of intermediate left clicks"*.

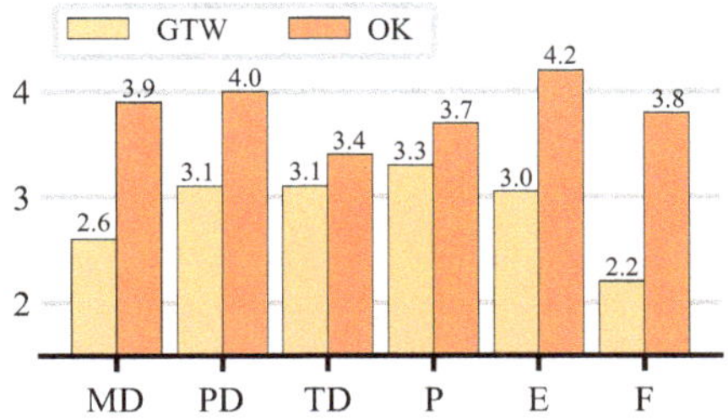

Figure 6.8 *Raw NASA-TLX scores. Lower scores signify lower perceived workload. MD, mental demand; PD, physical demand; TD, temporal demand; P, performance; E, effort; F, frustration*

Table 6.2 *Heuristics scores, higher scores are better*

Heuristics	**GTW**	**OK**
How was the *visibility* of the main interaction elements?	8.45	7.65
How comfortable was the *size* of the interaction elements?	7.95	6.95
How *intuitive* was the reading and scrolling experience?	8.00	6.05
How *easy* was handling the link navigation in the browser?	7.25	5.35
How easy was it to recover from *errors* made?	7.95	6.55
How *close* do you feel the interaction to conventional browsing?	7.90	6.40

6.3.2 Feasibility evaluation

The goal of a field study is to assess whether an eye-controlled interface is feasible in a real-world scenario. In this regard, we present results from the second phase trials of the MAMEM project, in which GazeTheWeb had been offered to people with motor impairment for home use. The system consisted of a laptop (with GazeTheWeb installed as startup program) and an eye tracking system, deployed to the homes of each of the 30 participants for one month.

6.3.2.1 Methodology

For the field study, GazeTheWeb system had been placed for one month at the homes of the participants at an accessible place within their homes. The major aim of the study was to assess the eye-controlled Web browsing behavior of the users in daily use environment. Hence, there were no usage guidelines, rewards, or requests to influence them for using the system. At the time of deployment, a person from the medical supervisory team gave an introduction and initial guidance through the system in native language to the participant and caretaker. After the deployment, participants were free to use the system as per their needs and preferences. The person from the medical supervisory team provided telephone support in native language, if there were any technical issues during the usage. For easier operation, GazeTheWeb started automatically after the startup of the laptop and offered the participant to perform a calibration of the eye tracker. Furthermore, if the participants' eyes could not be detected for 30 s, the participants have been offered a recalibration upon return. To record the behavioral pattern of field study, the system automatically logged the Web browsing activity to a custom Firebase[¶] real-time database, e.g., the system logged loaded webpages and times, clicks, amount of inserted text, and general interface use. Due to ethical reasons, we did not record the textual content itself but stored the string-edit distances. To assess the attitudinal aspect, we asked the participants to fill SUS questionnaire after one month usage, and for any open feedback about their overall experience, positive or negative aspects.

In total, 30 target-group users of eye tracking as an assistive technology were recruited for the field study. As per the recruitment process of participants, each of the candidates had been visited to verify if the eye-tracking system would work with them. Each associated clinical organization in the MAMEM project contacted participants of one potential target group. MDA Hellas, Greece, recruited ten participants (four females and six males, average age 31.5, sd = 4.8) with neuromuscular disease. These participants are referred to as MDA 1 to MDA 10 in the following. AUTH—School of Medicine, Greece, recruited ten participants (four females and six males, average age 55.6, sd = 7.3) with Parkinson's disease, referred to as AUTH 1 to AUTH 10 in the following. SHEBA—Academic Medical Center Hospital, Israel—also recruited ten participants (ten males, average age 38.1, sd = 10.8) with spinal cord injuries, referred to as SHEBA 1 to SHEBA 10 in the following. Most of the participants already had some computer accessories and assistive solutions available at their homes. Therefore, we did not expect that the participants would completely switch their existing means of interaction and start using GazeTheWeb excessively. However, the persistent usage of GazeTheWeb even by some participants would be relevant indicators on its functionality and effectiveness in supporting daily browsing activities. After one month usage, the system was collected back from the participant home, and the participants were asked to fill questionnaires, including an SUS, to quantify their gaze interaction experience.

As per ethics guidelines, the clinical protocol was followed after the ICH-GCP65 guidelines,[**] and an ethical approval (Helsinki Approval) had been obtained.

6.3.2.2 Measurements

GazeTheWeb worked successfully for the entire one month period, and we could observe the user behavior through the logged data on Firebase. The participants could visit and interact with variety of Web sites as per their needs and preferences, which signifies the usability of GazeTheWeb in supporting everyday browsing operations and handling dynamic webpages. An average SUS score of 73.2 among all participants indicates the general acceptability and satisfaction of GazeTheWeb.

For the field study, we collected data for total run-time of 186.24 h in Web browsing. Out of this, 118.93 h were categorized of active participation in front of

[¶] https://firebase.google.com
[**] http://www.ich.org/products/guidelines/efficacy/efficacy-single/article/good-clinical-practice.html

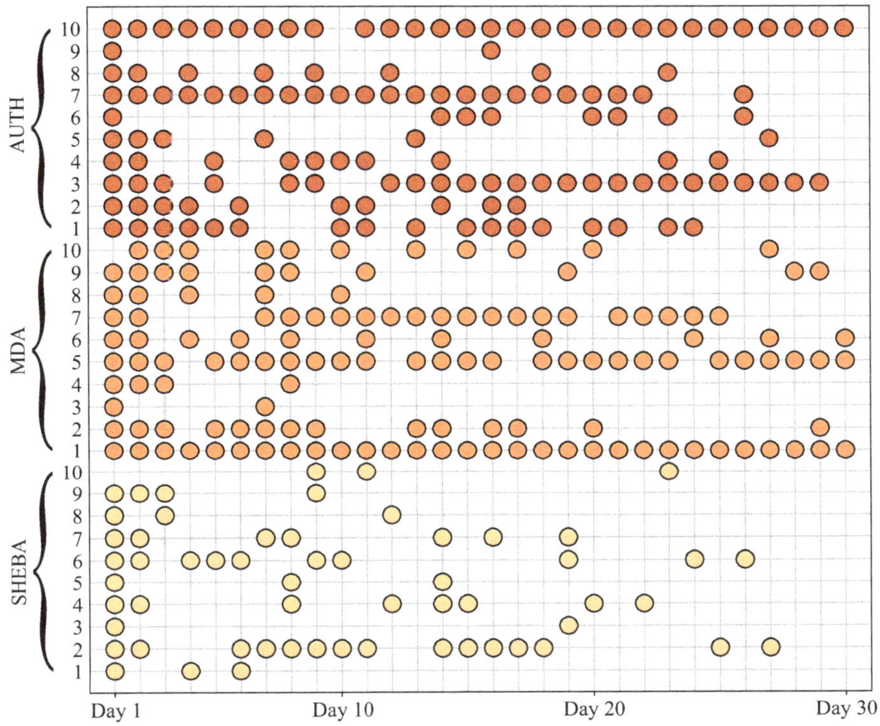

Figure 6.9 Daily use of GazeTheWeb in the field study. The vertical axis shows the participants. The horizontal axis displays the days since the setup of the system at the homes of the participants. Each dot signifies at least one start of the system by a participant on a specific day

the system, since we could check for the presence of participants from the gaze signals. There were regular use of the system by some of the participants, as shown in Figure 6.9. The participants have browsed to 456 unique domains, on which they have visited 8,415 Web pages. Table 6.3 indicates the top ten domains browsed by the participants. In total, there have been 8,027 clicks on Web pages and 22,811 characters have been entered in text inputs. Furthermore, participants have browsed to 498 URLs by typing and added 189 bookmarks. They also made the use of the multi-tab browsing through 857 tab switches.

The frequent usage of GazeTheWeb during the field study, including the visits of various websites, -pages, and the activities (e.g., clicks and text entry), indicates that users were able to interact with the variety of pages and to perform desired browsing operations. However, for the pages where the participants did not stay long or did not revisit, one could argue that interaction was difficult or not functional. For example, the gaming domains specifically received less interest among participants, such as

Table 6.3 Top 10 visited websites by participants. "Visits" are caused through URL input, navigation, bookmarks, or history use. "Stays" are visits that exceed an active time of one min, during which the participant had been registered by the eye-tracking system. "Pages" is the count of visited pages on the website. "Hours" is the time of browsing by a participant on the website. "duckduckgo.com" had been preset as search engine

Rank	Domain	Visits	Stays	Pages	Hours
1	facebook.com	369	229	1,208	24.3
2	youtube.com	271	203	1,396	26.5
3	duckduckgo.com	235	37	483	2.1
4	google.gr	123	24	270	1.6
5	mail.google.com	100	66	774	4.0
6	accounts.google.com	71	7	258	0.5
7	instagram.com	54	34	125	1.4
8	google.com	54	12	136	0.8
9	twitter.com	52	9	65	0.8
10	newsit.gr	50	46	257	3.7

"games.yo-yoo.co.il" with two visits and a stay duration of 14 min, "freegames.com" with one visit for 1.8 min, and "actiongame.com" with three visits and a stay duration of 12.2 min. Action games incorporate specialized controls for gameplay with high frequency of mouse and keyboard input, which would be nontrivial to adapt for gaze interaction in the proposed methodology of GazeTheWeb. There has been only one stay at "docs.google.com," with a duration of 6.2 min, and nobody has visited Google Maps. This indicated the limitations of the GazeTheWeb for specific websites and interesting future work direction for us.

Since the participants used the system for a month period, the longitudinal observations also provide specifics on their learning behavior and attitude about GazeTheWeb and feasibility of interaction with eye tracking in general. The usage behavior and feedback of participants, who used GazeTheWeb persistently over the course of a month, indicate that these participants get accustomed to the technology, and, hence, increase their usage and acceptance toward the system. For example, Figure 6.10 showcases the usage behavior of MDA 5, who consistently increased her activity over the time. The subjective feedback from MDA 5 also correlates with the pattern as she mentioned in her feedback that the *"usage was getting easier and faster as time went by. She used GazeTheWeb for 2 to 5 hours per day, while her previous use [of the Web] was almost zero. She reopened her Facebook accounts and Instagram, managed to communicate with a person abroad, [got] entertained by YouTube and was informed. She could organize a travel trip that took place through the use of GazeTheWeb. She would definitely use the eye tracker in the future."* More interestingly, while returning the equipment, she told us that she would miss it. Few weeks after the study she approaches us by phone to ask if she would again have

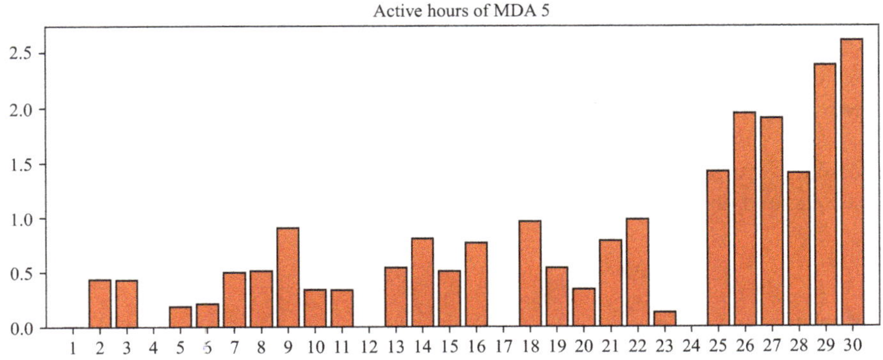

Figure 6.10 Daily use of GazeTheWeb for participant MDA 5 in terms of active hours in front of the system

the opportunity to use it through the MAMEM program or if she could buy it from somewhere.

Besides the positive signals, the longitudinal observation also reveals the challenges of eye tracking in daily usage. The repeated calibration requirement of eye tracker is one of such issues reported by most participants. The eye tracker gets often overheated, and sometimes it was required to unplug and restart the system again; hence, the participants needed assistance from their caretaker, which also can be considered current limitation for target-group feasibility. In summary, we attribute the infrequent usage of GazeTheWeb and eye-tracking technology in general to a multitude of factors as follows: (a) low motivation of participants in general; (b) low interest in using a technical system in general; (c) technical limitations of eye tracking calibration, precision, and accuracy; (d) the physical and cognitive demand of using an eye-tracking system; and (e) limitations of GazeTheWeb to perform some browsing activity or interact with specific websites.

6.4 Summary

Eye tracking has been explored as a computer input channel for several years. Various approaches have already been published with the focus on improving the performance and usability in interaction, and there is a continuous need to assess previous work and enhance the interaction methods further. For this purpose, a suitable evaluation methodology is an essential aspect for the innovation and advancement. Especially, for the research to be reproducible, it is highly recommended to clearly define all experimental variables as part of the evaluation methodology.

In this regard, this chapter elaborates on different evaluation methods used in gaze interaction research. We first discussed the common terminologies and guidelines for HCI experiments, and how they apply for eye tracking research. We summarized how the evaluation of atomic interactions like pointing, selection, and typing needs to be performed. Besides the micro-optimization for better pointing and typing, we argue the need to optimize and evaluate eye-controlled interfaces and their feasibility in daily use. Hence, we discussed the methodology to evaluate the eye-controlled application interfaces by means of comparative and feasibility evaluations of a browser application we had developed. We envision that these guidelines and examples would help the researchers to quickly recognize and solicit suitable study design, procedure, baseline, measures, and analysis methods to evaluate their ideas and approaches.

References

[1] Sengupta K, Sun J, Menges R, *et al.* Analyzing the Impact of Cognitive Load in Evaluating Gaze-Based Typing. In: 30th IEEE International Symposium on Computer-Based Medical Systems. vol. Special Track on Multimodal Interfaces for Natural Human–Computer Interaction: Theory and Applications. Thessaloniki: IEEE; 2017.

[2] Menges R, Kumar C, Staab S. Improving User Experience of Eye Tracking-Based Interaction: Introspecting and Adapting Interfaces. ACM Trans. Comput.-Hum. Interact. 2019. p. 46. DOI: https://doi.org/10.1145/3338844

[3] MacKenzie IS. Human–Computer Interaction: An Empirical Research Perspective (1st ed.). Morgan Kaufmann Publishers Inc., San Francisco, CA: Newnes; 2012.

[4] Martin DW. Doing Psychology Experiments (7th ed.). Belmont, CA: Wadsworth Cengage Learning; 2008.

[5] Menges R, Kumar C, Wechselberger U, *et al.* Schau genau! A gaze-controlled 3D game for entertainment and education. Journal of Eye Movement Research. 2017;10:220. Available from: https://bop.unibe.ch/JEMR/article/view/4182.

[6] Gao L. Latin Squares in Experimental Design Michigan State University. 2005.

[7] Kumar C, Hedeshy R, MacKenzie S, Staab S. TAGSwipe: Touch Assisted Gaze Swipe for Text Entry. In: Proceedings of the SIGCHI Conference on Human Factors in Computing Systems. CHI'20. New York, NY: ACM; 2020. Available from: http://dx.doi.org/10.1145/3313831.3376317

[8] Majaranta P, Aoki H, Donegan M, *et al.* Gaze Interaction and Applications of Eye Tracking: Advances in Assistive Technologies. 1st ed. Hershey, PA: IGI Global; 2011.

[9] Feit AM, Williams S, Toledo A, *et al.* Toward Everyday Gaze Input: Accuracy and Precision of Eye Tracking and Implications for Design. In: Proceedings of the 2017 CHI Conference on Human Factors in Computing Systems. CHI'17. New York, NY: ACM; 2017. p. 1118–1130. Available from: http://doi.acm.org/10.1145/3025453.3025599.

[10] MacKenzie IS, Tanaka-Ishii K. Text Entry Systems: Mobility, Accessibility, Universality. Amsterdam, Elsevier; 2010.

[11] MacKenzie IS. Evaluating eye tracking systems for computer input. In Gaze Interaction and Applications of Eye Tracking: Advances in Assistive Technologies. IGI Global; 2012. p. 205–225.

[12] Hornof AJ, Cavender A. EyeDraw: Enabling Children With Severe Motor Impairments to Draw With Their Eyes. In: Proceedings of the SIGCHI Conference on Human Factors in Computing Systems. CHI'05. New York, NY: ACM; 2005. p. 161–170. Available from: http://doi.acm.org/10.1145/1054972.1054995.

[13] World Medical Association. World Medical Association Declaration of Helsinki. Ethical principles for medical research involving human subjects. Bulletin of the World Health Organization. 2001;79(4):373.

[14] Kar A, Corcoran P. A review and analysis of eye-gaze estimation systems, algorithms and performance evaluation methods in consumer platforms. IEEE Access. 2017;5:16495–16519.

[15] Greenland S, Robins JM, Pearl J, *et al.* Confounding and collapsibility in causal inference. Statistical Science. 1999;14(1):29–46.

[16] Brooke J, SUS—A quick and dirty usability scale. Usability evaluation in industry. 1996;189(194):4–7.

[17] Hart SG, Staveland LE. Development of NASA-TLX (Task Load Index): Results of empirical and theoretical research. In: Advances in Psychology. vol. 52. Amsterdam: Elsevier; 1988. p. 139–183.

[18] Nielsen J, Molich R. Heuristic Evaluation of User Interfaces. In: Proceedings of the SIGCHI Conference on Human Factors in Computing Systems. CHI'90. New York, NY: ACM; 1990. p. 249–256. Available from: http://doi.acm.org/10.1145/97243.97281.

[19] Nielsen J. How to Conduct a Heuristic Evaluation. vol. 1. Nielsen Norman Group; 1995. p. 1–8.

[20] Sengupta K, Kumar C, Staab S. Usability Heuristics for Eye-Controlled User Interfaces. In: 19th European Conference on Eye Movements; 2017. Available from: http://cogain2017.cogain.org/camready/poster3-Sengupta.pdf.

[21] Jacob RJK. What You Look at is What You Get: Eye Movement-Based Interaction Techniques. In: Proceedings of the SIGCHI Conference on Human Factors in Computing Systems. CHI'90. New York, NY: ACM; 1990. p. 11–18. Available from: http://doi.acm.org/10.1145/97243.97246.

[22] MacKenzie IS. Fitts' Law As a Research and Design Tool in Human–Computer Interaction. Human–Computer Interaction. 1992;7(1):91–139. Available from: http://dx.doi.org/10.1207/s15327051hci0701_3.

[23] Zhang X, Ren X, Zha H. Modeling Dwell-Based Eye Pointing Target Acquisition. In: Proceedings of the SIGCHI Conference on Human Factors in Computing Systems. CHI'10. New York, NY: ACM; 2010. p. 2083–2092. Available from: http://doi.acm.org/10.1145/1753326.1753645.

[24] Blanch R, Guiard Y, Beaudouin-Lafon M. Semantic Pointing: Improving Target Acquisition With Control-Display Ratio Adaptation. In: Proceedings of the SIGCHI Conference on Human Factors in Computing Systems. CHI'04. New York, NY: ACM; 2004. p. 519–526. Available from: http://doi.acm.org/10.1145/985692.985758.

[25] Zhai S, Morimoto C, Ihde S. Manual and Gaze Input Cascaded (MAGIC) Pointing. In: Proceedings of the SIGCHI Conference on Human Factors

in Computing Systems. CHI'99. New York, NY: ACM; 1999. p. 246–253. Available from: http://doi.acm.org/10.1145/302979.303053.

[26] Grossman T, Balakrishnan R. The Bubble Cursor: Enhancing Target Acquisition by Dynamic Resizing of the Cursor's Activation Area. In: Proceedings of the SIGCHI Conference on Human Factors in Computing Systems. CHI'05. New York, NY: ACM; 2005. p. 281–290. Available from: http://doi.acm.org/10.1145/1054972.1055012.

[27] International Organization for Standardization. Ergonomics of Human–System Interaction (Subcommittee) ISE. Ergonomic Requirements for Office Work With Visual Display Terminals (VDTs): Guidance on Usability. ISO 9241-11:1998.

[28] Zhang X, MacKenzie IS. Evaluating Eye Tracking With ISO 9241—Part 9. In: International Conference on Human–Computer Interaction. Berlin: Springer; 2007. p. 779–788.

[29] MacKenzie IS, Isokoski P. Fitts' Throughput and the Speed-Accuracy Tradeoff. In: Proceedings of the SIGCHI Conference on Human Factors in Computing Systems. New York, NY: ACM; 2008. p. 1633–1636.

[30] Soukoreff RW, MacKenzie IS. Towards a standard for pointing device evaluation, perspectives on 27 years of Fitts' law research in HCI. International Journal of Human–Computer Studies. 2004;61(6):751–789.

[31] Rajanna V, Hammond T. A Fitts' Law Evaluation of Gaze Input on Large Displays Compared to Touch and Mouse Inputs. In: Proceedings of the Workshop on Communication by Gaze Interaction. New York, NY: ACM; 2018. p. 8.

[32] Hansen JP, Rajanna V, MacKenzie IS, *et al.* A Fitts' Law Study of Click and Dwell Interaction by Gaze, Head and Mouse With a Head-Mounted Display. In: Proceedings of the Workshop on Communication by Gaze Interaction. New York, NY: ACM; 2018. p. 7.

[33] Bates R, Istance H. Zooming Interfaces!: Enhancing the Performance of Eye Controlled Pointing Devices. In: Proceedings of the Fifth International ACM Conference on Assistive Technologies. Assets'02. New York, NY: ACM; 2002. p. 119–126. Available from: http://doi.acm.org/10.1145/638249.638272.

[34] Ashmore M, Duchowski AT, Shoemaker G. Efficient Eye Pointing With a Fisheye Lens. In: Proceedings of Graphics Interface 2005. GI'05. School of Computer Science, University of Waterloo, Waterloo, Ontario: Canadian Human–Computer Communications Society; 2005. p. 203–210. Available from: http://dl.acm.org/citation.cfm?id=1089508.1089542.

[35] Schenk S, Dreiser M, Rigoll G, *et al.* GazeEverywhere: Enabling Gaze-Only User Interaction on an Unmodified Desktop PC in Everyday Scenarios. In: Proceedings of the 2017 CHI Conference on Human Factors in Computing Systems. CHI'17. New York, NY: ACM; 2017. p. 3034–3044. Available from: http://doi.acm.org/10.1145/3025453.3025455.

[36] Dorr M, Böhme M, Martinetz T, *et al.* Gaze Beats Mouse: A Case Study. In: Proceedings of COGAIN; 2007. p. 16–19.

[37] Kumar M, Winograd T. GUIDe: Gaze-Enhanced UI Design. In: CHI'07 Extended Abstracts on Human Factors in Computing Systems. CHI EA'07. New York, NY: ACM; 2007. p. 1977–1982. Available from: http://doi.acm.org/10.1145/1240866.1240935.

[38] Diaz-Tula A, Morimoto CH. AugKey: Increasing Foveal Throughput in Eye Typing With Augmented Keys. In: Proceedings of the 2016 CHI Conference on Human Factors in Computing Systems. CHI'16. New York, NY: ACM; 2016. p. 3533–3544. Available from: http://doi.acm.org/10.1145/2858036.2858517.

[39] Kurauchi A, Feng W, Joshi A, *et al.* EyeSwipe: Dwell-Free Text Entry Using Gaze Paths. In: Proceedings of the 2016 CHI Conference on Human Factors in Computing Systems. CHI'16. New York, NY: ACM; 2016. p. 1952–1956. Available from: http://doi.acm.org/10.1145/2858036.2858335.

[40] Tuisku O, Majaranta P, Isokoski P, *et al.* Now Dasher! Dash Away!: Longitudinal Study of Fast Text Entry by Eye Gaze. In: Proceedings of the 2008 Symposium on Eye Tracking Research & Applications. New York, NY: ACM; 2008. p. 19–26.

[41] Wobbrock JO, Rubinstein J, Sawyer MW, *et al.* Longitudinal Evaluation of Discrete Consecutive Gaze Gestures for Text Entry. In: Proceedings of the 2008 Symposium on Eye Tracking Research & Applications. New York, NY: ACM; 2008. p. 11–18.

[42] Sengupta K, Menges R, Kumar C, *et al.* Impact of Variable Positioning of Text Prediction in Gaze-Based Text Entry. In: Proceedings of the 11th ACM Symposium on Eye Tracking Research & Applications. New York, NY: ACM; 2019. p. 74.

[43] Soukoreff RW, MacKenzie IS. Metrics for Text Entry Research: An Evaluation of MSD and KSPC, and a New Unified Error Metric. In: Proceedings of the SIGCHI Conference on Human Factors in Computing Systems. New York, NY: ACM; 2003. p. 113–120.

[44] MacKenzie IS. KSPC (Keystrokes per Character) as a Characteristic of Text Entry Techniques. In: International Conference on Mobile Human–Computer Interaction. Berlin: Springer; 2002. p. 195–210.

[45] Antonenko P, Paas F, Grabner R, *et al.* Using electroencephalography to measure cognitive load. Educational Psychology Review. 2010;22(4): 425–438.

[46] Başar E. Brain Function and Oscillations: Volume II: Integrative Brain Function. Neurophysiology and Cognitive Processes. Berlin: Springer Science & Business Media; 2012.

[47] Hwang T, Kim M, Hwangbo M, *et al.* Comparative Analysis of Cognitive Tasks for Modeling Mental Workload With Electroencephalogram. In: Engineering in Medicine and Biology Society (EMBC), 2014 36th Annual Int'l Conf. of the IEEE; 2014. p. 2661–2665.

[48] Fadzal CCW, Mansor W, Khuan L, *et al.* Short-Time Fourier Transform Analysis of EEG Signal From Writing. In: Signal Processing and Its Applications (CSPA), 2012 IEEE 8th Int'l Colloquium on; 2012. p. 525–527.

[49] Sengupta K, Menges R, Kumar C, *et al.* GazeTheKey: Interactive Keys to Integrate Word Predictions for Gaze-Based Text Entry. In: Proceedings of the 22Nd International Conference on Intelligent User Interfaces Companion. IUI'17 Companion. New York, NY: ACM; 2017. p. 121–124. Available from: http://doi.acm.org/10.1145/3030024.3038259.

[50] Heikkilä H. EyeSketch: A Drawing Application for Gaze Control. In: Proceedings of the 2013 Conference on Eye Tracking South Africa. ETSA'13. New York, NY: ACM; 2013. p. 71–74. Available from: http://doi.acm.org/10.1145/2509315.2509332.

[51] Abe K, Owada K, Ohi S, *et al.* A system for Web browsing by eye-gaze input. Electronics and Communications in Japan. 2008;91(5):11–18. Available from: http://dx.doi.org/10.1002/ecj.10110.

[52] Porta M, Ravelli A. WeyeB, an Eye-Controlled Web Browser for Hands-free Navigation. In: Proceedings of the 2nd Conference on Human System Interactions. HSI'09. Piscataway, NJ: IEEE Press; 2009. p. 207–212. Available from: http://dl.acm.org/citation.cfm?id=1689359.1689396.

[53] Kumar C, Menges R, Staab S. Assessing the Usability of Gaze-Adapted Interface Against Conventional Eye-Based Input Emulation. In: 2017 IEEE 30th International Symposium on Computer-Based Medical Systems (CBMS); 2017. p. 793–798.

[54] Menges R, Kumar C, Müller D, *et al.* GazeTheWeb: A Gaze-Controlled Web Browser. In: Proceedings of the 14th Web for All Conference. W4A'17. New York, NY: ACM; 2017. Available from: http://dx.doi.org/10.1145/3058555.3058582.

[55] Lankford C. Effective Eye-Gaze Input Into Windows. In: Proceedings of the 2000 Symposium on Eye Tracking Research & Applications. ETRA'00. New York, NY: ACM; 2000. p. 23–27. Available from: http://doi.acm.org/10.1145/355017.355021.

[56] Zhang X, Liu X, Yuan SM, *et al.* Eye Tracking Based Control System for Natural Human–Computer Interaction. Computational Intelligence and Neuroscience: CIN; 2017.

[57] Sweetland J. Optikey: Type, Click, Speak; 2016. https://github.com/OptiKey/OptiKey.

Chapter 7
Machine-learning techniques for EEG data

Vangelis P. Oikonomou[1], Spiros Nikolopoulos[1], and Ioannis Kompatsiaris[1]

In this chapter, we present an introductory overview of machine-learning techniques that can be used to recognize mental states from electroencephalogram (EEG) signals in brain–computer interfaces (BCIs). More particularly, we discuss how to extract relevant and robust information from noisy EEG signals. Due to the spatial properties of the EEG acquisition modality, learning robust spatial filters is a crucial step in the analysis of EEG signals. Optimal spatial filters will help us extract relevant and robust features, helping considerably the subsequent recognition of mental states. Also, a few classification algorithms are presented to assign this information into a mental state. Furthermore, particular care will be given on algorithms and techniques related to steady-state visual evoked potentials (SSVEPs) BCI and sensorimotor rhythms (SMRs) BCI systems. The overall objective of this chapter is to provide the reader with practical knowledge about how to analyze EEG signals.

7.1 Introduction

Machine learning has become a core component in any data analysis approach, including the analysis of EEGs. In this chapter, we will discuss the basic ideas of machine learning and how these ideas have been used to analyze EEG signals. More specifically, we will concentrate our efforts around supervised learning methodologies to design spatial filters and classifiers. This chapter serves as the introductory material for the various usages of EEG signal in the design of BCI systems, which are presented in the subsequent chapters. While this chapter introduces the main ideas of algorithms, we do not aim for a full treatment of the available literature. Rather, we present a somewhat biased view, mainly drawing from the authors' work.

7.1.1 What is the EEG signal?

Electroencephalography is a brain imaging technique that measures the brain's electrical activity by placing sensors on the head's surface. During the activation of a neuron,

[1]The Multimedia Knowledge and Social Media Analytics Laboratory, Information Technologies Institute, Centre for Research and Technology-Hellas (CERTH), Thessaloniki, Greece

a tiny weak electrical field is generated, but as neural activity becomes synchronous across thousands of neurons, the electrical activity is summed up and it is powerful enough to be measured outside the head. Current technological advancements give us the ability to measure accurately this activity, hence producing the EEG, a biomedical signal describing the status of the brain. The EEG signal is produced by performing recordings at multiple different brain sites (i.e., multichannels' recordings). It is a nonstationary high dimensional time series with temporal and spatial correlations. The EEG is used in many applications in brain studies for diagnosing brain disorders such as the epilepsy [1,2], Alzheimer's disease [3,4], the stroke [5], and Parkinson's disease [6]. Besides health-care applications, EEG is used as a communication channel in BCIs systems [7,8]. In this chapter, we will concentrate our study on using the EEG for BCI applications, especially in SSVEP BCI and SMR BCI.

7.1.2 EEG-based BCI paradigms

An EEG-based BCI system (Figure 7.1) translates the recorded electric brain activity to output commands. The input of a BCI system is the electrophysiological brain activity, while the output is the device commands. The brain activity is recorded through the use of an EEG system. After that, the analysis of EEG signals is performed in order to extract the intended commands of the user. A BCI system contains the following modules: (a) stimulator module (or BCI paradigm): this module is responsible to produce/enhance the desired brain activity; (b) signal acquisition module: which is responsible to acquire the EEG signals during the system operation; (c) signal processing module: which is responsible for the analysis of EEG signals and the translation/transformation of them into meaningful "codewords"; and (d) device commands module: which is appointed with the task to translate the "codewords"

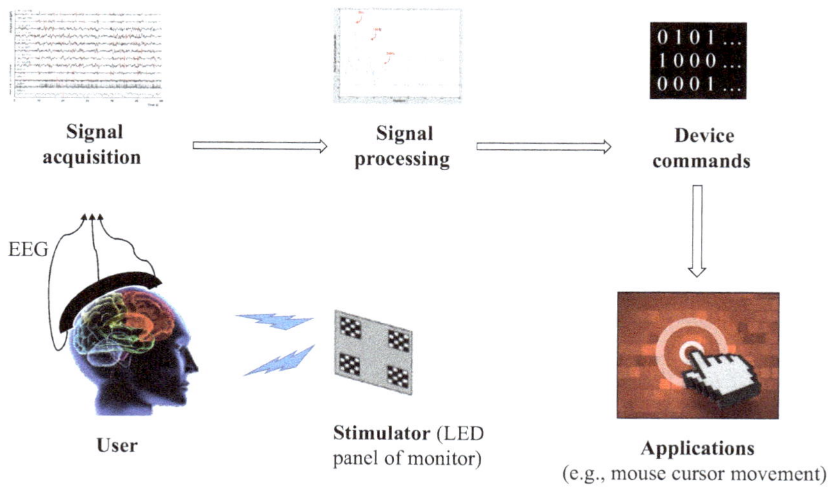

Figure 7.1 Basic parts of a BCI system

into interface commands according to the application setup. In the next paragraphs, we provide a short description of two basic BCI paradigms, the SSVEP paradigm and the SMR paradigm.

7.1.2.1 SSVEP paradigm

An SSVEP-based BCI system enables the user to select among several commands that depend on the application, e.g., directing a cursor on a computer screen. Each command is associated with a repetitive visual stimulus that has distinctive properties (e.g., frequency). The stimuli are simultaneously presented to the user who selects a command by focusing his/her attention on the corresponding stimulus. When the user focuses his/her attention on the stimulus, an SSVEP is produced that can be observed in the oscillatory components of the user's EEG signal, especially in the signals generated from the primary visual cortex. In these components, we can observe the frequency of the stimulus, as well as its harmonics. SSVEPs can be produced by repetitively applying visual stimuli to the user with frequencies higher to 6 Hz. Compared to other brain signals (e.g., P300 and SMRs) used for BCI approaches, SSVEP-based BCI systems have the advantage of achieving higher accuracy and higher information transfer rate. In addition, short/no training time is required by its users.

7.1.2.2 SMR paradigm

SMRs are brain waves that appear on EEG recordings from areas of the brain, which are associated with planning, control, and execution of voluntary movements. The execution of movements involves different brain regions, especially when the movement is related to the left or right hand. More specifically, the movement of the left hand reflects changes in EEG rhythms on the right part of the brain, while the right-hand movement reflects changes in EEG rhythms on the left part of the brain. Similar observations can be made when we imagine the corresponding movement. BCI systems based on SMRs try to exploit the previous observation and to find which brain area is activated.

To gather EEG data, an experimental protocol of two basic steps is applied. In the first step, the EEG data are acquired without showing any feedback to the user (calibration step), while at the second step, feedback is incorporated into the overall procedure (feedback step). In most cases, the feedback is the prediction of the BCI system. In both steps, a number of EEG trials are acquired for further processing. The trials acquired during the first step are gathered in order to train a classifier to be exploited in the subsequent feedback step, while the ones from the latter step are related to the training of the user. Thus, by providing feedback, the user is learning to adapt to the system by regulating appropriate brain rhythms. It is worth to point out here a significant difference between SSVEP and SMR BCI systems with respect to the produced brain waves. SSVEP signals are produced due to the existence of an external stimulus, while the SMRs are the results of an internal subject's process [8]. Furthermore, in SMR-based systems, a closed-loop connection is developed between the user and the system [8]. All the previously stated have considerable effects on the type of application that these two BCI systems could be applied.

7.1.2.3 General properties of an EEG dataset

EEG datasets are very complicated and difficult to analyze. The complexity and difficulty derive from the various properties of the acquisition process, the human brain physiology, and the nature of the collected data. During the acquisition process, we could collect data from 8 surface brain sites (channels) up to 256 sites. Furthermore, these data could be collected under various sampling rates ranging from 128–2,048 Hz. Also, during an experiment, we could collect EEG data from many (healthy or not) participants. Under the same experimental conditions, we expect some kinds of similarity between EEG data of different participants, but great diversity between them could be also present. In addition, EEG data are collected under various experimental conditions and many times for each condition resulting in EEG segments called trials. Furthermore, each EEG trial contains thousand of raw features. Most EEG studies result into a dataset with too many raw features (EEG samples) and very few trials making the data analysis procedure very difficult. Based on the previous short description, it is difficult to have a holistic approach for analyzing EEG data (i.e., an approach that works fair enough on all EEG datasets), in contradiction to other pattern recognition problems such as face recognition, speech analysis, and visual object recognition.

7.1.3 What is machine learning?

Learning is the process of acquiring new knowledge or skills through study, experience, or being taught. Humans learn before birth and continue until death as a consequence of ongoing interactions between people and their environment. As we see, in humans, learning involves the processing of information (or data) in order to produce new knowledge or skills. In analogy, machine learning involves an algorithm or a computer program that is able to learn from the data. At the very basic level, machine learning is about predicting the future based on the past. More specifically, we define a model with some parameters, and learning is the execution of an algorithm to optimize the parameters of the model using the training data or past experience. The model may be predictive, to make predictions about the future, or descriptive, to gain knowledge from data, or both.

Machine-learning methods are divided into two large approaches. In the supervised learning approach, the goal is to learn a mapping, $f(\cdot)$, from inputs $\mathbf{x}$ to outputs y, given a dataset. The dataset $\mathscr{D}$ is composed by pairs of inputs–outputs $\mathscr{D} = \{(\mathbf{x}_i, y_i)\}_{i=1}^{N}$, where N is the number of training examples. In the simplest case, each training input $\mathbf{x}_i$ is a D-dimensional vector of numbers, which is called *features*. However, in more general setting, each $\mathbf{x}_i$ can represent a more complex object such as images and EEG trials. The output (or response variable) y_i can in principle be anything, but two cases are of great interest for us. In the first case, each y_i takes value from some finite set, $y_i \in \{1, 2, \ldots, C\}$. This problem is known as classification. In the second case, y_i is a real-valued scalar. This problem is known as regression. The second approach of machine-learning methods is called unsupervised learning. In this type of method, we are only given the inputs, $\mathscr{D} = \{(\mathbf{x}_i)\}_{i=1}^{N}$ and the goal is to find valuable patterns in the data. This kind of learning includes clustering approaches and some forms of dimensionality reduction methods such as the principal component analysis (PCA) and the independent component analysis.

Due to the extensive use of supervised learning methods, in other chapters of this book, next we provide a more elaborate description of the supervised learning approach. The goal in supervised learning is to learn the mapping $f(\cdot)$ between the input and the output values (i.e., $y=f(\mathbf{x})$). In most cases, this mapping/function depends on model/function parameters $\boldsymbol{\theta}$ ($y=f(\mathbf{x};\boldsymbol{\theta})$). In order to find the optimal function, one needs to specify a suitable loss function that evaluates the goodness of the fitting $\ell(y_i, f(\mathbf{x}_i;\boldsymbol{\theta}))$ between true values and predicted values. Given a specific loss function, the best function $f(\cdot)$ is obtained when we minimize the expected risk (or generalization error), which cannot be minimized directly [9]. Therefore, we use an approximation of the expected risk, the averaged loss on the training samples: $(1/N)\sum_{i=1}^{N}\ell(y_i, f(\mathbf{x}_i;\boldsymbol{\theta}))$ [9]. In some cases, to avoid overfitting and account for model inconsistencies, we use the modified averaged loss function $\mathscr{L}(\boldsymbol{\theta}) = (1/N)\sum_{i=1}^{N}\ell(y_i, f(\mathbf{x}_i;\boldsymbol{\theta})) + \lambda\|Tf\|^2$, where the term $\lambda\|Tf\|^2$ is a regularization term over the properties of function $f(\cdot)$.

7.1.4 What do you want to learn in EEG analysis for BCI application?

In EEG analysis, machine-learning techniques are found on the signal processing module (Figure 7.2). In the typical case, the signal processing module consists of four submodules: (a) preprocessing, (b) feature extraction (FE), (c) feature selection (FS), and (d) classification. The first three submodules have the goal to make the data suitable for the classification process, which will give us the appropriate "codewords." In each of the previous submodules, we can find many machine-learning techniques.

After collecting the data (i.e., EEG trials), a critical step is the application of a dimensionality reduction method. It must be pointed out that in BCI applications, the raw data are of very large dimension. For example, in an SSVEP experiment, where we collect data from 256 channels for 4 s with a sampling frequency of 256 Hz, we obtain a feature vector of 262,144 features. Furthermore, the number of training samples is typically small, up to a few hundred samples. In order to make the classification feasible, the dimensionality of the input data needs to be significantly reduced, and informative features have to be extracted. Dimensionality reduction can be achieved

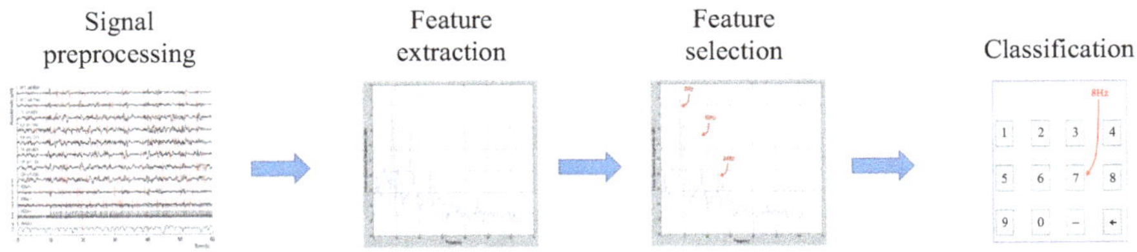

Figure 7.2 Basic parts of the signal processing module in a BCI system. Pattern recognition system of an EEG-based BCI system.

by performing FE or FS, or both. FE involves the transformation of the existing features into a lower dimensional space (i.e., PCA, common spatial pattern (CSP)) [10]. An FE approach closely related to BCI is the learning of optimal spatial filters. In contradiction to FE, FS is about selecting a subset of the existing features without a transformation. Although FS can be seen as a special case of FE, in practice it is quite different, since FS searches for a feature's subset that minimizes some cost function (i.e., student's t-statistics, biserial correlation) [11].

BCI experiments often aim at comparing specific brain states. Typically, we choose a neurophysiological paradigm that maximizes the contrast between various brain states. After recording brain imaging data, the goal of the analysis is to find significant differences in the spatial and the temporal characteristics of the data contrasting the different states as accurately as possible. The brain states are related to the type of BCI paradigm. For example, an SSVEP BCI system uses the fact that the visual cortex produces brain activity related to the visual stimulus frequency, while an SMR BCI system is based on the fact that the movement imagination of different parts of the human body (i.e., hands) activates distinct brain areas in motor cortex. As we can observe, the discrimination of brain states can be formulated as a classification problem (i.e., learning to discriminate the brain states). Widely used classifiers in the context of BCI are: the linear discriminant analysis (LDA) [12], the Bayesian LDA (BLDA) [13,14], and the support vector machines (SVMs) [12]. Concluding this section, we can mention that from a machine-learning perspective, we are interested, mainly, in learning good features to feed the classifier as well as to learn a boundary with good generalization capabilities (a classifier) in order to discriminate the brain states.

7.2 Basic tools of supervised learning in EEG analysis

In this section, a short description of three very important tools is provided. These tools are the generalized Rayleigh quotient function, the linear regression model, and the Bayesian framework. These tools can be found at the heart of most machine-learning algorithms, especially those related to EEG signal processing, as we will see in the next sections.

7.2.1 Generalized Rayleigh quotient function

In some problems, we want to maximize (or minimize) a cost function that is described as the ratio of two quantities. One significant cost function of this kind is the generalized Rayleigh quotient function given by

$$J(\mathbf{w}) = \frac{\mathbf{w}^T \mathbf{A} \mathbf{w}}{\mathbf{w}^T \mathbf{B} \mathbf{w}} \tag{7.1}$$

where $\mathbf{A} \in \Re^{N \times N}$ and $\mathbf{B} \in \Re^{N \times N}$ are symmetric matrices ($\mathbf{B}$ is also invertible) and $\mathbf{w} \in \Re^{N \times 1}$ is a vector (to be selected). We can observe that this function is a scalar and the value of $J(\mathbf{w})$ is irrelevant to the norm of $\mathbf{w}$, scaling $\mathbf{w}$ does not change the value of the function (i.e., $J(a\mathbf{w}) = J(\mathbf{w})$). Note here that when matrices **A** and **B** are also positive definite (i.e., $\mathbf{w}^T\mathbf{A}\mathbf{w} > 0$, $\mathbf{w}^T\mathbf{B}\mathbf{w} > 0$) then $J(\mathbf{w}) > 0$.

Maximizing the $J(\mathbf{w})$ is equivalent to the following constrained optimization problem:

$$\max_{\mathbf{w}} \mathbf{w}^T\mathbf{A}\mathbf{w} \text{ subject to } \mathbf{w}^T\mathbf{B}\mathbf{w} = 1 \tag{7.2}$$

Adopting the Lagrange multipliers approach, we can show that the optimal **w** is the eigenvector (corresponding to the largest eigenvalue, $\lambda_{\max}$) of the following generalized eigenvalue problem [15,16]:

$$\mathbf{A}\mathbf{w} = \lambda\mathbf{B}\mathbf{w}. \tag{7.3}$$

Also, we can observe that the function $J(\mathbf{w})$ is bounded below and above from the smallest and largest eigenvalue ($\lambda_{\min} \leq J(\mathbf{w}) \leq \lambda_{\max}$). Furthermore, the previous study can be extended to include more general cases when more than one eigenpairs are needed [16].

Finally, it is worth to note here that the previous function has found numerous application in machine-learning and signal processing communities. Close connections of the generalized Rayleigh quotient function can be found with the maximization of signal-to-noise ratio [15,17], the PCA [15], the canonical correlation analysis (CCA) [16], the partial least squares (PLS) [16], and the Fisher's LDA [18].

7.2.2 Linear regression modeling

One of the most important models to describe relations between input and output quantities is the linear regression model [18]. In this model, the output (observations) $\mathbf{y} = \{y_1, \ldots, y_N\}$ is described as a linear combination of the input (predictors) given by the following equation:

$$\mathbf{y} = \mathbf{\Phi}\mathbf{w} + \mathbf{e}, \tag{7.4}$$

where $\mathbf{\Phi}$ is the design matrix of size $N \times p$ and it is assumed to be known for the problem under study, **w** is the vector of weights of the linear combination and has size $p \times 1$, and **e** is the additive noise assumed to be zero mean and Gaussian distributed, $p(\mathbf{e}) = \mathcal{N}(0, \mathbf{C}_{\mathbf{e}}^{-1})$, where $\mathbf{C}_{\mathbf{e}}^{-1}$ is the inverse precision (covariance) matrix. The form of this matrix defines the properties of the additive noise. Usually, we assume that the error samples are independent and identically distributed; in that case, a simple approach is to assume $\mathbf{C}_{\mathbf{e}}^{-1} = \lambda\mathbf{I}$. Also, more general forms can be used such as a diagonal precision matrix, where we use for each observation y_n a separate precision λ_n. This form of the precision matrix helps to use in indirect way more useful distributions such as the student t distribution [19]. Finally, the autoregressive model can be alternatively used to describe the autocorrelation between the error samples, where it can be written in the general form of the additive noise [20].

In the next paragraphs, the role of the design matrix and the various forms of it will be described. The design matrix has the following general form:

$$\mathbf{\Phi} = \begin{bmatrix} \phi_1(\mathbf{x}_1) & \phi_2(\mathbf{x}_1) & \cdots & \phi_p(\mathbf{x}_1) \\ \phi_1(\mathbf{x}_2) & \phi_2(\mathbf{x}_2) & \cdots & \phi_p(\mathbf{x}_2) \\ \cdots & \cdots & \cdots & \cdots \\ \phi_1(\mathbf{x}_N) & \phi_2(\mathbf{x}_N) & \cdots & \phi_p(\mathbf{x}_N) \end{bmatrix}, \tag{7.5}$$

where $\{\mathbf{x}_n\}_{n=1}^{N}$ are the input variables and $\phi_j(\cdot), j = 1, \ldots, p$ are the basis functions, both, the input variables and the basis functions, are assumed to be known. According to the linear model described previously, each observation y_n is described as a linear combination of p basis functions:

$$y_n = \sum_{j=1}^{p} w_j \phi_j(\mathbf{x}_n) + e_n = \mathbf{w}^T \boldsymbol{\phi}(\mathbf{x}_n) + e_n. \tag{7.6}$$

We see that the basis functions describe the relationship between the observations and the input variables. In the literature, many forms for the basis functions have been proposed. In the case, where a linear relationship between the observations and the input variables is assumed, then the basis functions take the form $\boldsymbol{\phi}(\mathbf{x}_n) = \mathbf{x}_n$. It is important to observe here that by using nonlinear functions, we allow the model to be also nonlinear to the input variables, while we keep the linearity with respect to the weights. One possible choice is to use polynomial basis functions where the basis function has the form of powers of the input variables, i.e., $\phi_j(x) = x^j$. Other choices of basis functions are: the Gaussian basis functions, the logistic sigmoid functions, the Fourier basis functions, and the wavelet basis functions.

7.2.3 Maximum likelihood (ML) parameter estimation

Assuming that the noise follows white Gaussian distribution, i.e., $\mathbf{e} \sim N(0, \lambda \mathbf{I})$, then the likelihood of the observations $\mathbf{y}$ is given by

$$p(\mathbf{y}; \mathbf{w}, \lambda) = \left(\frac{\lambda}{2\pi}\right)^{N/2} \exp\left\{-\frac{\lambda}{2}\|\mathbf{y} - \mathbf{\Phi}\mathbf{w}\|^2\right\} \tag{7.7}$$

Based on the previous formulation, the learning of the linear regression model becomes a maximum likelihood (ML) estimation problem for the regression model parameters $\Theta = \{\mathbf{w}\}$, in the sense of maximizing the log-likelihood function given by

$$L_{ML}(\Theta) = \log p(\mathbf{y}; \mathbf{w}) = \left\{\frac{N}{2}\log\lambda - \frac{\lambda}{2}\|\mathbf{y} - \mathbf{\Phi}\mathbf{w}\|^2\right\}. \tag{7.8}$$

Setting the partial derivatives of the previous function with respect to the parameters, equal to zero, the following update rules for the model parameters are obtained

$$\hat{\mathbf{w}} = (\mathbf{\Phi}^T\mathbf{\Phi})^{-1}\mathbf{\Phi}^T\mathbf{y}. \tag{7.9}$$

We want to mention here that we have made the assumption that the matrix $\mathbf{\Phi}^T\mathbf{\Phi}$ is invertible. Furthermore, we can observe that the previous solution can be, also, obtained if we adopt the LS formulation [21].

7.2.4 Bayesian modeling of machine learning

Bayesian inference provides a mathematical framework that can be used for modeling, where the uncertainties of the system are taken into account and the decisions are made according to logical principles. These main tools are random variables, the probability distributions, and the rules of probability calculus. The Bayesian framework is based on Bayes' rule given by

$$p(\mathbf{w}|\mathbf{y}) = \frac{p(\mathbf{y}|\mathbf{w})p(\mathbf{w})}{p(\mathbf{y})} \tag{7.10}$$

An important property of this framework is that the model parameters $\mathbf{w}$ are treated as random variables with a probability distribution being assigned over them.

The basic components of the Bayesian framework is described next.

Prior distribution The prior information consists of beliefs about the possible and impossible parameters' values and their relative likelihoods before anything has been seen. The prior distribution is a mathematical representation of this information:

$$p(\mathbf{w}) = \text{Information on parameters } \mathbf{w} \text{ before arises any observations.} \tag{7.11}$$

The lack of prior information can be expressed by using a non-informative prior [18,22]. Besides incorporating prior knowledge into our problem, from a model perspective view, prior distribution introduces four significant properties:

- Avoids overfitting since restricts parameters to fit completely to the data.
- Provide generalization capabilities due to the supressness of overfitting.
- Avoid numerical instabilities (or model inaccuracies) since places constraints onto the likelihood.
- Specific priors, such as sparse priors, favor simpler models to explain the data (Occam's razor).

Likelihood function Between the measurements and the parameters, there is a noisy or inaccurate relationship. This relationship is modeled using the likelihood distribution:

$$p(\mathbf{y}|\mathbf{w}) = \text{Distribution of observation } \mathbf{y} \text{ given the parameter } \mathbf{w}. \tag{7.12}$$

Using only the likelihood function, to learn the parameters, could result in overfitting, since the parameters will learn very accurately the current observed data, so accurately that will overfit to the observed data losing the ability to generalize in new data.

Posterior Posterior distribution is the conditional distribution of parameters given the observation $\mathbf{y}$ and represents the information that we have after the observation $\mathbf{y}$ has been obtained. It can be computed by using Bayes' rule:

$$p(\mathbf{w}|\mathbf{y}) = \frac{p(\mathbf{y}|\mathbf{w})p(\mathbf{w})}{p(\mathbf{y})} \tag{7.13}$$

where the normalization constant is given by

$$p(\mathbf{y}) = \int p(\mathbf{y}|\mathbf{w})p(\mathbf{w})d\mathbf{w}. \tag{7.14}$$

Predictive posterior distribution The predictive distribution is the distribution of the new observation y_{new}:

$$p(y_{new}|\mathbf{y}) = \int p(y_{new}|\mathbf{w})p(\mathbf{w}|\mathbf{y})d\mathbf{w}. \tag{7.15}$$

The predictive distribution can be used for computing the probability distribution of the new observation, which has not been observed yet.

7.3 Learning of spatial filters

Spatial filtering is the process of filtering the signals by using information from the spatial domain. More specifically, in the spatial filtering, "new" channels are created as a combination of the original ones. In EEG analysis, well-known spatial filters are the bipolar and Laplacian which are local spatial filters [23]. A bipolar filter is defined as the difference between two neighboring channels, while a Laplacian filter is defined as four times the value of a central channel minus the values of the four neighboring channels around. The previous spatial filters are defined a priori, i.e., the filter coefficients are known and fixed. There are different ways to define spatial filters. In particular, the filter coefficients (or weights) can be fixed in advance or they can be data-driven, i.e., the weights are obtained during a learning procedure. More formal, the above procedure can be described by the following linear model:

$$\mathbf{x}^{(new)} = \mathbf{W}\mathbf{x}, \tag{7.16}$$

where $\mathbf{x}^{(new)}$ is the spatially filtered EEG, $\mathbf{x}$ is the original EEG, and $\mathbf{W}$ is a matrix contains the set of spatial filters. In the next subsections, we describe two methods for learning the set of spatial filters, the CCA and the CSPs. We will see both methods are special cases of the generalized Rayleigh quotient function and, hence, they present great similarity with the well-known PCA method.

7.3.1 Canonical correlation analysis

CCA is a multivariate statistical method, where the goal is to find the underlying correlations between two sets of data [16,24]. The basic assumption of this approach is that these two sets of data are only a different view (or representation) of the same original (hidden) data. More specifically, a linear projection is computed for each representation such that they are maximally correlated in the dimensionally reduced (hidden) space. Let us assume that the two different views of the hidden data can be represented by two matrices $\mathbf{X} \in \Re^{KxP}$ and $\mathbf{Y} \in \Re^{HxP}$. Formally, CCA approach seeks to find two vectors $\mathbf{w} \in \Re^{K}$ and $\mathbf{v} \in \Re^{H}$ in order to maximize the linear correlation between the projections $\mathbf{w}^T\mathbf{X}$ and $\mathbf{v}^T\mathbf{Y}$. This is achieved by solving the following optimization problem:

$$\max \rho = \max_{\mathbf{w},\mathbf{v}} \frac{\mathbf{w}^T\mathbf{X}\mathbf{Y}^T\mathbf{v}}{\sqrt{\mathbf{w}^T\mathbf{X}\mathbf{X}^T\mathbf{w}\mathbf{v}^T\mathbf{Y}\mathbf{Y}^T\mathbf{v}}} \tag{7.17}$$

Since ρ is invariant to the scaling of $\mathbf{w}$ and $\mathbf{v}$, the previous optimization problem can be also formulated as

$$\max_{\mathbf{w},\mathbf{v}} \mathbf{w}^T\mathbf{X}\mathbf{Y}^T\mathbf{v} \text{ subject to } \mathbf{w}^T\mathbf{X}\mathbf{X}^T\mathbf{w} = 1 \text{ and } \mathbf{v}^T\mathbf{Y}\mathbf{Y}^T\mathbf{v} = 1$$

Assuming that $\mathbf{Y}\mathbf{Y}^T$ is nonsingular, it can be shown that $\mathbf{w}$ can be obtained by solving the following optimization problem

$$\max_{\mathbf{w}} \mathbf{w}^T\mathbf{X}\mathbf{Y}^T(\mathbf{Y}\mathbf{Y}^T)^{-1}\mathbf{Y}\mathbf{X}^T\mathbf{w} \text{ subject to } \mathbf{w}^T\mathbf{X}\mathbf{X}^T\mathbf{w} = 1 \tag{7.18}$$

The previous formulations attempt to find the eigenvectors corresponding to top eigenvalues of the following generalized eigenvalue problem:

$$\mathbf{X}\mathbf{Y}^T(\mathbf{Y}\mathbf{Y}^T)^{-1}\mathbf{Y}\mathbf{X}^T\mathbf{w} = \lambda\mathbf{X}\mathbf{X}^T\mathbf{w} \tag{7.19}$$

where λ is the eigenvalue corresponding to the eigenvector $\mathbf{w}$. Finally, we collect all the eigenvectors (or spatial filters) in one matrix, $\mathbf{W}_x$. By applying the transformation $\mathbf{W}_x$ to data $\mathbf{X}$, we obtain the spatial filtered data.

In EEG signal processing, $\mathbf{X}$ refers to the set of multichannel EEG signals and $\mathbf{Y}$ refers to the set of reference signals. The idea behind using CCA for spatial filtering is to find a spatial filter that maximizes the correlation between the spatially filtered signal and the reference signals, hence reducing noise in EEG signals. CCA has been used in numerous occasions to analyze EEG signals such as P300 signals [25] and SSVEP signals [26]. Also, CCA can be used in various ways in EEG-based BCI community, depending on the construction of the two matrices. For example, in [25], a number of multichannel EEG trials have been concatenated into one large matrix $\mathbf{X}$, while matrix $\mathbf{Y}$ has been constructed by using averaged EEG trials (i.e., averaged evoked responses). Using CCA, a set of spatial filters was obtained. A slightly different approach was adopted in SSVEP BCI studies [26], where the general idea is to find the canonical correlation coefficients between a multichannel EEG trial, $\mathbf{X}$, and various multichannel reference signals, $\mathbf{Y}_i$. These coefficients were used as features for the subsequent classification. Furthermore, in SSVEP BCI systems, extensions of basic CCA algorithm have been proposed based on the exploitation of filterbanks and on the advanced construction of the reference signals [27,28]. Also, in [29], CCA was used as a kernel function to describe similarities between the multichannel EEG trials. In addition to BCI applications, CCA has been applied to remove artifacts from EEG signals [30,31] and to evaluate the correlation between EEG complexity and cognitive dysfunction (as measured by the Neuropsychiatric Inventory scores) in AD [32].

7.3.2 Common spatial patterns

The CSP algorithm is an algorithm that provides us with a set of spatial filters. These filters are obtained after performing a learning procedure, during which the variance of the spatially filtered signals is maximized for one class (e.g., one mental imagery task) and minimized for the other class. At the beginning, the CSP algorithm was applied on multichannel data from two classes/conditions [33]. However, extensions of the algorithm to handle multi-class problems have been proposed [34,35].

The CSP algorithm performs a decomposition of the signal through the matrix $\mathbf{W}$, which contains the spatial filters. More specifically, this algorithm transforms the EEG signal from the original into a new domain which is occupied by the "new" channels

$$\mathbf{x}^{(CSP)} = \mathbf{W}\mathbf{x}, \tag{7.20}$$

where $\mathbf{x} \in \Re^{C\times 1}$ is the EEG signal at time point t, $\mathbf{x}^{(CSP)} \in \Re^{C\times 1}$ is the decomposed "new" EEG signal and $\mathbf{W} \in \Re^{C\times C}$ is the matrix with the spatial filters $\mathbf{w}_i$, $i = 1, \ldots, C$, and C is the number of channels. The spatial filters are obtained by maximizing (or extremizing) the following function [36]:

$$J(\mathbf{w}) = \frac{\mathbf{w}^T\mathbf{C}_1\mathbf{w}}{\mathbf{w}^T\mathbf{C}_2\mathbf{w}} \tag{7.21}$$

where T denotes transpose, and $\mathbf{C}_i$ is the covariance matrix of ith class. The previous maximization problem is equal to maximize $\mathbf{w}^T\mathbf{C}_1\mathbf{w}$ subject to the constraints $\mathbf{w}^T\mathbf{C}_2\mathbf{w} = 1$. The last problem is equivalent to the generalized eigenvalue problem $\mathbf{C}_1\mathbf{w} = \lambda\mathbf{C}_2\mathbf{w}$. So, the spatial filters $\mathbf{w}_i$ are the generalized eigenvectors of the previous problem. It is worth to note here that in most cases after the application of CSP algorithm for spatial filtering, an additional step is performed in order to extract CSP-related features[36]. Once the spatial filters $\mathbf{w}_i$ are obtained, CSP FE consists in filtering the EEG signals using the $\mathbf{w}_i$ and then computing the resulting signals variance. As reported in [36] it is common to select three pairs of CSP spatial filters, corresponding to the three largest and smallest eigenvalues, hence resulting in a trial being described by six CSP features. However, the previous heuristic choice of the number of spatial filters depends heavily on the nature of the data (i.e., number of channels) as well as on the data analysis perspective (i.e., using filterbanks or not). Nevertheless, by choosing $2p(< C)$ spatial filters, corresponding to the p largest and smallest eigenvalues, the CSP algorithm is transformed into a spatial dimensionality reduction method (i.e., reduce the number of EEG channels). CSP algorithm was applied at first in BCI applications [33] with substantial results, and as expected, it has found many other applications in EEG analysis such as the prediction of epileptic seizure [37] and the recognition of AD [3].

7.4 Classification algorithms

In the next sections, we will describe classifiers that use linear functions of features to distinguish between classes. Linear classifiers are the most popular algorithms for BCI applications. Furthermore, these algorithms can be easily extended to cover nonlinear cases.

7.4.1 Linear discriminant analysis

Given a dataset $\mathscr{D} = \{(\mathbf{x}_i, y_i)\}_{i=1}^N$, where $\mathbf{x}_i$ are feature vectors and $y_i \in \{1(\omega_1), -1(\omega_2)\}$ are the class labels, we seek to find a weight vector $\mathbf{w}$ and a threshold w_0 such that: assign $\mathbf{x}$ to ω_1 if $f(\mathbf{x}) = \mathbf{w}^T\mathbf{x} + w_0 > 0$, otherwise to ω_2. Through $f(\mathbf{x})$, we want to project the D-dimensional feature vector into a scalar value (i.e., in 1-dimension). Much effort in machine-learning community is consumed to find an optimal, in some sense, projection (i.e., an optimal weight vector $\mathbf{w}$). An obvious choice is to select a projection that maximizes the class separation. Under this view, a measure of the separation of the classes is the separation of the projected class means, resulting in a weight vector that is equal to the difference of means from each class. The previous approach does not take into account the variability of the data resulting in significant overlap between classes. The idea proposed by Fisher is to find a projection that will give a large separation between the projected class means while also giving a small variance within each class, thereby minimizing the class overlap.

The Fisher criterion is defined to be the ratio of the between-class variance to the within-class variance and is given by

$$J(\mathbf{w}) = \frac{\mathbf{w}^T\mathbf{S}_B\mathbf{w}}{\mathbf{w}^T\mathbf{S}_W\mathbf{w}} \tag{7.22}$$

where $\mathbf{S}_B$ is the between-class covariance matrix and $\mathbf{S}_W$ is the total within class covariance matrix [18]. Maximizing the previous function with respect to $\mathbf{w}$, we obtain

$$\mathbf{w} = \mathbf{S}_W^{-1}(\mathbf{m}_{\omega_2} - \mathbf{m}_{\omega_1}) \tag{7.23}$$

Now, it remains to determine the threshold w_0. Under normality assumptions, the optimal threshold w_0 is given on closed form [38]. However, in non-normal situations, a different threshold maybe more appropriate. In that case, we can choose the threshold that minimizes misclassification error over the training data [38]. It must be noted here the similarities between Fisher's LDA with the CSP and the CCA. All the previous methods are special cases of the generalized Rayleigh quotient function.

LDA has found numerous applications on EEG analysis ranging from medical/health-care studies to a more general case such as BCI application. For example, in [39], it was used to classify the various EEG-based sleep stages in healthy subjects and patients, while in [37], it was used for epileptic seizure prediction. Furthermore, in [3], LDA was used to distinguish AD patients from healthy control subjects. Finally, in [11], it is was used to discriminate SMR rhythms, while in [40] to recognize the various SSVEP responses.

7.4.2 Least squares classifier

Given a dataset $\mathscr{D} = \{(\mathbf{x}_i, y_i)\}_{i=1}^N$, where $\mathbf{x}_i$ are feature vectors of size $p \times 1$ and $y_i = t_1$ if $x_i \in \omega_1$ or $y_i = t_2$ if $x_2 \in \omega_2$. Collecting all the features vectors in a matrix, $\mathbf{X} \in \Re^{N \times p}$, then the LS function is given by

$$J(\mathbf{w}) = \|\mathbf{y} - \mathbf{X}\mathbf{w}\|^2 \tag{7.24}$$

Maximizing the previous function with respect to $\mathbf{w}$, we obtain the LS solution (assuming that $\mathbf{X}^T\mathbf{X}$ is invertible):

$$\mathbf{w}_{LS} = (\mathbf{X}^T\mathbf{X})^{-1}\mathbf{X}^T\mathbf{y} \tag{7.25}$$

When a new feature vector, $\mathbf{x}$, arrives the quantity $\mathbf{x}^T\mathbf{w}_{LS}$ is compared to a threshold t_0 in order to decide its class. Depending on the data, different threshold values maybe used.

It is interesting to make some comments here with respect to the quantity $\mathbf{X}^T\mathbf{y}$, which can be written as

$$\mathbf{X}^T\mathbf{y} = \sum_{i=1}^{N} y_i \cdot \mathbf{x}_i = \sum_{i:\mathbf{x}_i \in \omega_1} t_1\mathbf{x}_i + \sum_{i:\mathbf{x}_i \in \omega_2} t_2\mathbf{x}_i = t_1N_1\mathbf{m}_1 + t_2N_2\mathbf{m}_2 \tag{7.26}$$

where N is the number of total samples, N_1 is the number of samples belonging to class 1, and N_2 is the number of samples belonging to class 2. We see that the vector $\mathbf{z} = \mathbf{X}^T\mathbf{y}$ is a weighted average of mean vectors of each class. For the sake of exposition, let us assume that $N_1 = N_2 = N/2$ then we have $\mathbf{z} = (N/2)(t_1\mathbf{m}_1 + t_2\mathbf{m}_2)$.

Using the target values $t_1 = 1$ and $t_2 = 0$, we obtain $\mathbf{z} = (N/2)\mathbf{m}_1$; hence, when a test feature vector arrives, we calculate the prediction score $\mathbf{x}^T\mathbf{w}_{LS} = \mathbf{x}^T(\mathbf{X}^T\mathbf{X})^{-1}(N/2)\mathbf{m}_1$ and compared it with a threshold to perform the decision. The prediction score can be further analyzed. More specifically, we can see that we project the new test feature vector into the space of training feature vectors (row space of matrix $\mathbf{X}$) and then we calculate the similarity between this projection and the mean of class 1. Finally, we compare the similarity with the threshold, t_0, in order to decide about the class. Loosely speaking, we can say that the decision rule checks how close to class 1 is the new (projected) feature vector.

Let us use the following target values $t_1 = 1$ and $t_2 = -1$ then we have $\mathbf{z} = (N/2)(\mathbf{m}_1 - \mathbf{m}_2)$. When a new (test) feature vector arrives, the prediction score is calculated according to $\mathbf{x}^T\mathbf{w}_{LS} = \mathbf{x}^T(\mathbf{X}^T\mathbf{X})^{-1}(N/2)(\mathbf{m}_1 - \mathbf{m}_2)$. Again, we project the new test feature vector into the space of training feature vectors but, now, we calculate the similarities between the projected feature vector and the two class means. Then, we compare the difference between the two similarities with the threshold, t_0, to decide about the class. At this case, we can say that the decision rule checks in which class the new (projected) feature vector "is closer." It is tempting at this point someone to assume that when $\mathbf{m}_2 = 0$, the two cases are identical. However, this is not really the case, since no such assumption is made in the first case. More specifically, in the first case, the term $t_2N_2\mathbf{m}_2$ goes to zero due to the given target value in $t_2(=0)$, and not due to $\mathbf{m}_2 = 0$. While the previous statements do not have huge practical implications, it is helpful to know the underlying assumptions that lead us to specific algorithms (or decision rules).

In the next paragraphs, we will show that under some assumptions on the values of class labels, the LDA classifier is a special case of LS classifier [18,38]. A more thorough analysis on this subject can be found in [18,38]. In order to be consistent with the theory and the assumptions of Section 7.4.1, the following LS function is used:

$$J(\mathbf{w}, w_0) = \|\mathbf{y} - \mathbf{X}\mathbf{w} - w_0\|^2 \tag{7.27}$$

Maximizing the function $J(\mathbf{w}, w_0)$ with respect to w_0 we obtain

$$\hat{w}_0 = \bar{y} - \mathbf{w}^T\mathbf{m} \tag{7.28}$$

where $\bar{y} = (1/N)\sum_{i=1}^{N} y_i$ and $\mathbf{m} = (1/N)\sum_{i=1}^{N} \mathbf{x}_i$. Replacing w_0 in (7.27) with the estimated quantity $\hat{w}_0$ and maximizing $J(\mathbf{w}, \hat{w}_0)$ with respect to $\mathbf{w}$ we obtain

$$\left(N\mathbf{S}_W + \frac{N_1N_2}{N}\mathbf{S}_B\right)\mathbf{w} = \mathbf{X}^T\mathbf{y} - N\bar{y}\mathbf{m} \tag{7.29}$$

where N is the number of total samples, N_1 is the number of samples belonging to class 1, and N_2 is the number of samples belonging to class 2. Now, it remains to define the values y_i that "we will regress." By defining $y_i = (N/N_1)$ if $\mathbf{x}_i \in \omega_1$ and $y_i = -(N/N_2)$ if $\mathbf{x}_i \in \omega_2$, it can be shown that $\bar{y} = 0$ and $\mathbf{X}^T\mathbf{y} = N(\mathbf{m}_1 - \mathbf{m}_2)$. Therefore, we obtain

$$\hat{w}_0 = -\mathbf{w}^T\mathbf{m} \tag{7.30}$$

$$\left(N\mathbf{S}_W + \frac{N_1N_2}{N}\mathbf{S}_B\right)\mathbf{w} = N(\mathbf{m}_1 - \mathbf{m}_2) \tag{7.31}$$

where $\mathbf{S}_W$ is the total within-class covariance matrix and $\mathbf{S}_B$ is the between-class covariance matrix[18,38]. In (7.31), we can observe that $\mathbf{S}_B\mathbf{w}$ lies in the direction of $\mathbf{m}_1 - \mathbf{m}_2$; hence, the final estimation for $\mathbf{w}$ is given by

$$\mathbf{w} \propto \mathbf{S}_W^{-1}(\mathbf{m}_{\omega_2} - \mathbf{m}_{\omega_1}) \tag{7.32}$$

We see that under some assumptions, the LS classifier is similar to Fisher LDA. This similarity gives us the opportunity to provide connections with the Bayesian framework in the following section.

7.4.3 Bayesian LDA

Given a dataset $\mathscr{D} = \{(\mathbf{x}_i, y_i)\}_{i=1}^{N}$, where $\mathbf{x}_i$ are feature vectors of size $p \times 1$ and y_i the target values. Collecting all the features vectors in a matrix, $\mathbf{X} \in \Re^{N\times p}$, we can consider the following linear model:

$$\mathbf{y} = \mathbf{X}\mathbf{w} + \mathbf{e} \tag{7.33}$$

where $\mathbf{e}$ is the noise and it follows a Gaussian distribution with zero mean and inverse variance (or precision) β. In that case, the likelihood function of parameters $\mathbf{w}$ is given by

$$p(\mathbf{y}|\mathbf{w}, \beta) = \frac{\beta}{2\pi}\exp\left\{-\frac{\beta}{2}\|\mathbf{y} - \mathbf{X}\mathbf{w}\|^2\right\} \tag{7.34}$$

Furthermore, we constrain the parameters $\mathbf{w}$ to follow a Gaussian distribution with zero mean and precision α before we observe the data (or class labels in our case). This prior distribution is given by

$$p(\mathbf{w}|\alpha) = \mathcal{N}(0, \alpha\mathbf{I}) = \frac{\alpha}{2\pi}\exp\left\{-\frac{\alpha}{2}\|\mathbf{w}\|^2\right\} \tag{7.35}$$

Then, according to the Bayes theorem, the posterior distribution is also a Gaussian distribution and it is given by

$$p(\mathbf{w}|\mathbf{y}, \alpha, \beta) = \mathcal{N}(\boldsymbol{\mu}, \boldsymbol{\Sigma}) \tag{7.36}$$

where

$$\boldsymbol{\Sigma} = (\beta\mathbf{X}^T\mathbf{X} + \alpha\mathbf{I})^{-1} \tag{7.37}$$

$$\boldsymbol{m} = \beta\boldsymbol{\Sigma}\mathbf{X}^T\mathbf{y} \tag{7.38}$$

The predictive distribution (the probability distribution over target values conditioned on an input vector) is given by

$$\begin{aligned} p(y_{new}|\mathbf{y}, \alpha, \beta, \mathbf{x}_{new}) &= \int p(y_{new}|\alpha, \beta, \mathbf{x}_{new}, \mathbf{w})p(\mathbf{w}|\alpha, \beta, \mathbf{y})d\mathbf{w} \\ &= \mathcal{N}(\mu, \Sigma) \end{aligned} \tag{7.39}$$

where

$$\Sigma = \frac{1}{\beta} + \mathbf{x}_{new}^T\boldsymbol{\Sigma}\mathbf{x}_{new} \tag{7.40}$$

$$\mu = \boldsymbol{\mu}^T\mathbf{x}_{new} \tag{7.41}$$

We observe that in this case, the predictive distribution is also a Gaussian. To decide about the class, a threshold is applied on the predictive mean μ [13,14]. In the previous simplified Bayesian-based analysis, we assume that we know a priori the parameters α and β. However, in most cases, these quantities are unknown. In order to define these quantities, either a cross validation approach may be adopted or we can extend the Bayesian framework by incorporating priors (and hyperpriors) over these parameters. In the case that priors are used, we obtain various iterative algorithms for the estimation of $\mathbf{w}$, α, and β [14,41]. Furthermore, in EEG signal analysis community, various priors have been explored over the parameters $\mathbf{w}$ [42]. Many variants of BLDA have been used for SMR rhythms discrimination [14], SSVEP responses recognition [41], epileptic seizure detection [43], and ERP classification [13]. Also the usage of Bayesian linear model is not restricted in EEG signals classification but it has found many other applications on EEG analysis. In [44], it was used to compress sensed EEG signals, while in [45], it was used to solve the inverse EEG problem and provide estimates of EEG sources.

7.4.4 Support vector machines

The most popular classification algorithm is the SVMs, which aims to find the optimal hyperplane that separates the positive class from the negative one by maximizing the margin between the two classes. This hyperplane, in its basic linear form, is represented by its normal vector $\mathbf{w}$ and a bias parameter w_0. These two terms are the parameters that are learnt during the training phase. Assuming that the data are linearly separable, there exist multiple hyperplanes that can separate the two classes, and hence, solve the classification problem. SVMs choose the one that maximizes the margin, assuming that this will generalize better to new unseen data. This hyperplane is found by solving the following minimization problem:

$$\min_{\mathbf{w},w_0} \frac{1}{2}\|\mathbf{w}\|^2 + C\sum_{i=1}^{N}\xi_i$$

$$\text{s.t.:} \quad y_i(\mathbf{w}^T\mathbf{x} + w_0) \geq 1 - \xi_i, \xi_i \geq 0, \quad i = 1, \ldots, N \tag{7.42}$$

where ξ_i are slack variables relaxing the constraints of perfect separation of the two classes and C is a regularization parameter controlling the trade-off between the simplicity of the model and its ability to better separate the two classes. SVM has been used in numerous applications with respect to EEG analysis. It has been used to solve various medical-based classification problems related, among others, to the epilepsy [46], to Alzheimer's disease [47], and to the recognition of sleep stage in patients [39]. Furthermore, it has found extensive use in BCI-related problems such as the discrimination of SMR rhythms [48–50] and the recognition of SSVEP responses [51,52].

7.4.5 Kernel-based classifier

In all aforementioned machine-learning techniques, we assume that the mapping/connection between the target values and the features was linear. However, easily the previous algorithms can be extended to include (nonlinear) feature space transformation $\phi(\cdot)$. In that case, the decision is based on the output of $f(\mathbf{x}) = \mathbf{w}^T\phi(\mathbf{x}) + w_0$, where $\phi(\mathbf{x})$ denotes the possibly nonlinear, feature-space transformation. Note here that assuming $\phi(\mathbf{x}) = \mathbf{x}$, we obtain the basic form of the decision function $f(\mathbf{x})$. A careful investigation of the aforementioned algorithms reveals that a significant quantity is the kernel function given by $k(\mathbf{x}_i, \mathbf{x}_j) = \phi(\mathbf{x}_i)^T\phi(\mathbf{x}_j)$, where $\mathbf{x}_i$ and $\mathbf{x}_j$ are two points of the training dataset. This function describes the similarity between these two points. By adopting the dual representation [53], we can see that the prediction can be made in terms of similarities between feature vectors instead of features. In addition, the kernel function $k(\mathbf{x}_i, \mathbf{x}_j)$ must be evaluated for all possible pairs of $\mathbf{x}_i$ and $\mathbf{x}_j$. The previous fact affects considerably the computation time of any kernel-based algorithm. Kernel-based algorithms avoid the problem of large dimensionality of features vectors but encounter problem when we have a large training set (large number of trials). A useful tactic to the previous problem is to favor kernel-based algorithms that have sparse solutions such as the SVM and the RVM [53].

It is worth to note here that the regression models of (7.33) can be easily kernelized [53]. Instead of working on the original feature space described from the following equation $\mathbf{y} = \mathbf{X}\mathbf{w} + \mathbf{e} = \sum_{n=1}^{D} w_n\mathbf{x}_n + \mathbf{e}$, we can work on kernel feature space by applying the kernel trick. In that case, the regression model is described by $\mathbf{y} = \sum_{n=1}^{N} w'_n k(\mathbf{x}, \mathbf{x}_n) + \mathbf{e} = \mathbf{X}'\mathbf{w}' + \mathbf{e}$, where the matrix $\mathbf{X}'$ is an $N \times N$ symmetric matrix with elements $X_{nm} = k(\mathbf{x}_n, \mathbf{x}_m)$, $k(\cdot)$ is the kernel function, and $\mathbf{w}'_k \in \Re^N$ is the new vector of regression coefficients. It is worth to note here that the kernel method can be useful in high dimensional settings, even if we only use a linear kernel. More specifically, to compute the regression coefficients into the original feature space (primal variables), the computational cost is $O(D^3)$, while in the kernel feature space is $O(N^3)$[53]. When $D \gg N$, as it is the case for the many problems in EEG analysis, the computational cost of working into the original feature space is considerable compared to the computational cost of kernel feature space. Finally, we must point out here that in most cases, an EEG dataset contains much more raw features than trials, and this fact must be taken into account when a machine-learning algorithm is used to process the EEG data.

7.5 Future directions and other issues

In the previous sections, we described the basic concept underlying the supervised learning task and how this task is applied to EEG analysis to learn optimal spatial filters and classifiers. However, other more specialized concepts of learning could be applied in order to obtain a model with better performance or a model that best suited in our needs. In this section, we will provide a brief description of these learning tasks.

7.5.1 Adaptive learning

In the previous sections, we described methods on how to learn optimal spatial filters and classifiers. However, we can observe that when the weights of the model are learned, they cannot change with the passage of time. In situations such as EEG analysis, this can be problematic, since the EEG signal is a nonstationary signal and its statistical properties can change with time. To attack this problem, variants of reported algorithms have been proposed where the weights of models are changing with time. Under this perspective, adaptive spatial filters have been proposed in [54,55]. Furthermore, adaptive classifiers have been applied on analyzing EEG data, see [56]. Finally, it is worth to point here that the Kalman filter algorithm is a very useful starting point for adaptive learning, especially, in applications related to the EEG data analysis [57].

7.5.2 Transfer learning and multitask learning

Transfer learning (TL) is the process of transferring the knowledge obtained during the learning of a task to learn another similar new task. The basic assumption on this type of learning is that the two tasks have a common underlying structure. A useful characteristic of the TL with respect to BCI applications is that it could reduce significantly the calibration time of a BCI system [58]. On the other side, multitask learning (MTL) is the process of learning, simultaneously, a number of tasks that share a common structure. While MTL and TL are very similar approaches, there is a critical difference. Assuming that we have two tasks, MTL tries to learn both tasks simultaneously, while TL tries to transfer the learning experience from the first task to learn the second one. As we see that MTL does not make distinctions between tasks[59]. With respect to BCI, MTL is trying to combine information from different tasks[60]. The *task* can be defined in terms of subjects, of sessions, and of channels. Furthermore, MTL and TL show us the road for subject-independent BCI systems.

7.5.3 Deep learning

The term *deep learning* is referred to a set of methodologies that are used to train convolutional neural networks of several dozen, or more, hidden layers. Deep models often have millions of parameters, due to their large and complicated structure, and we need enough labeled data to train such models. Assuming that enough data are acquired for a particular problem, then deep models can be trained to learn simultaneously the features and the classifier. In EEG analysis, especially in BCI concept, many attempts have been performed to use deep models [61,62]; however, their success is limited [56]. This behavior of deep models is expected to some degree, since the current EEG datasets are of small size; hence, there is not enough data to effectively train them. Finally, another basic criticism over deep models in EEG analysis is that "they are difficult to interpret in terms of the EEG problem"[56]. However, this is closely related to our overall understanding of the human brain. Clearly, we know and understand much better the basic principles of image processing and analysis than how the human brain is functioning.

7.6 Summary

In this chapter, we provided basic information about the EEG signal and how this biomedical signal can be used in the context of noninvasive BCI systems. In addition, we described well-known machine-learning techniques that are used in EEG analysis. More specifically, at first, we provided a short description of the generalized Rayleigh quotient function, the linear regression model, and the Bayesian framework. These approaches can be found at the heart of many supervised learning algorithms related to the EEG analysis. In addition, we described basic ideas of widely used learning methods in the spectrum of EEG analysis, although being biased towards the BCI applications. These learning methods are related to the finding of optimal spatial filters and classifiers. Furthermore, we provided additional information about various connections between these methods. Then, we close this chapter by describing subcategories of the general supervised learning task, how these subcategories could be used in EEG analysis, and what are the benefits of using them.

References

[1] Acharya UR, Sree SV, Swapna G, *et al.* Automated EEG analysis of epilepsy: A review. Knowledge-Based Systems. 2013;45:147–165.

[2] Oikonomou VP, Tzallas AT, Fotiadis DI. A Kalman filter based methodology for EEG spike enhancement. Computer Methods and Programs in Biomedicine. 2007;85(2):101–108.

[3] Woon WL, Cichocki A, Vialatte F, *et al.* Techniques for early detection of Alzheimer's disease using spontaneous EEG recordings. Physiological Measurement. 2007;28(4):335–347.

[4] Tsolaki A, Kazis D, Kompatsiaris I, *et al.* Electroencephalogram and Alzheimer's disease: Clinical and research approaches. International Journal of Alzheimer's Disease. 2014;1–10. DOI: 10.1155/2014/349249.

[5] Tangwiriyasakul C, Mocioiu V, van Putten MJAM, *et al.* Classification of motor imagery performance in acute stroke. Journal of Neural Engineering. 2014;11(3):036001.

[6] Handojoseno AMA, Shine JM, Nguyen TN, *et al.* Analysis and prediction of the freezing of gait using EEG brain dynamics. IEEE Transactions on Neural Systems and Rehabilitation Engineering. 2015;23(5):887–896.

[7] Wolpaw JR, Birbaumer N, McFarland DJ, *et al.* Brain–computer interfaces for communication and control. Clinical Neurophysiology. 2002;113(6): 767–791.

[8] Graimann B, Allison BZ, Pfurtscheller G. Brain–Computer Interfaces: Revolutionizing Human–Computer Interaction. Berlin, Heidelberg: Springer Publishing Company, Incorporated; 2013.

[9] Lemm S, Blankertz B, Dickhaus T, *et al.* Introduction to machine learning for brain imaging. NeuroImage. 2011;56(2):387–399.

[10] Cunningham JP, Ghahramani Z. Linear dimensionality reduction: Survey, insights, and generalizations. Journal of Machine Learning Research. 2015; 16:2859–2900. Available from: http://jmlr.org/papers/v16/cunningham15a.html.

[11] Muller KR, Krauledat M, Dornhege G, *et al.* Machine learning techniques for brain–computer interfaces. Biomedical Engineering. 2004;49:11–22.

[12] Lotte F, Congedo M, Lécuyer A, *et al.* A review of classification algorithms for EEG-based brain–computer interfaces. Journal of Neural Engineering. 2007;4(2):R1–R13.

[13] Hoffmann U, Vesin JM, Ebrahimi T, *et al.* An efficient P300-based brain–computer interface for disabled subjects. Journal of Neuroscience Methods. 2008;167(1):115–125.

[14] Zhang Y, Zhou G, Jin J, *et al.* Sparse Bayesian classification of EEG for brain–computer interface. IEEE Transactions on Neural Networks and Learning Systems. 2015;27(99):2256–2267.

[15] Moon TD, Stirling WC. Mathematical Methods and Algorithms for Signal Processing. NJ: Prentice-Hall; 2000.

[16] Sun L, Ji S, Ye J. Multi-Label Dimensionality Reduction. Boca Raton. FL: CRC Press, Taylor and Francis Group; 2014.

[17] Borloz B, Xerri B. Subspace signal-to-noise ratio maximization: The constrained stochastic matched filter. In: Proceedings of the Eighth International Symposium on Signal Processing and Its Applications, 2005. vol. 2; 2005. p. 735–738.

[18] Bishop CM. Pattern Recognition and Machine Learning (Information Science and Statistics). NY: Springer; 2007.

[19] Tipping ME, Lawrence ND. Variational inference for Student-*t* models: Robust Bayesian interpolation and generalised component analysis. Neurocomputing. 2005;69(1):123–141.

[20] Oikonomou VP, Blekas K, Astrakas L. A sparse and spatially constrained generative regression model for fMRI data analysis. IEEE Transactions on Biomedical Engineering. 2012;59(1):58–67.

[21] Kay SM. Fundamentals of Statistical Signal Processing: Estimation Theory. Upper Saddle River, NJ, USA: Prentice-Hall; 1993.

[22] Barber D. Bayesian Reasoning and Machine Learning. New York, NY, USA: Cambridge University Press; 2012.

[23] Sanei S and Chambers JA. EEG Signal Processing. Hoboken: Wiley; 2007.

[24] Hardoon DR, Szedmak SR, Shawe-Taylor JR. Canonical correlation analysis: An overview with application to learning methods. Neural Computation. 2004;16(12):2639–2664.

[25] Spüler M, Walter A, Rosenstiel W, *et al.* Spatial filtering based on canonical correlation analysis for classification of evoked or event-related potentials in EEG data. IEEE Transactions on Neural Systems and Rehabilitation Engineering. 2014;22(6):1097–1103.

[26] Lin Z, Zhang C, Wu W, *et al.* Frequency recognition based on canonical correlation analysis for SSVEP-based BCIs. IEEE Transactions on Biomedical Engineering. 2006;53(12):2610–2614.

[27] Nakanishi M, Wang Y, Wang YT, *et al.* A comparison study of canonical correlation analysis based methods for detecting steady-state visual evoked potentials. PLoS One. 2015;10:e0140703.

[28] Zhang Y, Zhou G, Jin J, *et al.* L1-regularized multiway canonical correlation analysis for SSVEP-based BCI. IEEE Transactions on Neural Systems and Rehabilitation Engineering. 2013;21(6):887–896.

[29] Oikonomou VP, Nikolopoulos S, Kompatsiaris I. A Bayesian multiple kernel learning algorithm for SSVEP BCI detection. IEEE Journal of Biomedical and Health Informatics. 2018;23:1990–2001.

[30] Sweeney KT, McLoone SF, Ward TE. The use of ensemble empirical mode decomposition with canonical correlation analysis as a novel artifact removal technique. IEEE Transactions on Biomedical Engineering. 2013;60(1): 97–105.

[31] De Clercq W, Vergult A, Vanrumste B, *et al.* Canonical correlation analysis applied to remove muscle artifacts from the electroencephalogram. IEEE Transactions on Biomedical Engineering. 2006;53(12):2583–2587.

[32] Fan M, Yang AC, Albert C, *et al.* Topological pattern recognition of severe Alzheimer's disease via regularized supervised learning of EEG complexity. Frontiers in Neuroscience. 2018;12. Available from: https://www.frontiersin.org/article/10.3389/fnins.2018.00685.

[33] Blankertz B, Tomioka R, Lemm S, *et al.* Optimizing spatial filters for robust EEG single-trial analysis. IEEE Signal Processing Magazine. 2008;25(1): 41–56.

[34] Chin ZY, Ang KK, Wang C, *et al.* Multi-class filter bank common spatial pattern for four-class motor imagery BCI. In: 2009 Annual International Conference of the IEEE Engineering in Medicine and Biology Society; 2009. p. 571–574.

[35] Grosse-Wentrup M, Buss M. Multiclass common spatial patterns and information theoretic feature extraction. IEEE Transactions on Biomedical Engineering. 2008;55(8):1991–2000.

[36] Lotte F, Guan C. Regularizing common spatial patterns to improve BCI designs: Unified theory and new algorithms. IEEE Transactions on Biomedical Engineering. 2011;58(2):355–362.

[37] Alotaiby TN, Alshebeili SA, Alotaibi FM, *et al.* Epileptic seizure prediction using CSP and LDA for scalp EEG signals. Computational Intelligence and Neuroscience. 2017;1–11. DOI: 10.1155/2017/1240323

[38] Webb AR. Statistical Pattern Recognition, Second Edition. West Sussex, England: John Wiley and Sons, Ltd.; 2002.

[39] Boostani R, Karimzadeh F, Nami M. A comparative review on sleep stage classification methods in patients and healthy individuals. Computer Methods and Programs in Biomedicine. 2017;140:77–91.

[40] Maronidis A, Oikonomou VP, Liaros G, *et al.* Steady state visual evoked potential detection using Subclass Marginal Fisher Analysis. In: 2017 8th International IEEE/EMBS Conference on Neural Engineering (NER); 2017. p. 37–41.

[41] Oikonomou VP, Maronidis A, Liaros G, *et al.* Sparse Bayesian learning for subject independent classification with application to SSVEP-BCI. In: 2017 8th International IEEE/EMBS Conference on Neural Engineering (NER); 2017. p. 600–604.

[42] Oikonomou VP, Nikolopoulos S, Petrantonakis P, *et al.* Sparse Kernel machines for motor imagery EEG classification. In: 2018 40th Annual International Conference of the IEEE Engineering in Medicine and Biology Society (EMBC); 2018. p. 207–210.

[43] Zhou W, Liu Y, Yuan Q, *et al.* Epileptic seizure detection using lacunarity and Bayesian linear discriminant analysis in intracranial EEG. IEEE Transactions on Biomedical Engineering. 2013;60(12):3375–3381.

[44] Zhang Z, Jung T, Makeig S, *et al.* Spatiotemporal sparse Bayesian learning with applications to compressed sensing of multichannel physiological signals. IEEE Transactions on Neural Systems and Rehabilitation Engineering. 2014;22(6):1186–1197.

[45] Hashemi A, Haufe S. Improving EEG source localization through spatio-temporal sparse Bayesian learning. In: 2018 26th European Signal Processing Conference (EUSIPCO); 2018. p. 1935–1939.

[46] Subasi A, Gursoy MI. EEG signal classification using PCA, ICA, LDA and support vector machines. Expert Systems with Applications. 2010;37(12):8659–8666.

[47] Lehmann C, Koenig T, Jelic V, *et al.* Application and comparison of classification algorithms for recognition of Alzheimer's disease in electrical brain activity (EEG). Journal of Neuroscience Methods. 2007;161(2):342–350.

[48] Herman P, Prasad G, McGinnity TM, *et al.* Comparative analysis of spectral approaches to feature extraction for EEG-based motor imagery classification. IEEE Transactions on Neural Systems and Rehabilitation Engineering. 2008;16(4):317–326.

[49] Brodu N, Lotte F, Lécuyer A. Comparative study of band-power extraction techniques for motor imagery classification. In: 2011 IEEE Symposium on Computational Intelligence, Cognitive Algorithms, Mind, and Brain (CCMB); 2011. p. 1–6.

[50] Oikonomou VP, Georgiadis K, Liaros G, *et al.* A comparison study on EEG signal processing techniques using motor imagery EEG data. In: 2017 IEEE 30th International Symposium on Computer-Based Medical Systems (CBMS); 2017. p. 781–786.

[51] Oikonomou VP, Liaros G, Georgiadis K, *et al.* Comparative evaluation of state-of-the-art algorithms for SSVEP-based BCIs; 2016. arXiv:1602.00904.

[52] Chatzilari E, Liaros G, Georgiadis K, *et al.* Combining the benefits of CCA and SVMs for SSVEP-based BCIs in real-world conditions. In: Proceedings of the 2nd International Workshop on Multimedia for Personal Health and Health Care. MMHealth'17. New York, NY, USA: ACM; 2017. p. 3–10. Available from: http://doi.acm.org/10.1145/3132635.3132636.

[53] Murphy KP. Machine Learning: A Probabilistic Perspective. Cambridge, Massachusetts: MIT Press; 2012.

[54] Woehrle H, Krell MM, Straube S, *et al.* An adaptive spatial filter for user-independent single trial detection of event-related potentials. IEEE Transactions on Biomedical Engineering. 2015;62(7):1696–1705.

[55] Meng J, Sheng X, Zhang D, *et al.* Improved semisupervised adaptation for a small training dataset in the brain–computer interface. IEEE Journal of Biomedical and Health Informatics. 2014;18(4):1461–1472.

[56] Lotte F, Bougrain L, Cichocki A, *et al.* A review of classification algorithms for EEG-based brain–computer interfaces: A 10 year update. Journal of Neural Engineering. 2018;15(3):031005.

[57] Oikonomou VP, Tzallas AT, Konitsiotis S, *et al.* The use of Kalman filter in biomedical signal processing. In: Moreno VM, Pigazo A, editors. Kalman Filter: Recent Advances and Applications. Rijeka: IntechOpen; 2009. Available from: https://doi.org/10.5772/6805.

[58] Jayaram V, Alamgir M, Altun Y, *et al.* Transfer learning in brain–computer interfaces. IEEE Computational Intelligence Magazine. 2016;11(1):20–31.

[59] Yang Q, Zhang Y. An overview of multi-task learning. National Science Review. 2017;5(1):30–43.

[60] Alamgir M, Grosse-Wentrup M, Altun Y. Multitask learning for brain–computer interfaces. In: Teh YW, Titterington M, editors. Proceedings of the Thirteenth International Conference on Artificial Intelligence and Statistics. vol. 9 of Proceedings of Machine Learning Research. Chia Laguna Resort, Sardinia, Italy; 2010. p. 17–24.

[61] Cecotti H, Gräser A. Convolutional neural networks for P300 detection with application to brain–computer interfaces. IEEE Transactions on Pattern Analysis and Machine Intelligence. 2011;33(3):433–445.

[62] Kwak NS, Müller KR, Lee SW. A convolutional neural network for steady state visual evoked potential classification under ambulatory environment. PLoS One. 2017;12(2):1–20. Available from: https://doi.org/10.1371/journal.pone.0172578.

Chapter 8
BCIs using steady-state visual-evoked potentials

Vangelis P. Oikonomou[1], Elisavet Chatzilari[1], Georgios Liaros[1], Spiros Nikolopoulos[1], and Ioannis Kompatsiaris[1]

Brain–computer interfaces (BCIs) have been gaining momentum in making human–computer interaction more natural, especially for people with neuromuscular disabilities. Among the existing solutions, the systems relying on electroencephalograms (EEGs) occupy the most prominent place due to their noninvasiveness. However, the process of translating EEG signals into computer commands is far from trivial, since it requires the optimization of many different parameters that need to be tuned jointly. In this chapter, we focus on the category of EEG-based BCIs that rely on steady-state-visual-evoked potentials (SSVEPs) and perform a comparative evaluation of the most promising algorithms existing in the literature. Moreover, we will also describe four novel approaches that are able to improve the accuracy of the interaction under different operational context.

8.1 Introduction

The EEG represents the mean electrical activity of the brain cells in different locations of the head. We acquire the brain activity by placing scalp electrodes on the surface of the head. To ensure reproducibility among studies, an international system for electrode placement, the 10–20 international system, has been defined. In this system, the electrodes' locations are related to specific brain areas. For example, electrodes O1, O2, and Oz are above the visual cortex. Each EEG signal can therefore be related to an underlying brain area. However, this is a broad approximation that depends strongly on the placement of the electrodes.

The EEG is a valuable tool in the diagnosis of numerous brain disorders. Nowadays, the recording of EEG signals is a routine clinical procedure and it is widely regarded as the physiological "gold standard" to monitor and quantify electric brain activity. The electric activity of the brain is usually divided into three categories: (1) bioelectric events produced by single neurons, (2) spontaneous activity, and

[1]The Multimedia Knowledge and Social Media Analytics Laboratory, Information Technologies Institute, Centre for Research and Technology-Hellas (CERTH), Thessaloniki, Greece

(3) EPs. EEG spontaneous activity is measured on the scalp or on the brain. Clinically, meaningful frequencies lie between 0.1 and 100 Hz. Event-related potentials (ERPs) are the changes of spontaneous EEG activity related to a specific event. ERPs triggered by specific stimuli, VEP, auditory, or somatosensory are called EPs. It is assumed that ERPs are generated by the activation of specific neural populations, time-locked to the stimulus, or that they occur as the result of the reorganization in ongoing EEG activity. The basic problem in the analysis of ERPs is their successful detection within the larger EEG activity, since ERP amplitudes are an order of magnitude smaller than that of the rest EEG components.

VEPs have attracted special interest due to their usage on the construction of BCIs systems. When the stimulation frequency is at low rate (<4 Hz), the potentials are called transient VEPs, while stimulation on higher rate (>6 Hz) produces SSVEPs [1]. When identical stimuli are presented at high frequency (e.g., 8 Hz), the visual system stops producing transient responses and enters into a steady state where the visual system resonates at the stimulus frequency. In other words, when the human eye is excited by a visual stimulus, the brain generates electrical activity at the same (or multiples of) frequency of the visual stimulus. Besides, the significance of SSVEPs in clinical studies, their employment as a basic building block of BCIs makes them a very important tool. An SSVEP-based BCI gives the user the ability to select among several commands, where each command is associated with a repetitive visual stimulus that has distinctive properties (e.g., frequency). The general architecture of an SSVEP BCI system contains the following modules: (1) stimulator module: this module is responsible for the production of the visual stimuli, and, in most cases, it is an LED panel or a monitor; (2) signal acquisition module: it is responsible for the acquisition of SSVEP signals; (3) signal processing module: it is responsible for the analysis of SSVEP signals and the translation/transformation of them into meaningful “codewords”; and (4) device commands module: it is appointed with the task to translate the “codewords” into interface commands according to the application setup. In this chapter, our interest lies in the signal processing module.

SSVEP is the brain response evoked in the occipital and occipital–parietal areas of the brain by a visual stimulus flashing at a fixed frequency [2] and includes the fundamental frequency of the visual stimulus that has generated this response, as well as its harmonics. Due to this property, methods based on power spectrum density analysis (PSDA) (i.e., fast Fourier transform) have been extensively used [3,4]. However, PSDA methods are sensitive to noise and they need a large data window to estimate the frequency spectrum with sufficient resolution [5,6]. To overcome the previously mentioned problems, approaches based on spatial filtering have been proposed. These approaches are based on reference templates and they solve an optimization problem on multichannel SSVEP data for obtaining the optimal spatial filters [5–7].

A well-known method of spatial filtering is the canonical correlation analysis (CCA) [5] where the correlation between the reference templates and the acquired EEG signal (SSVEP responses) is used to detect the target frequency. However, one major drawback of CCA is that the templates are based on sine–cosine, resulting in inaccurate representation of the underlying information due to overfitting [8].

Extensions of basic CCA method have been proposed to alleviate this problem, including a multi-way extension of CCA (MCCA) and its L1-regularized version (L1MCCA) [8], itCCA that averages over multiple training trials for constructing individual templates[9], as well as a combination of standard CCA and itCCA [10]. Recently, the multivariate synchronization index [11] has been also proposed to analyze SSVEP data by using time-delayed versions of them. CCA method has proven very useful method for the analysis of SSVEP. In addition, CCA is a special case of the generalized eigenvalue problem; hence, it is natural to pay close attention to various others special cases of the generalized eigenvalue problem. Recently, methods based on the generalized eigenvalue problem have been proposed in [12–14]. More specifically, in task-related component analysis [14] and in correlated component analysis [13], spatial filters were learned in order to increase the discrimination ability of the subsequent classification scheme, while in the Subclass Marginal Fisher Analysis (SMFA) [12], discriminative features were learned. SMFA belongs to a general category of techniques, known as subspace learning, which is the process of reducing the dimensionality of the raw data while retaining as much discriminant information as possible.

Furthermore, methods that do not make the use of any reference templates, based on deep learning techniques, have been proposed in [15,16]. However, these methods have been tested with a limited number (four or five) of stimulus frequencies and with SSVEP responses segmented into overlapping segments of (at least) 1 s, resulting in BCI systems with low ITR values. Furthermore, deep learning methods (due to complex structure of the underlying model) require large enough datasets and are computationally expensive with limited expressiveness in BCI applications [17]. Finally, in [18], the use of multivariate linear regression (MLR) was proposed to learn discriminative features for improving SSVEP classification using only four stimulus frequencies. Extending the work in [18], linear regression models under the Bayesian framework have been used in [19,20]. Furthermore, in [21], different kernel spaces under a multiple kernel scheme are used to improve the classification performance. The idea of using multiple kernels is motivated by the need to combine different views of the data and our expectation on improving the performance of SSVEP classification stems from the fact that it has been already proposed for the classification of P300 signals [22] and the clustering of fMRI time series [23]. In this chapter, we will present various regression-based SSVEP recognition systems reported in the literature based on the conception of synchronous mode [24].

8.2 Regression-based SSVEP recognition systems

The input to signal processing module is a set of EEG trials $\mathscr{X}_i \in \Re^{M \times P}, i = 1, \ldots, N$. Another form of representation of the ith EEG trial is that of vector (instead of matrix) $\mathbf{x}_i = vec(\mathscr{X}_i)$, where vec represents the procedure which converts a matrix into a vector (concatenation of P temporal points from M channels). Now, the set of EEG trials is $\mathbf{x}_1, \mathbf{x}_2, \ldots, \mathbf{x}_N \in \Re^D$ (feature vectors), where each vector has dimension $D = M \times P$ and N is the number of training samples. The classes are represented by adopting the 1-of-K coding scheme, where K is the number of classes. More specifically, for a training sample $\mathbf{x}_i$ belonging to class m, its label is specified as

$$\mathbf{y}_i = [y_1, y_2, \ldots, y_K], \text{ where } y_j = \begin{cases} 1, & \text{if } j = m \\ 0, & \text{otherwise} \end{cases} \tag{8.1}$$

The previous formulation provides us with the indicator matrix $\mathbf{Y} = [\mathbf{y}_1, \mathbf{y}_2, \ldots, \mathbf{y}_N]^T, \in \Re^{N\times K}$. Assuming that each column of matrix $\mathbf{Y}$ can be expressed as a linear combination of feature vector, we obtain the following K regression models:

$$\mathbf{y}_k = \mathbf{X}\mathbf{w}_k + \mathbf{e}_k, \quad k = 1, \ldots, K \tag{8.2}$$

The previous assumption leads us to K regression models, where each regression model is trying to learn the labels of one class versus the rest. To obtain an estimate for the model parameters $\mathbf{w}_k$, we will describe two frameworks, the least squares and the sparse Bayesian learning. But before that, it is needed to provide relevant information related to (8.2). The vector $\mathbf{y}_k \in \Re^N$ contains 0s and 1s, with the nth element being 1 if the nth feature vector belongs to class k. The matrix $\mathbf{X} \in \Re^{N\times D}$ contains the EEG samples (feature vectors) $\mathbf{x}_i$, $i = 1, \ldots, N$ and $\mathbf{e}_k$ denotes the noise of the model following a Gaussian distribution with zero mean and precision (inverse variance) β_k. Finally, the $\mathbf{w}_k \in \Re^D$ is a vector containing the model parameters.

8.2.1 Multivariate linear regression (MLR) for SSVEP

The MLR classification scheme was first appeared for the analysis of SSVEP data in [18]. This scheme combines into one general framework/algorithm—a feature-extraction-based method and a classifier. In this section, we provide a brief description of the MLR classification scheme.

In regression, given a training data set comprising of N observations $\mathbf{x}_n \in \Re^D$ where $n = 1, \ldots, N$, together with corresponding target values $\mathbf{y}_n$, the goal is to predict the value of target for a new value of observations. The observations and the targets are connected through a projection matrix $\mathbf{W}$, which is unknown in our case and needs to be estimated. In this case, the least squares method can be applied to compute the projection matrix $\mathbf{W} \in \Re^{D\times K}$:

$$\mathbf{W}_{MLR} = (\mathbf{X}\mathbf{X}^T)^{\dagger}\mathbf{X}^T\mathbf{Y} \tag{8.3}$$

where $(\mathbf{X}\mathbf{X}^T)^{\dagger}$ denotes the pseudo-inverse of $(\mathbf{X}\mathbf{X}^T)$. The least squares formulation can also be applied for classification problems by treating accordingly the target values. More specifically, in classification problems, the target values contain information related to the class labels by using various coding schemes [25]. In our analysis, the 1-of-K coding scheme has been adopted.

After learning the projection matrix via MLR between training samples and multi-class label matrix, the training data are projected onto a lower dimensional space expanded by columns of $\mathbf{W}$, which represent the features of training data. Hence, sample features $\mathbf{Z} \in \Re^{N\times K}$ can be extracted as $\mathbf{Z} = \mathbf{X}\mathbf{W}_{MLR}$, and $\mathbf{z}_i \in \Re^K$ denotes features corresponding to EEG sample $\mathbf{x}_i$. For subsequent analysis of test data, we project them to the space learned by MLR and as a final step, the k-nearest-neighbor (kNN) algorithm is adopted to classify the subspace features extracted by the MLR.

8.2.2 Sparse Bayesian LDA for SSVEP

An alternative approach to learning the projection matrix $\mathbf{W}$ is through the Bayesian framework. The Bayesian framework has been extensively used to analyze EEG for BCI applications. More specifically, the Bayesian Linear Discriminant Analysis (BLDA) with sparse priors has been used for sensorimotor rhythms discrimination [26] and for SSVEP responses recognition [21]. In this chapter, we describe the general idea of the previous works restricting ourselves to the analysis of SSVEP data.

The overall model is constituted by K regression models (see (8.2)); hence, we can work on each model separately. Also the subscript k has been removed in the remaining of this subsection to facilitate the exposition of the algorithm. Crucial part in a Bayesian setting plays the priors and hyperparameters of the probabilistic model and how we are handling these two issues. In the literature, two well-known methodologies are widely used: the Bayesian evidence framework and the variational Bayesian (VB) framework [25]. In our study, we adopt the VB framework since it provides us the ability to use prior (and hyperprior over hyperparameters) distributions overall model parameters. A suitable choice for the prior distribution is the Automatic Relevance Determination (ARD) prior [27,28] since provides us with simpler models. This type of sparse prior is based on a hierarchical modeling approach, where the parameter vector $\mathbf{w}$ is treated as a random variable with Gaussian prior of zero mean and variance a_i^{-1} for each element in the vector $\mathbf{w}$:

$$p(\mathbf{w}|\mathbf{a}) = \prod_{i=1}^{D} N(0, a_i^{-1}), \tag{8.4}$$

where D is the length of the vector $\mathbf{w}$. Also each parameter a_i, which controls the prior distribution of the parameters $\mathbf{w}$, follows a Gamma distribution, so the overall prior over all a_i is a product of Gamma distributions given by $p(\mathbf{a}) = \prod_{i=1}^{D} Gamma(a_i; b_a, c_a)$.

Besides the prior and hyperparameters over $\mathbf{w}$, it is needed to deal with the noise (more specifically with its variance), since in our study, it is an unknown quantity. The overall precision (inverse variance) β of the noise follows a Gamma distribution: $p(\beta) = Gamma(\beta; b, c) = (1/\Gamma(c))(\beta^{(c-1)}/b^c)\exp\{-(\beta/b)\}$, where b and c are the scale and the shape of the Gamma distribution, respectively.

Now, the prior governing all model parameters $\{\mathbf{w}, \mathbf{a}, \beta\}$ is given by $p(\mathbf{w}, \mathbf{a}, \beta) = p(\mathbf{w}|\mathbf{a}) \prod_{i=1}^{D} p(a_i)p(\beta)$. In comparison with the BLDA, described in the previous chapter in this book, we can observe here that a different prior is used over $\mathbf{w}$. Also the noise precision as well as the parameters of the prior are treated as unknown probabilistic quantities and they must be estimated together with parameters vector $\mathbf{w}$. The likelihood of the data is given by

$$p(\mathbf{y}|\mathbf{w}, \beta) = \frac{\beta^{N/2}}{(2\pi)^{N/2}} \cdot \exp\left\{-\frac{\beta}{2}(\mathbf{y} - \mathbf{X}\mathbf{w})^T(\mathbf{y} - \mathbf{X}\mathbf{w})\right\} \tag{8.5}$$

The VB methodology [25] dictates that we need an approximate distribution of the true distribution in order to facilitate the computations. In our analysis, we define an approximate posterior based on one factorization over the parameters $\{\mathbf{w}, \mathbf{a}, \beta\}$. More specifically, we choose the following factorization:

$$q(\mathbf{w}, \mathbf{a}, \beta) = q(\mathbf{w}|\mathbf{a}) \prod_{i=1}^{D} q(a_i)q(\beta),$$

which means that we assume posteriori independence between model parameters. Applying the VB methodology, and taking into account the previous factorization, the following posteriors are obtained:

$$q(\mathbf{w}) = N(\hat{\mathbf{w}}, \mathbf{C}_{\mathbf{w}}), \tag{8.6}$$

$$q(\beta) = Gamma(\beta; b', c'), \tag{8.7}$$

$$q(\mathbf{a}) = \prod_{i=1}^{D} Gamma(a_i; b'_{a_i}, c'_{a_i}), \tag{8.8}$$

where

$$\mathbf{C}_{\mathbf{w}} = (\hat{\beta}\mathbf{X}^T\mathbf{X} + \hat{\mathbf{A}})^{-1}, \tag{8.9}$$

$$\hat{\mathbf{w}} = (\hat{\beta}\mathbf{X}^T\mathbf{X} + \hat{\mathbf{A}})^{-1}\hat{\beta}\mathbf{X}^T\mathbf{y}, \tag{8.10}$$

$$\frac{1}{b'_{a_i}} = \frac{1}{2}(\hat{w}_i^2 + \mathbf{C}_{\mathbf{w}}(i,i)) + \frac{1}{b_a}, \tag{8.11}$$

$$c'_{a_i} = \frac{1}{2} + c_a, \tag{8.12}$$

$$\hat{a}_i = b'_{a_i}c'_{a_i}, \tag{8.13}$$

$$\frac{1}{b'_\beta} = \frac{1}{2}(\mathbf{y} - \mathbf{X}\mathbf{w})^T(\mathbf{y} - \mathbf{X}\mathbf{w}) + tr(\mathbf{X}^T\mathbf{X}\mathbf{C}_{\mathbf{w}}) + \frac{1}{b}, \tag{8.14}$$

$$c'_\beta = \frac{N}{2} + c, \tag{8.15}$$

$$\hat{\beta} = b'_\beta c'_\beta. \tag{8.16}$$

In the previous equations, the matrix $\hat{\mathbf{A}}$ is a diagonal matrix with the mean of parameters a_i in its main diagonal. Equations (8.9)–(8.16) are applied iteratively until convergence. The previous algorithm is applied for each regression model resulting in K classification scores. When a feature vector is provided and we want to classify it into one of the K classes, the following procedure is adopted:

- Calculate the K classification scores using the predictive means.
- Use the kNN algorithm to classify the classification scores.

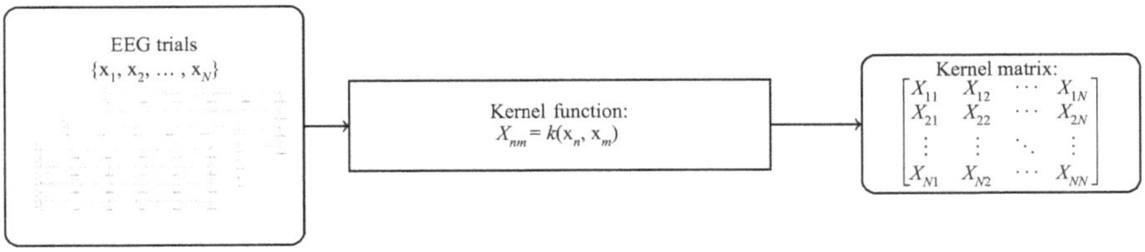

Figure 8.1 Graphic representation of kernel transformation

8.2.3 Kernel-based BLDA for SSVEP (linear kernel)

Regression models of (8.2) can be easily kernelized [29]. By applying the kernel trick, the regression models of feature space $\mathbf{y}_k = \mathbf{X}\mathbf{w}_k + \mathbf{e}_k = \sum_{n=1}^{D} w_{kn}\mathbf{x}_n + \mathbf{e}_k$ are transformed into $\mathbf{y}_k = \sum_{n=1}^{N} w'_{kn}k(\mathbf{x}, \mathbf{x}_n) + \mathbf{e}_k = \mathbf{X}'\mathbf{w}'_k + \mathbf{e}_k$, where the matrix $\mathbf{X}'$ is a $N \times N$ symmetric matrix with elements $X_{nm} = k(\mathbf{x}_n, \mathbf{x}_m)$, $k(\cdot)$ is the kernel function, and $\mathbf{w}'_k \in \Re^N$ is the new vector of regression coefficients. In supervised learning settings, feature-based classifiers are trying to predict the class label by giving optimal (in some sense) weights in each feature. From the other side, kernel-based classifiers are trying to predict the class label by giving weights to the similarities between the training samples (see Figure 8.1). We can observe that the kernel-based transformation is useful in high dimensional settings, even if the linear kernel is used. More specifically, to compute the regression coefficients $\mathbf{w}_k$ into the original feature space, the computational cost is $O(D^3)$, while in the kernel space is $O(N^3)$[29]. When $D \gg N$, as it is the case for the SSVEP analysis, the computational cost of working into the original feature space is considerable compared to the computational cost of kernel space.

8.2.4 Kernels for SSVEP

In the literature, various kernel functions have been proposed such as the linear and Gaussian kernels. In this section, we describe two kernel functions appropriate for multichannel data (and hence SSVEP data). More specifically, the two kernels are based on CCA and Partial Least Squares (PLS) methods.

8.2.4.1 CCA-based kernel

CCA [30] seeks to find the underlying correlations between two sets of data (paired data). Let us assume that $\mathbf{a} \in \Re^{D_a}$ and $\mathbf{b} \in \Re^{D_b}$ are two multivariate variables and their corresponding (paired) sets are $S_a = \{\mathbf{a}_1, \mathbf{a}_2, \ldots, \mathbf{a}_n\}$, $S_b = \{\mathbf{b}_1, \mathbf{b}_2, \ldots, \mathbf{b}_n\}$. Furthermore, let us denote with $f^{CCA}(S_a, S_b)$ the process (or function) of calculating the maximum canonical correlation between the two sets S_a and S_b. The function $f^{CCA}(\cdot, \cdot)$ has two important properties: it is symmetric ($f^{CCA}(S_a, S_b) = f^{CCA}(S_b, S_a)$) and nonnegative ($f^{CCA}(S_a, S_b) \geq 0$). Hence, the previous function can be interpreted as a measure of similarity and it can be used to produce a valid kernel matrix [29]. For SSVEP analysis, we can create a kernel matrix $\mathbf{\Phi}_{CCA}$ based on CCA by defining the sets S_a and S_b to be the multichannel EEG trials:

$$\mathbf{\Phi}_{CCA} = \begin{bmatrix} f^{CCA}(\mathscr{X}_1, \mathscr{X}_1) & f^{CCA}(\mathscr{X}_1, \mathscr{X}_2) & \cdots & f^{CCA}(\mathscr{X}_1, \mathscr{X}_N) \\ f^{CCA}(\mathscr{X}_2, \mathscr{X}_1) & f^{CCA}(\mathscr{X}_2, \mathscr{X}_2) & \cdots & f^{CCA}(\mathscr{X}_2, \mathscr{X}_N) \\ \vdots & \vdots & \ddots & \vdots \\ f^{CCA}(\mathscr{X}_N, \mathscr{X}_1) & f^{CCA}(\mathscr{X}_N, \mathscr{X}_2) & \cdots & f^{CCA}(\mathscr{X}_N, \mathscr{X}_N) \end{bmatrix}$$

Finally, it must be noted that in our approach, we do not propose a kernelized version of CCA such as in [31], but CCA is used as the function to construct the kernel matrix.

8.2.4.2 PLS kernel

Assuming two EEG trials $\mathscr{X}_i$ and $\mathscr{X}_j$. The linear regression of $\mathscr{X}_j$ on $\mathscr{X}_i$ is defined as $\mathscr{X}_j = \mathbf{C}\mathscr{X}_i + \mathbf{V}$, where $\mathbf{C}$ and $\mathbf{V}$ are coefficient and noise matrices. Note here that in this subsection, the terms "noise" and "coefficient" are not the same as these of the linear regression model described previously. PLS method decomposes the two trials into the following form:

$$\mathscr{X}_i = \mathbf{T}\mathbf{P}^T + \mathbf{E} \tag{8.17}$$

$$\mathscr{X}_j = \mathbf{U}\mathbf{Q}^T + \mathbf{F} \tag{8.18}$$

where $\mathbf{T}$ and $\mathbf{U}$ are matrices containing the latent vectors, $\mathbf{P}$ and $\mathbf{Q}$ are the loading matrices, and, $\mathbf{E}$ and $\mathbf{F}$ are the residual matrices. The decomposition of trial $\mathscr{X}_j$ can be expressed by using the trial $\mathscr{X}_i$ by

$$\mathscr{X}_j = \mathbf{B}\mathscr{X}_i + \mathbf{F}^* \tag{8.19}$$

where $\mathbf{F}^*$ is the residual matrix and the matrix $\mathbf{B}$ is given by

$$\mathbf{B} = \mathscr{X}_i\mathbf{U}(\mathbf{T}\mathscr{X}_i\mathscr{X}_i^T\mathbf{U})^{-1}\mathbf{T}^T\mathscr{X}_j \tag{8.20}$$

In our study, we define as the kernel function $\kappa(\cdot,\cdot)$ between $\mathscr{X}_i$ and $\mathscr{X}_j$ the trace of matrix $\mathbf{B}$, i.e., $\kappa(\mathscr{X}_i, \mathscr{X}_j) = trace(\mathbf{B})$. It is worth to note here that the previous construction approach does not produce a Mercer kernel, and approaches like the support vector machines cannot be applied[29]. In this particular analysis, the term "kernel function" is used to describe the relationship between two objects in a broader sense than the classical case, where the kernel function is defined in terms of similarity or positive definiteness of the Gramian matrix (Mercer kernel). In order to have a positive definite kernel, we could use an eigendecomposition of PLS kernel matrix and remove the negative eigenvalues, as suggested in [32].

8.2.5 Multiple kernel approach

In the previous section, we provided information about three significant kernels in SSVEP analysis, the linear kernel, the CCA-based kernel, and the PLS-based kernel. In this section, we will provide information on how we could combine these kernels under one general framework. Multiple kernel learning (MKL) is one of the most promising kernel optimization approaches, which aims at simultaneously learning a combined kernel and the associated predictor in supervised learning settings [33–35]. More specifically, the combined kernel is modeled as a convex sum of base kernels as follows: $k_{\eta_k}(\mathbf{x},\mathbf{x}_n) = \sum_{\ell=1}^{L} \eta_{\ell k} k_\ell(\mathbf{x},\mathbf{x}_n), \sum_{\ell=1}^{L} \eta_\ell = 1, \eta_\ell \geq 0$; where L is the number of kernels. Furthermore, the combined kernel matrix $\mathbf{\Phi}_{\eta_k}$ is provided by $\mathbf{\Phi}_{\eta_k} = \sum_{\ell=1}^{L} \eta_{\ell k}\mathbf{\Phi}_\ell, \sum_{\ell=1}^{L} \eta_{\ell k} = 1, \eta_{\ell k} \geq 0$. Also we can extend the linear regression model in order to take into account the combined kernel $\mathbf{\Phi}_{\eta_k}$:

$$\mathbf{y}_k = \sum_{n=1}^{N} w_{nk} \sum_{\ell=1}^{L} \eta_{\ell k} k_\ell(\mathbf{x},\mathbf{x}_n) + \mathbf{e}_k = \mathbf{\Phi}_{\eta_k}\mathbf{w}_k + \mathbf{e}_k. \tag{8.21}$$

The previous equation can be written slightly different by interchanging the sum operators. In that case, we have a new representation of the linear regression model given by

$$\begin{aligned}\mathbf{y}_k &= \sum_{\ell=1}^{L} \eta_{\ell k} \sum_{n=1}^{N} w_{nk} k_\ell(\mathbf{x},\mathbf{x}_n) + \mathbf{e}_k \\ &= \sum_{\ell=1}^{L} \eta_{\ell k}(\mathbf{\Phi}_\ell \mathbf{w}_k) + \mathbf{e}_k = \mathbf{\Phi}_{\mathbf{w}_k}\boldsymbol{\eta}_k + \mathbf{e}_k,\end{aligned} \tag{8.22}$$

where the matrix $\mathbf{\Phi}_{\mathbf{w}_k} \in \Re^{N\times L}$ and each column of this matrix is constructed by weighting each base kernel matrix with the regression coefficients $\mathbf{w}_k$. As we can see the class labels $\mathbf{y}_k$ are described by (8.21) or (8.22). Thus, by knowing the regression coefficients $\mathbf{w}_k$ and the mixing coefficients $\boldsymbol{\eta}_k$, it is possible to provide predictions. A graphic representation of the MKL scheme is provided in Figure 8.2. In [21], a methodology for combining the MKL scheme with the Sparse Bayesian Learning (SBL) framework is provided and we will adopt this procedure in our current experimental analysis.

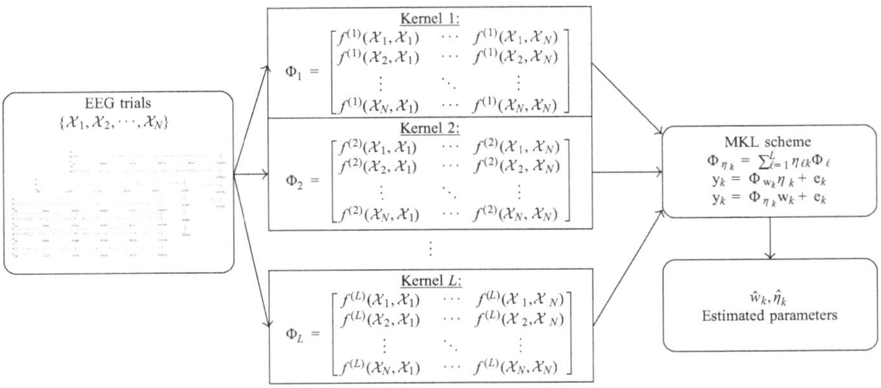

Figure 8.2 Graphic representation of multiple kernel learning approach. In this figure, symbol $f^{(\cdot)}(\cdot,\cdot)$ represents a general kernel function © 2019. Reprinted, with permission, from Reference [21]

8.3 Results

In order to validate the performance of the described/reviewed regression-based pattern recognition system, we use the EEG dataset described in [10]. In this dataset, 12-target visual stimuli were presented on a 27-in. LCD monitor. Ten healthy subjects with normal or corrected-to-normal vision participated in this study. EEG data were recorded with eight electrodes covering the occipital area. For each subject, the experiment consisted of 15 blocks. In each block, subjects were asked to gaze at one of the visual stimuli indicated by the stimulus program in a random order for 4s, and complete 12 trials corresponding to all 12 targets. Data epochs, comprising eight-channel SSVEPs, were extracted according to event triggers generated by the stimulus program. All data epochs were down-sampled to 256 Hz. The EEG data have been band-pass filtered from 6 to 80 Hz with an infinite impulse response filter using the filtfilt() function in MATLAB. As indicated in [10], a latency delay of 0.135 ms in the visual system is considered.

For evaluation purposes, we have used the classification accuracy. The classification is defined as the ratio of the number of correctly classified trials to the total number of trials. In our analysis, the classification accuracy is estimated using a leave-one-block-out cross validation. In each of 15 rounds, cross-validation was performed using 14 blocks for training and 1 block for testing. The previous procedure is the same as that described in [10,18]. Finally, the accuracy has been calculated at various time windows ranging from 0.5 to 4 s.

The goal of an SSVEP pattern recognition algorithm is to take as input one EEG trial, $\mathscr{X}$, and assign it into one of $K(=12)$ classes where each class corresponds to a stimulation frequency $f_k, k = 1, \ldots, K$. In this section, we provide results that show the effectiveness of various kernels to discriminate various SSVEP responses. More specifically, the linear, the CCA, and the PLS kernel are used. Also the previous kernels are combined by adopting the MKL scheme [21]. As a baseline method, we use the MLR approach and the BLDA with linear kernel. Next, we provide a short explanation of the methods' acronym.

- MLR is the acronym for the MLR method.
- LIN is the acronym of the basic version of BLDA where the linear kernel is used.
- PLS is the acronym of BLDA when the PLS kernel is used.
- MKL_CCA_LIN is the acronym of BLDA when we combine the CCA kernel with the linear kernel under the MKL scheme.
- MKL_PLS_CCA is the acronym of BLDA when we combine the CCA kernel with the PLS kernel under the MKL scheme.

The results are shown in Figure 8.3. The LIN method presents the worst performance compared to MLR, especially in small time windows. However, when this kernel is combined by the CCA kernel in the MKL scheme (MKL_CCA_LIN method), the performance becomes much better than MLR. But it cannot provide us with classification rates such high as the PLS method. The PLS method provides the most promising results, which can be augmented slightly if this kernel is combined with the CCA kernel under the MKL scheme (MKL_PLS_CCA method). In Figure 8.4,

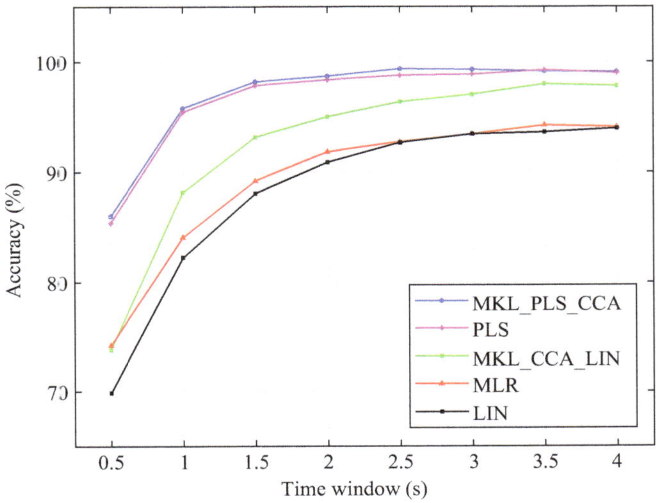

Figure 8.3 Average accuracy overall subjects for time windows from 0.5 to 4 s (0.5 s interval)

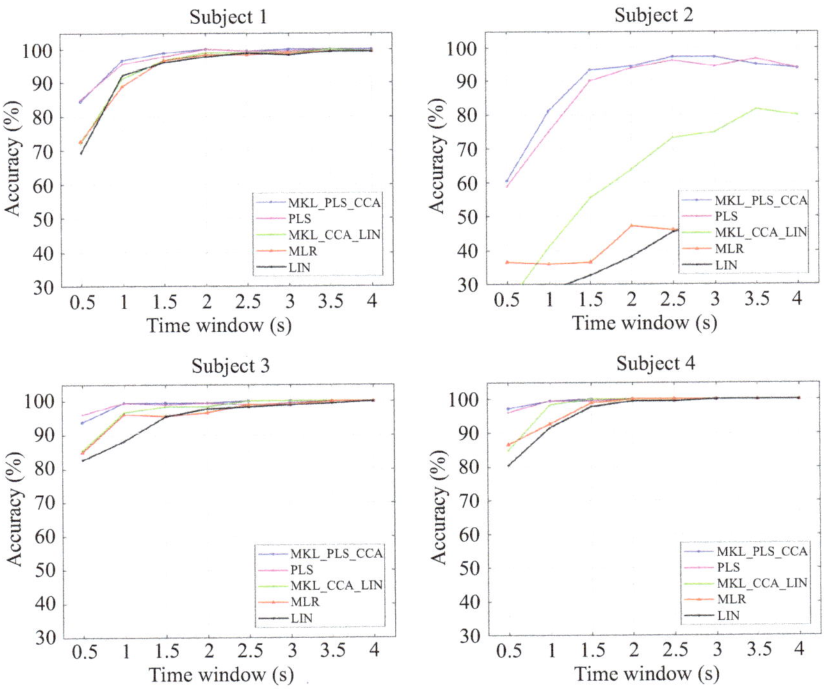

Figure 8.4 Average accuracy per subject for time windows from 0.5 to 4 s (0.5 s interval)

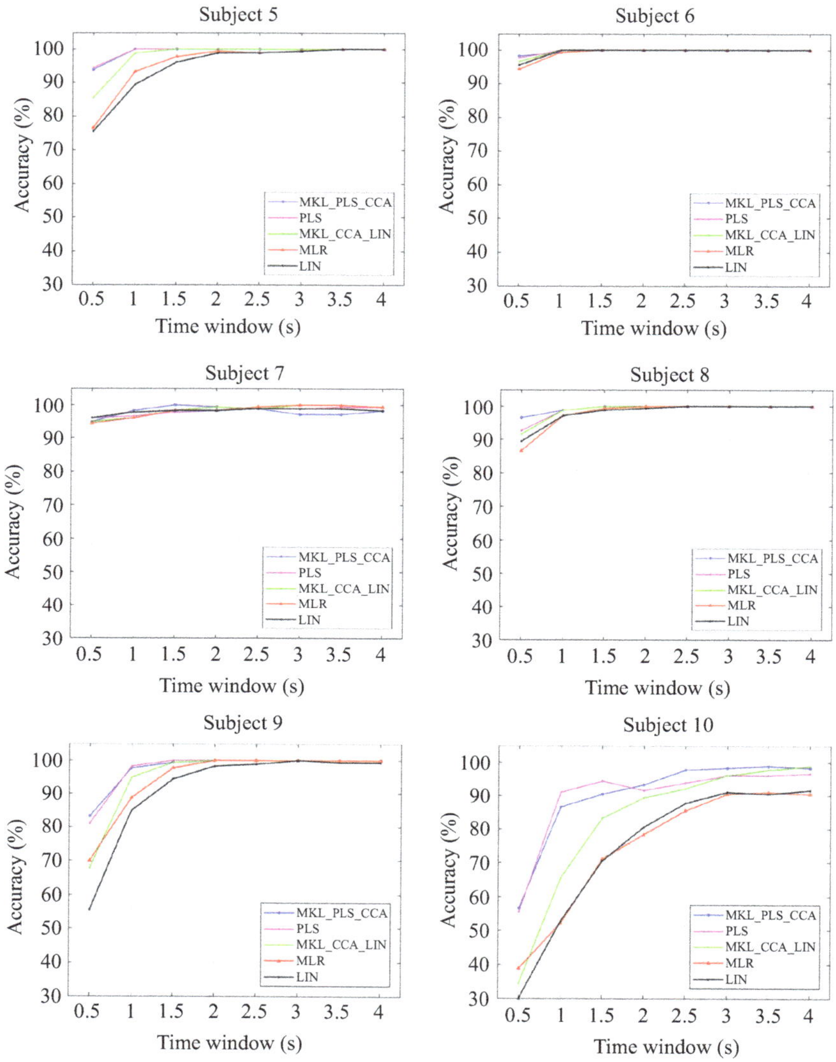

Figure 8.4 (Continued)

the accuracy for each subject is provided for all methods. We can observe that the combined use of PLS and CCA provides the best performance for each subject. In this figure, it can also be observed that there exist a significant number of subjects that all methods present similar performance (Subjects 6–8). Furthermore, we can observe that Subjects 2 and 10 are the subjects that contribute most in the observed differences between the PLS-related methods and the rest methods. Especially, in Subject 2, the MLR and LIN methods fail to provide accurate performance.

8.4 Summary

In this chapter, we presented existing methods for the discrimination of SSVEP responses based on the linear regression model. At first, we presented the MLR method where massive linear regression models were used to find an optimal linear projection between the EEG trials and the corresponding labels. This projection gave us the opportunity to extract new discriminative features. The performance of MLR, presented here and elsewhere [18,21], suggests that the linear-regression-based techniques are valuable tools in the BCI context. The general key steps of MLR method have been extended in more complicated models by adopting the Bayesian framework. Due to large number of raw EEG features, the MLR approach uses, as a preprocessing step, a dimensionality reduction technique based on principal component analysis. To alleviate the problem of large number of raw features, we used the BLDA method with sparse priors resulting into an algorithm that chooses the most important raw features. However, the use of BLDA on the original feature domain does not solve the problem of computational cost, which we alleviated by adopting the idea of kernels. Furthermore, we presented two recently used kernels in SSVEP analysis and an extension of kernel-idea by using multiple kernels under the same framework. The results have shown the usefulness of all methods. A careful investigation and comparisons with state-of-the-art methods of the literature reveal that all methods provide us with state-of-the-art performance. Finally, the performance of the presented methods (in this chapter) could be enhanced by adopting a filter bank analysis or an ensemble strategy such as those presented in [14].

References

[1] American Clinical Neurophysiology Society. Guideline 9B: Recommended Standards for Visual Evoked Potentials; 2008. https://www.acns.org/pdf/guidelines/Guideline-9B.pdf.

[2] Gao S, Wang Y, Gao X, *et al.* Visual and Auditory Brain–Computer Interfaces. IEEE Transactions on Biomedical Engineering. 2014;61(5):1436–1447.

[3] Cheng M, Gao X, Gao S, *et al.* Design and Implementation of a Brain–Computer Interface With High Transfer Rates. IEEE Transactions on Biomedical Engineering. 2002;49(10):1181–1186.

[4] Wang H, Li T, Huang Z. Remote Control of an Electrical Car With SSVEP-Based BCI. In: 2010 IEEE International Conference on Information Theory and Information Security, Beijing, China; 2010. p. 837–840.

[5] Lin Z, Zhang C, Wu W, *et al.* Frequency Recognition Based on Canonical Correlation Analysis for SSVEP-Based BCIs. IEEE Transactions on Biomedical Engineering. 2006;53(12):2610–2614.

[6] Friman O, Volosyak I, Graser A. Multiple Channel Detection of Steady-State Visual Evoked Potentials for Brain–Computer Interfaces. IEEE Transactions on Biomedical Engineering. 2007;54(4):742–750.

[7] Ge S, Wang R, Leng Y, *et al.* A Double-Partial Least-Squares Model for the Detection of Steady-State Visual Evoked Potentials. IEEE Journal of Biomedical and Health Informatics. 2017;21(4):897–903.

[8] Zhang Y, Zhou G, Jin J, *et al.* L1-Regularized Multiway Canonical Correlation Analysis for SSVEP-Based BCI. IEEE Transactions on Neural Systems and Rehabilitation Engineering. 2013;21(6):887–896.

[9] Bin G, Gao X, Wang Y, *et al.* A High-Speed BCI Based on Code Modulation VEP. Journal of Neural Engineering. 2011;8(2):025015. Available from: http://stacks.iop.org/1741-2552/8/i=2/a=025015.

[10] Nakanishi M, Wang Y, Wang YT, *et al.* A Comparison Study of Canonical Correlation Analysis Based Methods for Detecting Steady-State Visual Evoked Potentials. PLoS One. 2015;10:e0140703.

[11] Zhang Y, Guo D, Yao D, *et al.* The Extension of Multivariate Synchronization Index Method for SSVEP-Based BCI. Neurocomputing. 2017;269:226–231.

[12] Maronidis A, Oikonomou VP, Nikolopoulos S, *et al.* Steady State Visual Evoked Potential Detection Using Subclass Marginal Fisher Analysis. In: 8th International IEEE EMBS Conference on Neural Engineering, Shangai, China; 2017: p. 37–41.

[13] Zhang Y, Guo D, Li F, *et al.* Correlated Component Analysis for Enhancing the Performance of SSVEP-Based Brain–Computer Interface. IEEE Transactions on Neural Systems and Rehabilitation Engineering. 2018;26(5):948–956.

[14] Nakanishi M, Wang Y, Chen X, *et al.* Enhancing Detection of SSVEPs for a High-Speed Brain Speller Using Task-Related Component Analysis. IEEE Transactions on Biomedical Engineering. 2018;65(1):104–112.

[15] Cecotti H. A Time-Frequency Convolutional Neural Network for the Offline Classification of Steady-State Visual Evoked Potential Responses. Pattern Recognition Letters. 2011;32(8):1145–1153.

[16] Kwak NS, Muller KR, Lee SW. A Convolutional Neural Network for Steady State Visual Evoked Potential Classification Under Ambulatory Environment. PLoS One. 2017;12(2):1–20.

[17] Lotte F, Bougrain L, Cichocki A, *et al.* A Review of Classification Algorithms for EEG-Based Brain–Computer Interfaces: A 10 Year Update. Journal of Neural Engineering. 2018;15(3):031005.

[18] Wang H, Zhang Y, Waytowich NR, *et al.* Discriminative Feature Extraction via Multivariate Linear Regression for SSVEP-Based BCI. IEEE Transactions on Neural Systems and Rehabilitation Engineering. 2016;24(5):532–541.

[19] Oikonomou VP, Liaros G, Nikolopoulos S, *et al.* Sparse Bayesian Learning for Multiclass Classification With Application to SSVEP-BCI. In: 7th Graz Brain–Computer Interface Conference, Graz, Austria; 2017.

[20] Oikonomou VP, Maronidis A, Liaros G, *et al.* Sparse Bayesian Learning for Subject Independent Classification With Application to SSVEP-BCI. In: 8th International IEEE EMBS Conference on Neural Engineering, Shanghai, China; 2017: p. 600–604.

[21] Oikonomou VP, Nikolopoulos S, Kompatsiaris I. A Bayesian Multiple Kernel Learning Algorithm for SSVEP BCI Detection. IEEE Journal of Biomedical and Health Informatics. 2018;23:1990–2001.

[22] Yoon K, Kim K. Multiple Kernel Learning Based on Three Discriminant Features for a P300 Speller BCI. Neurocomputing. 2017;237:133–144.
[23] Oikonomou VP, Blekas K. An Adaptive Regression Mixture Model for fMRI Cluster Analysis. IEEE Transactions on Medical Imaging. 2013;32:649–660.
[24] Georgiadis K, Laskaris N, Nikolopoulos S, *et al.* Discriminative Codewaves: A Symbolic Dynamics Approach to SSVEP Recognition for Asynchronous BCI. Journal of Neural Engineering. 2018;15(2):026008.
[25] Bishop CM. Pattern Recognition and Machine Learning (Information Science and Statistics). NY: Springer; 2007.
[26] Zhang Y, Zhou G, Jin J, *et al.* Sparse Bayesian Classification of EEG for Brain–Computer Interface. IEEE Transactions on Neural Networks and Learning Systems. 2015;27:2256–2267.
[27] MacKay DJ. Bayesian Interpolation. Neural Computation. 1992;4:415–447.
[28] Tipping ME. Sparse Bayesian Learning and the Relevance Vector Machine. Journal of Mach Learn Research. 2001;1:211–244.
[29] Murphy KP. Machine Learning: A Probabilistic Perspective. Cambridge, MA: MIT Press; 2012.
[30] Hardoon DR, Szedmak SR, Shawe-Taylor JR. Canonical Correlation Analysis: An Overview With Application to Learning Methods. Neural Computation. 2004;16(12):2639–2664.
[31] Shawe-Taylor J, Cristianini N. Kernel Methods for Pattern Analysis. New York, NY: Cambridge University Press; 2004.
[32] Vert JP, Tsuda K, Schölkopf B. A Primer on Kernel Methods. Cambridge, MA: MIT Press; 2004. p. 35–70.
[33] Rakotomamonjy A, Bach FR, Canu S, *et al.* SimpleMKL. Journal of Machine Learning Research. 2008;9:2491–2521.
[34] Gönen M, Alpaydin E. Multiple Kernel Learning Algorithms. Journal of Machine Learning Research. 2011;12:2211–2268.
[35] Gönen M. Bayesian Efficient Multiple Kernel Learning. In: ICML 2012: Proceedings of the 29th International Conference on Machine Learning. New York, NY: ACM; 2012. p. 241–248.

Chapter 9

BCIs using motor imagery and sensorimotor rhythms

Kostas Georgiadis[1,2], *Nikos A. Laskaris*[2,3], *Spiros Nikolopoulos*[1], *and Ioannis Kompatsiaris*[1]

Motor imagery (MI) Brain Computer Interfaces (BCI) are considered the most prominent paradigms of endogenous BCIs, as they comply with the requirements of asynchronous implementations. As MI BCIs can be operated via the movement imagination of one limb (e.g., left hand), after a training period, the user can harness such an interface without the aid of external cue(s) that are considered ideal for self-paced implementations. MI BCIs have been employed in several cases as a means of both communication restoration and neurorehabilitation. Neuromuscular disease (NMD), although rarely studied within the context of MI BCIs, presents significant interest mainly due to the disease's progressive nature and the impact it has on each patient's brain reorganization.

9.1 Introduction to sensorimotor rhythm (SMR)

Endogenous BCIs receive continuous attention from the neuroscientific community as there is no need for external stimulation for them to be operated by the user. Their experimental design encapsulates the perspective of asynchronous (i.e., self-paced) BCIs, and this is the main reason why endogenous BCIs currently receive significant attention, even though a considerable training period that can last from a couple of days to several months is required for the user before harnessing such a system. Endogenous BCIs include the slow cortical potential techniques, like the thought translation device [1] and the MI paradigms, that require the user to perform a mental task, including planning, imagination, and execution of a movement (e.g., one hand [2,3], left or right hand [4], feet [5], or even a combination of limbs and tongue [6]). Besides the two aforementioned approaches, more recently introduced alternatives within the same concept, incorporate speech imagination [7] and mental arithmetic [8].

[1] Information Technologies Institute, CERTH, Thermi-Thessaloniki, Greece
[2] Department of Informatics, Aristotle University of Thessaloniki, Thessaloniki, Greece
[3] Neuroinformatics Group, Aristotle University of Thessaloniki, Thessaloniki, Greece

BCIs based on movement imagination are considered the most prominent paradigms of endogenous BCIs as they provide the most natural path to adopt in a real-life setting where simple commands like "turn left" or "turn right" need to be realized by an actuator. MI BCIs brain decoding usually relies on the sensorimotor rhythm (SMR), oscillations detected in the Electroencephalogram (EEG) signal captured from the electrodes placed on the sensorimotor cortex, the part of the brain that is associated with planning, control, and execution of voluntary movements [9]. MI-related modulations in brain activity are usually associated with both α and β rhythms over the sensorimotor areas. Nevertheless, there are also cases that SMR activity can be associated with different brain rhythms, due to mainly the wide subject variability encountered on EEG measurements.

9.2 Common processing practices

The early studies exploited the power alteration identified whenever a movement is imagined or planned or once it is completed. More specifically, there is a noticeable power decrease in μ-band whenever a movement imagination task is performed known as event-related desynchronization, while once the mental task has been terminated, there is a power increase in β-band, usually referred to as event-related synchronization or beta-rebound [10]. Within the same context, several band-power-based approaches have also emerged [11].

A second popular approach is the technique of common spatial patterns (CSPs) [12], where spatial filtering is combined with classification so as to decode the intended movement, based on the entropy minimization for the one mental task and the maximization for the other. As the core CSP algorithm is designed only for binary-related MI problems, several modifications have been proposed to accommodate the existence of more than two classes (e.g., filter bank CSP (FBCSP) [13] or wavelet CSP [14]). A recent comparison study [15] between power-based approaches, the CSP and FBCSP algorithms, demonstrated that there are no significant differences in terms of performance among them given that an appropriate classification scheme is employed.

Besides the previously mentioned time-domain-based approaches, the detection of discriminative patterns of functional connectivity or covariation, as these emerge in the sensor or source space, has also been attempted [16,17]. Recent approaches include concepts from complex network theory [18] and Riemannian geometry [19], as a means to classify the multichannel EEG-signals based on distinct estimates of connectivity pattern and covariance matrix, respectively. Phase synchrony has also recently entered into the picture and led to novel alternative ways in decoding an indented movement by describing the functional inter-areal interactions during MI [20]. The metric of phase locking value (PLV) is usually employed and features from either the static or dynamic connectivity patterns, as they emerge over the sensor space, have been demonstrated to facilitate the effective decoding of user's intentions [21].

Finally, recent approaches based on the graph signal-processing notion have entered the picture of MI decoding [22–26], where spectral graph theory is blended with signal-processing techniques offering a unified framework to handle signals over irregular domains (as is the case of EEG traces recorder via a given sensor array).

9.3 MI BCIs for patients with motor disabilities

As BCIs, regardless of the selected experimental strategy, can be operated by exploiting the user's brain activity without requiring any physical activity, they are considered ideal assistive mechanisms for people with partial or complete loss of their fine motor skills [27], as they usually cannot use conventional communication channels. As a result, BCIs can provide an alternative communication pathway to people suffering from motor disabilities, therefore, restoring the communication of patients with their environment. Moreover, BCIs have also proven to be extremely effective in terms of rehabilitation, known as neurorehabilitation, by promoting functional recovery providing the patients the opportunity to relearn a number of their motor functions. Beyond their initial motivation, BCI applications have been recently involved in novel approaches extended beyond rehabilitation or communication restoration, including gaming [28] and mental workload monitoring [29].

Depending on the type of disability, different brain regions are affected or even damaged, resulting in the obstruction or the cessation of various motor functions. Furthermore, the loss of motor skills may be gradual or sudden (e.g., chronic strokes) affecting the patient's brain reorganization in different ways. In this direction, several studies have examined the potential use of MI BCIs for enabling the communication or aiding in terms of rehabilitation in several motor disabilities with diversified severity.

9.3.1 MI BCIs for patients with sudden loss of motor functions

Spinal cord injuries (SCIs) (e.g., due to car accident) can result in sudden and complete loss of the affected individuals' motor skills, leading to paraparesis or paraplegia. Similarly, an ischemic stroke (IS) can cause (temporary) paralysis of arms and legs and can even lead to speaking difficulties. As rehabilitation is an integral factor in both cases (i.e., SCI and IS), there are several MI-based studies that examine neurorehabilitation therapies, indicating that BCIs can be used as a means of recovery [30,31]. Besides rehabilitation, MI BCIs have been successfully used, for both conditions, as communication tools in several studies [32,33].

9.3.2 MI BCIs for patients with gradual loss of motor functions

Kübler *et al.* [34] examined patients suffering from amyotrophic lateral sclerosis (ALS) and their ability to operate an MI BCI by mentally controlling (e.g., left-hand movement imagination) the horizontal movement of a mouse cursor, proving that all participants could improve their ability to harness the interface after several training sessions. In the same direction, Bai *et al.* [35], including also primary lateral sclerosis patients in their study, expanded the use of MI BCI beyond the binary cursor control to a four-directional cursor control. Similar trends were observed in multiple sclerosis

(MS) patients [36] that could operate MI BCIs when external cues are provided to them.

9.4 MI BCIs for NMD patients

9.4.1 Condition description

The abovementioned MI-BCI approaches have been investigated in several studies with participants suffering from motor disabilities, ALS, SCI, MS, and CS. However, only a limited number of studies have been done on people suffering from NMD [37].

In contrast with SCI and CS, NMD is a progressive condition that often initiates with the affection of specific group of muscles and finally spreads to many other groups, resulting in gradual loss of a patient's fine motor skills. Therefore, significant mental effort is required by the patients to make a move or even attempt to move their limbs in their everyday life for several years, prior to the complete loss of their movement control. In this direction, the initial motivation of this study was to examine how NMD-patients, as novice BCI users, would perform in simple MI tasks (imagination of left-/right-hand movement) without any training and/or feedback. We hypothesized that, due to long-lasting self-organization, phase synchrony would govern their reconfigured brain networks and could be detected in the sensor space when they were cued to imagine a limb movement (which for them is almost equivalent to try to realize the same movement).

In this direction, it is important to examine whether NMD patients are characterized by increased level of both phase-synchrony and brain network organization as hypothesized, by comparing NMD patients with health control population [38].

9.4.2 Experimental design

9.4.2.1 Participants

Twelve individuals, separated into two groups (seven males and five females, aged 36.08 ± 6.45), participated in the experimental procedure. More specifically, the first group consists of six people suffering from various types of NMD and the second of six able-bodied with a matching sociodemographic profile. Table 9.1 provides information about each participant, while a more detailed description (e.g., inclusion criteria and clinical characteristics) is available in [39]. All subjects had normal or corrected-to-normal vision and none of them had taken any psychoactive or psychotropic substance. Participants were BCI-naive users, as they had no prior experience with BCIs. Prior to the experimental session, subjects and their caretakers were informed about the experimental procedure and a consent form (approved by the Ethical Committee of MDA HELLAS—protocol no. TH.COM-23, 26/01/2017) was signed by the participants or in the cases of inability by their caretakers.

9.4.2.2 Experimental environment

During the experimental procedure, participants were seated in a comfortable armchair (or in their wheelchair) placed 50 cm from a 22-in. liquid crystal display with

Table 9.1 Subject demographics

Participant ID	Gender	Age	Participant ID	Gender	Age	Condition
S1	F	46	P1	M	35	SMA III
S2	F	31	P2	M	44	Muscular dystrophy
S3	M	40	P3	M	32	Muscular dystrophy type II
S4	M	43	P4	F	36	Tunisian muscular dystrophy
S5	F	39	P5	M	25	Duchenne muscular dystrophy

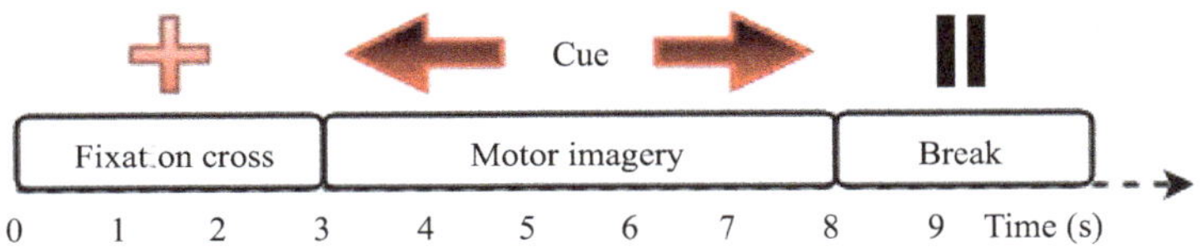

Figure 9.1 The timeline of the experimental procedure (depicted for a single trial)

the EEG cap attached on their scalp. Throughout the entire process, participants were required to place their hands in the armrests and to minimize any kind of upper limb movement in order to minimize the recorded artifactual activity.

9.4.2.3 Experimental design

Prior to the initiation of the MI task, where subjects were asked to imagine the movement of their left or right hand, resting state for 3 min was recorded. The cue for the initiation of movement imagination was given by a red arrow (onset), appearing either on the left or right side of the screen, pointing in the same direction and indicating the corresponding imagery movement. The arrow remained on the screen for approximately 5 s, indicating the continuation of movement imagination to the subject. Once the arrow disappeared from the screen, subjects could rest and prepare themselves for the next arrow appearance. Prior to the arrow presentation, a fixation cross was displayed on the screen for 3 s, indicating the beginning of a new trial. Figure 9.1 presents the sequence of events on a single-trial level. Each subject participated in two identical sessions, each one consisting of 20 random arrow appearances, equally distributed among the two classes, resulting in 40 trials (20 for each imagery movement class). The brain activity was recorded using the BePlusLTM Bioelectric Signal Amplifier,* an EEG headset with 61 + 2 (ground and reference) electrodes placed according to the 10–10 international system with a sampling frequency of 256 Hz and the impedance was set below 10 K before beginning the recording in every session.

*http://www.ebneuro.biz/en/neurology/ebneuro/galileo-suite/be-plus-ltm

9.4.2.4 Preprocessing

During the offline processing and prior to the trial segmentation so as to avoid edge effects, a third-order bandpass filter (0.5–45 Hz) was applied to the EEG signals. Furthermore, "bad" sensors were identified and excluded from further analysis using a spectral analysis based scheme. The remainder "good" sensors, denoted hereby as N_{sensor}, were employed in the subsequent average re-reference procedure.

Independent component (IC) analysis (ICA) was then used to reduce physiological artifacts. Using a semi-supervised procedure that employed the ranking of ICs, based on kurtosis/skewness and the visual inspection of their spectra and topographies, artifactual components were identified and removed before reconstructing the multichannel single-trial data.

Seven commonly used EEG frequency bands were defined: (1–4); (4–8); 1 (8–10); 2 (10–13); 1 (13–20); 2 (20–30); (30–45) Hz and the neural activity of each brain rhythm was examined independently. Once again, band-pass filtering was implemented via third-order Butterworth filters, applied in zero-phase mode.

9.4.2.5 PLV measurements and functional connectivity patterns

Phase synchronization is a well-established concept for describing the coordinated function of distinct neural assemblies based on the recorded signals. When studied at the level of sensor space, the brain signals recorded at distinct sites are used (by one of the available estimators) to detect whether the relative phases of the underlying oscillatory processes bear any systematic relation across time.

The PLV measurement, introduced by Lachaux *et al.* [40], is a very popular estimator of phase synchrony, with the great advantage of computational simplicity that motivated its use in the context of MI-BCIs. Considered a function, PLV receives two signal traces and produces a scalar ranging between 0 and 1, with 1 indicating a "perfect" coupling between the brain areas and 0 indicating functional independence. For a pair of single-trial signals $x_k(t)$ $xr(t)$, with $k, r = 1 \ldots N_{sensor}$ and $t = t_1 \ldots t_2$, from distinct recording sites, PLV is estimated as follows:

$$PLV(x_k, x_r) = \left(\frac{1}{t_2 - t_1} \left| \sum_{t_1}^{t_2} \exp(i(\Delta\phi(t))) \right| \right) \tag{9.1}$$

with $\Delta\phi(t) = \phi_k(t) - \phi_r(t)$ denoting the difference between the instantaneous phases of the two processes and discrete time parameter t running along the latencies of interest (for instance the 5 s interval during the presentation of an arrow on the screen). Hilbert transform is applied to the corresponding band-limited brain activity $x_k(t)$ to produce each phase signal $\phi_k(t)$. By efficiently parallelizing the computations implied by (9.1), the PLV computations extend to all sensor couples.

In this way, for each frequency band, an $[N_{sensor} \times N_{sensor}]$ matrix is formed with entries $W_{kr} = PLV(x_k, x_r)$ and is treated as a weighted adjacency matrix $\mathbf{W}$, including the connectivity pattern of a graph that spans the sensor space and reflects the brain's functional organization. Considering the symmetry in PLV measurements, $PLV(x_k, x_r) = PLV(x_r, x_k)$ and the fact that all diagonal elements W_{kk} equal to 1, it is easy to realize that a more economical description of a connectivity pattern can be obtained by vectorizing the upper triangular part of W, i.e., gathering all $(N_{sensor}xN_{sensor}-1)/2$ elements W_{kr} with $r < k$ in a single vector, denoted as $vec(\mathbf{W})$.

9.4.2.6 Network metrics

The functional connectivity graph defined by **W** matrix, with nodes the recording sites and edges the links between the sites weighed by the associated pairwise PLV values, can be characterized based on network topology metrics [41]. In this chapter, weighted graphs were selected aiming to reveal the self-organization tendencies of the underlying cortical network and provide a comparative analysis of them between healthy and NMD condition. Toward this end, the following three well-established evaluation metrics were employed to compare the recording conditions and the physiological states.

Strength equals to the sum of connectivity weights is attached to a given node:

$$S_k = \sum_{r \neq k} W_{kr} \tag{9.2}$$

Global efficiency (GE)/Local efficiency (LE) is a metric which expresses how efficiently information is transferred via the network, at a global/local level. Network's efficiency is directly linked with the concept of the shortest paths, which in our case were estimated after turning the functional coupling strengths w_{kr} to pairwise distances $d_{kr} = 1 - w_{kr}$ and applying Dijkstra's algorithm. Adopting the formulation of GE as defined in [42]:

$$GE = \frac{1}{N_{sensors}(N_{sensors}-1)} \sum_{k,r \neq k} \frac{1}{l_{rk}} \tag{9.3}$$

with l_{rk} denoting the length of the shortest path between nodes (i.e., sensors) r and k.

LE requires the confinement of the previous calculations to each subgraph G_k and then integrating across nodes:

$$LE = \frac{1}{N_{sensors}} \sum_{r \neq k} LE(k) = \frac{1}{N_{sensors}} \sum_{r \neq k} \frac{1}{N_{Gk}(N_{Gk}-1)} \sum_{i,j \in G_k} \frac{1}{l_{ij}} \tag{9.4}$$

9.4.2.7 Time-indexed patterns of functional connectivity

In an attempt to track more precisely the dynamics of cortical self-organization during MI, we formulated several instantiations of the connectivity pattern for all single trials, by means of a stepping window that confined the integration in (9.1) within successive (overlapping) temporal segments. The "cycle-criterion" [43,44], which adapts the temporal resolution so as three cycles from the lowest frequency of the band-limited brain signal to be included at each step along the time-axis, was adapted to define the window length. In this way, a sequence $vec(W[\tau])$, $\tau = 1, 2, \ldots, N_\tau$ encaptulating the evolving functional connectivity during the hand movement imagination was generated.

9.4.2.8 Feature screening

As the number of features was extremely high (i.e., 1,830 in the case of static PLVs, with this number being multiplied by the selected number of steps in the time-indexed

PLVs) it was clear that, besides the "curse of dimensionality issues," it would not be possible for all couplings to contribute highly discriminative information for the decoding task. In this direction, a "filter" approach (using MATLAB rankfeatures command accompanied by the "Wilcoxon" criterion), including feature ranking and selection, was employed to identify the most reliable coupling in the designing process of a classifier. More specifically, in the case of static connectivity patterns of this operation are estimated as follows:

$$Score(r) = rankfeatures(\{vec(\mathbf{W}^{i}_{left})\}_{i=1:N_{trials}}, \{vec(\mathbf{W}^{j}_{right})\}_{i=1:N_{trials}}), \quad r = 1, 2 \ldots, N_{pairs} \tag{9.5}$$

resulted in a vector of scores reflecting the relative discriminative power of each coupling. Feature selection was accomplished by identifying the set of ten most discriminative couplings.

For the case of time-indexed connectivity, the previous command was applied repeatedly at every latency of the stepping window resulting in a time-indexed score

$$Score(r, \tau) = rankfeatures(\{vec(\mathbf{W}[t]^{i}_{left})\}_{i=1:N_{trials}}, \{vec(\mathbf{W}[t]^{j}_{right})\}_{i=1:N_{trials}}), \quad \tau = 1, 2 \ldots, N_{\tau} \tag{9.6}$$

To identify the most important features among the $(N_{pairs} \cdot N_{\tau})$ available ones, a permutation test was applied, with the connectivity patterns being randomly partitioned, several times, into two groups and the computations implied by (9.6) being repeated for every random splitting. The computed $randScore(r, \tau)1 : N_{rand}$ measurements were used to form a "baseline" distribution of scores associated with the random case (no differences between the two imagery types would be detectable). From the formed distribution, the value of score index corresponding to the margin of 99.9% was identified and used as a threshold, $thr_{99.9\%}$, that was applied to the actual $Score(r, \tau)$ measurements so as to keep only the statistical significant couplings ($p < 0.001$).

After this trimming step that zeroed most of the measurements, a sparse matrix appeared that contained some spurious entries (associated with couplings that occasionally become significant for short lasting intervals). An additional data-sieving step (based on simple rowwise median filtering) was applied that eliminated most of them. The rationale behind this last step was the detection of couplings that could be considered both "useful" and "stable" regarding their discriminatory power.

A pair-dependent profile was derived by the sequence of these operations as shown next, where the operator $H(\cdot)$ denotes Heaviside step function operator and 1_N is column-vector of N ones.

$$\begin{aligned} I(r,t) &= H(Score(r,\tau) - thr_{99.9\%}), r = 1, 2, \ldots, N_{pairs}, t = 1, 2, \ldots, N_t \\ \hat{\mathbf{I}}_{[N_{pairs} x N_{\tau}]} &= runningMedian_{rowwise}(\mathbf{I}) \\ Profile(r) &= \hat{\mathbf{I}} \cdot 1^{N\tau} \end{aligned} \tag{9.7}$$

Finally, feature selection was accomplished by detecting the nonzero entries in this profile. A demonstration of this sequence of algorithmic steps can be seen in Figure 9.12.

9.4.2.9 SVM classifiers as MI-direction decoders

Support vector machines (SVMs) constitute a family of well-established classification algorithms. In the basic binary formulation, the training algorithm of SVM is designed to determine the optimal hyperplane that separates two classes while maximizing the margin between them. The selected hyperplane guarantees optimal generalization, meaning that it can cope better with unseen data. The class of an unseen pattern is determined based on its relative position with respect to the learned hyperplane and the confidence regarding the decision can be estimated using its distance to the hyperplane.

For the purposes of this chapter, a linear hyperplane was selected for the MI-direction decoding as it provided satisfactory results at low computational load (a combination of high importance for online implementations). In all cases reported next, SVM classification had been employed in a "personalized" mode. This trend, which ultimately led to subject-specific brain decoding, was initiated very early during the stage of feature selection. For each trial of an MI movement, the selected (discriminative) features (depending on subject and brain rhythm) were used to form the input pattern to be used in SVM training and validation.

The performance of the SVM-based binary classification ("left" vs "right") was measured, for each subject independently, under the two different feature-screening procedures, which in turn led to two distinct classification scenarios: one based on static and one based on "instantaneous" connectivity (sub)patterns. Classification performance was expressed in terms of accuracy and carefully validated using a cross-validation scheme that was dependent on the scenario.

The validation and testing procedure was performed on a single-subject basis. In the reported results, a leave-one-out-cross-validation (LOOCV) scheme had been selected to validate the accuracy of the proposed methodology. In the LOOCV, one trial is left out for testing purposes, while the remainder was used in the training process, with the procedure being repeated for all trials.

Having in mind to establish a procedure that could also be employed in a potential implementation of a personalized BCI, in which only a small training dataset would be available for crafting the decision function and the overall training should be completed within a reasonable time before the actual use of the BCI system, we proceeded as follows. We repeatedly form (by sampling with replacement) 30 sets of $2N_{trials}$, and the procedure described in (9.7) was applied to every bootstrap-resample resulting in an ensemble of curves $boot\_iProfile(r)_{boot\_i=1:30}$. Feature selection was accomplished by averaging these profiles and thresholding the obtained average curve.

9.4.2.10 Group analysis of pairwise couplings

The first part of the analysis was devoted to confirming the hypothesis that there were significant differences between NMD patients and controls regarding the strength of functional couplings. To this end, a single connectivity pattern was first derived (by trial-averaging) for each subject, brain rhythm and recording condition (i.e., "rest," "left," and "right"). A statistical comparison of the group medians in every pairwise coupling of the connectivity patterns was performed. The Wilcoxon rank sum test was repeatedly applied and the results were corrected for multiple testing, by means of false discovery rate (FDR = 0.05) [45].

Figure 9.2 illustrates the results for all brain rhythms and recording conditions (i.e., "left," "right," and "rest"). The statistically significant ($p < 0.05$) pairs stand out as colored entries in the shown matrices, with the color reflecting the sign of the observed differences. It was computed based on the medians of the groups ($med(PLV(\cdot)_{NMD})–med(PLV(\cdot)_{Control})$) and clearly indicates (since only red hue is observed) an increased coupling in the patients group compared to the control group, mostly in low and high brain rhythms. It is important to mention here that increased functional couplings were found in all frequency bands, although not clearly observed when a common color code was used.

The topological representation of the statistically significant functional couplings is provided in Figure 9.3, with the edge-width reflecting the difference in strength between the group-level medians of each pairwise coupling and the node-size the number of edges that have survived the statistical test ($p < 0.05$) and are incident to the node. It is clear that the NMD group is characterized by enhanced connectivity in

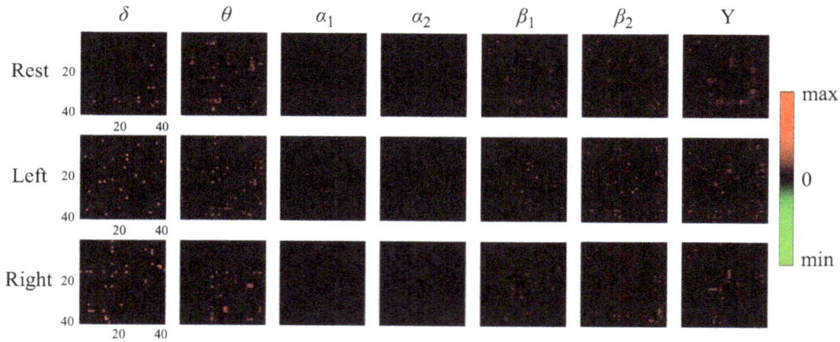

Figure 9.2 The results from the statistical comparison (group-level analysis) of averaged connectivity patterns between patients and controls. Each pairwise coupling was compared independently, for every band and recording condition, by means of Wilcoxon rank sum and the significant ones ($p < 0.05$; corrected for multiple comparisons) are indicated as nonzeroed entries of a "connectivity matrix," with a color code that encapsulates the difference in strength (of the median values in the corresponding groups). Red hue has to be interpreted as higher coupling in patients and green hue as higher coupling in controls, while color intensity reflects the strength of this effect. The absence of green hue in the diagrams clearly indicates the increased coupling in patients group compared to the control group

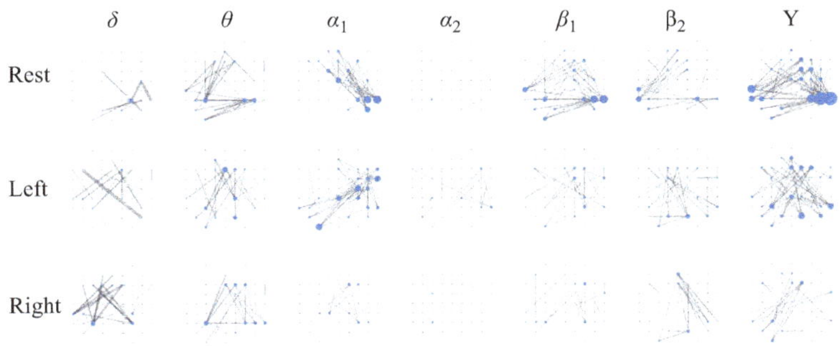

Figure 9.3 Topographical representation of the statistically significant functional couplings (shown in Figure 9.2). In the emerging graphs, the edge-width reflects the strength of the coupling and the node-size the number of edges incident to that node. The shown results correspond to group-level analysis and reflect higher connectivity in the NMD patients

all recording conditions. In the two MI-conditions, the majority of nodes being part of the statistical significant couplings, follow a distributed pattern that occasionally includes the primary and supplementary sensorimotor area (e.g., in "left": α_1, β_1 and γ rhythms).

9.4.2.11 Group analysis of network metrics

Next, the network organization associated with the functional connectivity patterns was compared. The three metrics (i.e., strength, GE, and LE) were first applied at the single-trial level (to "static" **W**s) and then averaged to derive a triad of measurements for each subject, brain rhythm and experimental condition were used to further justify the observed differences among the two groups.

Figure 9.4 compares these measurements, after deriving group-medians, with the stars in patients' bars indicating the statistically significant differences ($p < 0.01$, bonferroni-corrected) in the level of network-metrics, which resulted from the group-analysis of the corresponding measurements (NMD patients vs controls) performed using the Wilcoxon rank sum test. It is easy to observe that despite the lack of statistically significant differences in the case of strength (which practically corresponds to integrating the coupling strength across sensors), the network efficiency metrics (i.e., GE and LE) pose significant differences for rhythms faster than 8 Hz, where MI spectral activity is expected to be found. The observed differences in these two topological metrics (which reflect how efficiently the information flows within the brain network) are related to brain coordination and, hence, can be attributed to the NMD condition itself and the way it affects the patient's brain reorganization during its progression. Interestingly, differences in network organization during MI tasks were detected in $\alpha 2$ rhythm, even though the pairwise couplings did not show, individually, any difference between groups based on their PLV levels (see Figures 9.2 and 9.3).

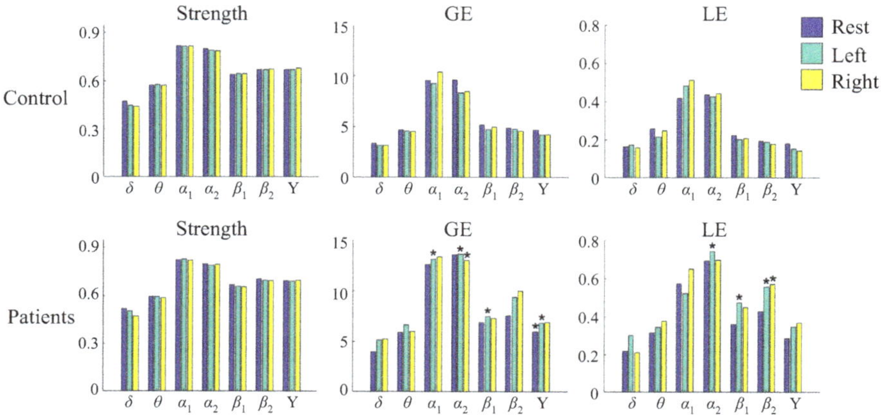

Figure 9.4 Contrasting the functional network organization between patients and controls using the standard networks metrics of strength, global efficiency (GE) and local efficiency (LE). The median values have been computed across the subjects of each group and presented for all brain rhythms. Statistically significant differences between the two groups have also been detected (using Wilcoxon rank sum test) and indicated with a star symbol in the corresponding bar of the patients' bar plot

9.4.2.12 Personalized MI decoding—SVM classification based on static patterns

The third stage of the analysis consists of an attempt to decode the MI direction based on the single-trial functional connectivity patterns and provides a comparative analysis of the two cohorts in terms of performance. A linear SVM in conjunction with standard, statistical, feature screening was used, with the screening being performed to confine the SVM design within the space spanned by the ten most informative functional couplings. Moreover, to reduce the possibility of overfitting, this step was incorporated in the LOOCV scheme (i.e., it was performed every time an SVM was about to be designed from the set of trials that had been reserved for training).

The acquired classification scores of the decoding task for each subject and all brain rhythms are illustrated in Figure 9.5. It is evident that the accuracy levels for the patients group are significantly higher, with five out of six subjects exceeding 75% accuracy, while even the participant with the lowest accuracy (i.e., P3) for this group reaches 65%. It is also interesting to notice that for the NMD group, the highest accuracy is associated with a common frequency band (i.e., β1) in four subjects (i.e., P1, P2, P4, and P6). On the contrary, only half of the subjects belonging to the control population exceed the level of 60% accuracy in any of the frequency bands, with subject S5 being considered the "best" subject for the control group, as it is the only case where 80% of the trials were correctly classified. To confirm rigorously the hypothesis that BCI-naive patients can perform better than controls in the employed MI tasks, we gathered the highest performance level from each individual in two sets of accuracies, $\text{Accuracies}_{i=1:6}^{NMD}$ /$\text{Accuracies}_{i=1:6}^{controls}$, and applied the Wilcoxon rank sum test that revealed a statistical significant difference ($p < 0.05$, one-tailed). For comparison purposes, PSD and CSP-based measurements were estimated (see Figures 9.6 and 9.7). Overall, the decoding performance stays below

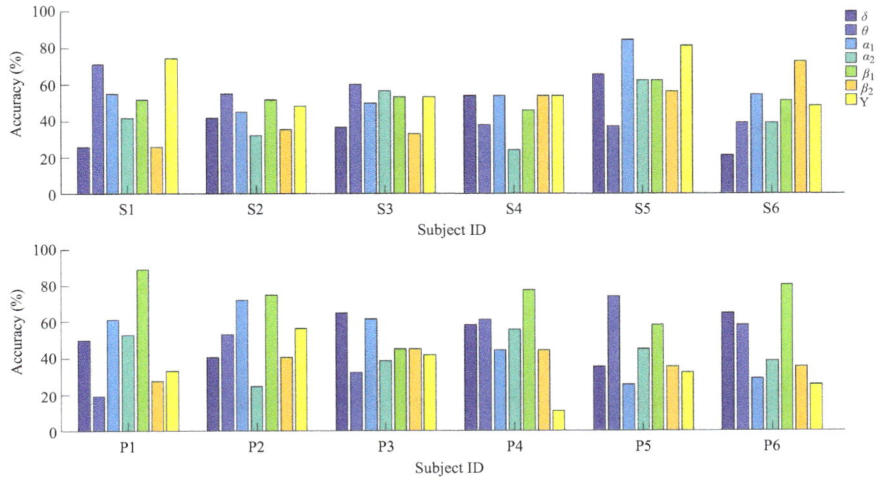

Figure 9.5 *The classification performance in the state discrimination task ("left" vs "right" hand movement imagery) when elements from the static connectivity patterns are used*

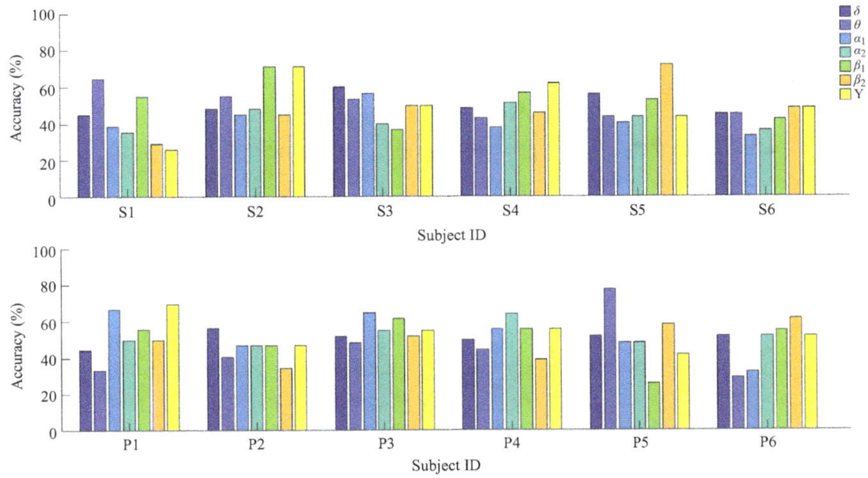

Figure 9.6 *The classification performance in the state discrimination task ("left" vs "right") when band-specific power-spectral density estimates are employed*

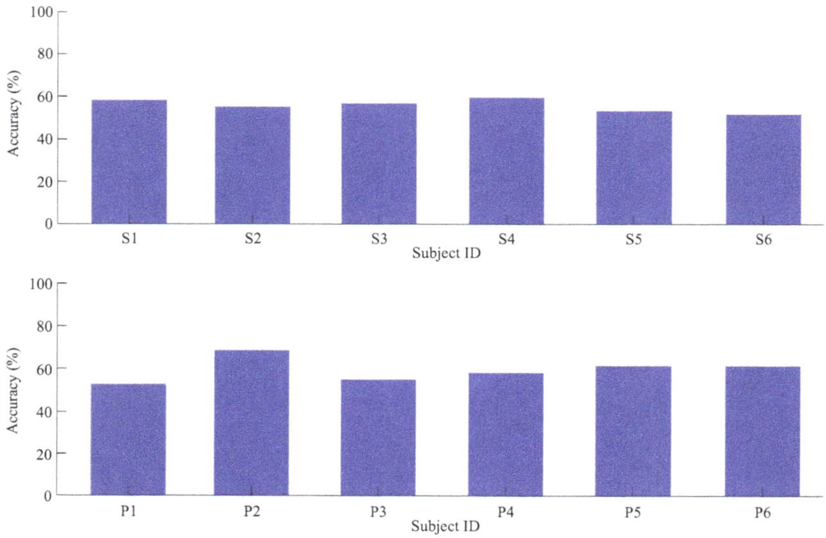

Figure 9.7 The classification performance in the state discrimination task ("left" vs "right") when the common spatial pattern algorithm is employed in the 8–30 Hz frequency band

75% (except for patient P5), i.e., lower than in the case of PLV measurements. In addition, there is no statistically significant advantage for the NMD group over the controls considering the highest performance level from each individual ($p = 0.43$, one-tailed).

9.4.2.13 Personalized MI decoding—SVM classification based on time-varying patterns

At the expense of increased computations and algorithmic complexity, the decoding of MI direction from time-varying connectivity patterns for the NMD patients was then explored. Both the beneficial phase-synchrony based representation, for the brain activity in this clinical group, and the fact that MI-BCIs have remained largely unexplored for NMD patients led to a deeper study regarding the relevant dynamic patterns of connectivity.

Supporting evidence, regarding the dynamic nature of the underlying phenomena, was provided by the feature screening procedures, since the obtained scores for the dynamic patterns were often higher than the ones for the static patterns. Working at a personalized level, the set of functional couplings was spotted, illustrated in Figure 9.8 that showed a stable and highly discriminative behavior (using bootstrapping and (9.7)). The fixed set of selected entries was extracted (in single-trial level) from the time-indexed connectivity patterns, $vec(\mathbf{W}[\tau])$, which had been computed with a time-step of 350 ms. The vectors were used to design and evaluate an "instantaneous" SVM (i.e., SVM^{τ}) that corresponds to each latency and also follows an

LOOCV scheme. The performances of this decoder were estimated by comparing the time-indexed predictions with the class labels of the trials and integrating the results across trials.

Figure 9.9 shows the corresponding performance curves for the "instantaneous" SVM classification scheme. The variability among subjects is evident, as subjects

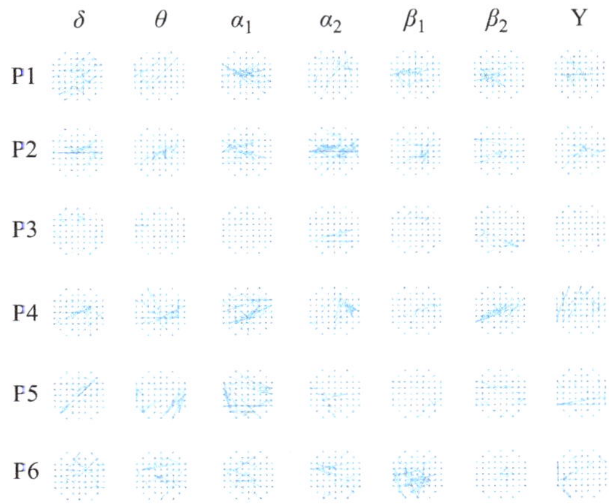

Figure 9.8 The statistically significant and temporally consistent couplings as detected by means of a permutation test (random re-labeling of trials) applied for each patient and brain rhythm independently

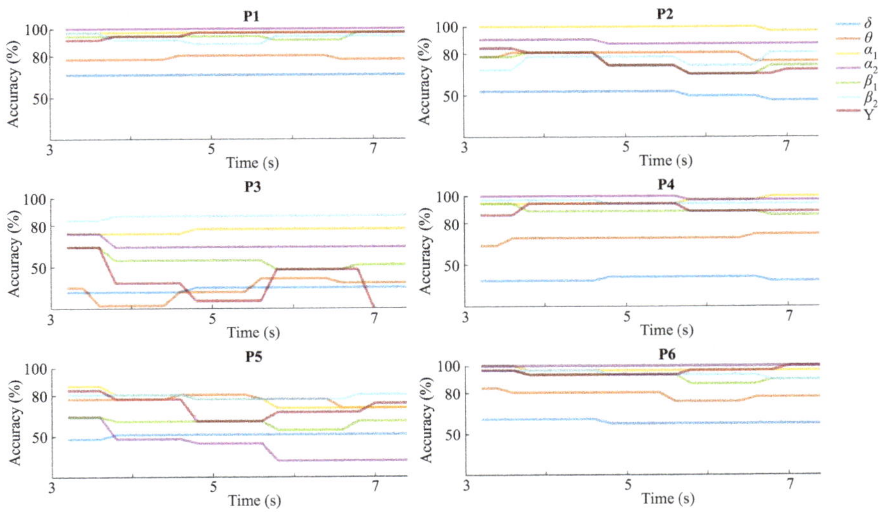

Figure 9.9 The classification performance in the "left" vs "right" task when the selected couplings (shown in Figure 9.8) are used to form multiple time-resolved patterns associated with each trial

P1, P3, and P6 reach the highest accuracy within the first second and maintain the high performance for the full trial length, while subjects P2 and P4 similarly achieve the highest performance levels within the first second but there is a decrease in the performance as the trial evolves, a trend that can be interpreted as declining engagement to the task. Finally, only subject P5 showed deterioration in performance after the first second. The observed variability can be attributed to the subject's devotion to the task, how he/she performed it, and possibly to NMD condition. Overall, this classification scheme appears to lead to optimal performance earlier in time.

9.5 Toward a self-paced implementation

9.5.1 *Related work*

Providing the BCI-user with the ability to operate a BCI at his/her own will is considered one of the fundamental aspects of BCIs and its significance for people with motor disabilities is even greater. In the related literature of MI-BCIs, there are only a few approaches that present a solution regarding a self-initiated motion. In two of them, [46,47], a two-stage classification scheme is adopted, with the first identifying the onset of an MI event, and the second determining the direction of the imagery movement. Additionally, a "brain switch" has been implemented based on the rhythm rebound (i.e., ERS). Either a simple thresholding scheme [48] or linear discriminant analysis [49] is employed to flag a significant departure from the ongoing activity that corresponds to an "idling" (baseline) state.

9.5.2 *An SVM-ensemble for self-paced MI decoding*

The high performance of the SVM decoders working with time-resolved connectivity patterns, $vec(\mathbf{W}[\tau])$, motivated the search for a decoding scheme that could operate without requiring an external trigger to define the trial initiation. The original idea was that a "local" SVM tailored to deal with patterns from latency τ_{sel} would show a high confidence level about its prediction only within a time-interval around that latency. While this scenario seemed to work well upon trial-averaging, it could not cope well with available MI trials, as it had the tendency to produce false-positive (FP) detections (see Figure 9.10(b)). As a consequence, a sequence of SVMi, $i = \tau_{sel1}, \tau_{sel2}, \ldots, \tau_{selM}$, was employed with the scope of making more stringent the decision about detecting an MI event. Assuming a trigger-agnostic scenario, these SVMs will run in parallel resulting in a time-indexed vector $Z(\tau) = [z^1(\tau), z^2(\tau), \ldots, z^M(\tau)]^T$, with entries

$$z^i(\tau) = SVM^i(FeatureExtraction(vec(\mathbf{W}[\tau]))) \qquad i = \tau_{sel1}, \tau_{sel1}, \ldots, \tau_{selM} \tag{9.8}$$

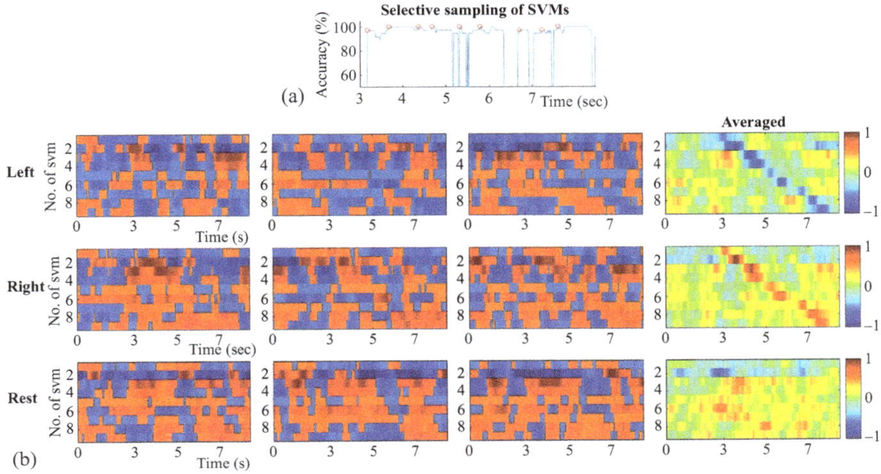

Figure 9.10 SVM-ensemble formation: (a) a set of consecutive but not "colliding" SVMs are combined in order to form an ensemble that will process, in parallel, the streaming coupling measurements and derive for each latency a vector of classification grades. (b) The latency-resolved multitude of instantaneous SVM predictions is shown for three exemplar single trials (first three columns in every row) along with the corresponding pattern resulted from averaging the individual ST profiles across all trials (rightmost column). Each SVM outputs a classification score ranging within [−1 1], with the sign indicating the movement side and the magnitude reflecting its confidence

Each entry z^i denotes the multiplication of classifier's confidence with the sign of its prediction (+/− is associated with "right"/"left" movement), i.e., a real number within [−1 1]. Deviating from the standard approaches for combining classifiers (e.g., voting), in the proposed scheme, the classifiers' outputs are combined based on temporal patterning (that reflects their relative positioning in time, which is associated with the optimal performance in the cued trials). An "instantaneous" classification index is derived by averaging the individual signed confidences after imposing the predefined lags

$$z_{ensemble}(t) = \frac{1}{M}\sum_{i}^{M} z^i(t + \tau_{seli}) \tag{9.9}$$

It is important to notice, here, that such an SVM-ensemble formation is feasible and computational tractable, due to the prior selection of a unique set of "stable" couplings. The proposed SVM-ensemble scheme is supported by two experimental observations. First, the time-indexed accuracy of the locally defined SVMs showed multiple, easy-detectable peaks (e.g., Figure 9.10(a)), which led to an automation process for selecting the SVMs. Second, there was no pair of SVMs among the selected "local" ones in the ensemble that demonstrated significant similarity. The latter fact means that all the selected SVMs were defining different separating-hyperplanes in the space of common features.

9.5.3 In quest of self-paced MI decoding

Finally, the possibility of decoding phase-connectivity patterns in a way that could be used in a future implementation of a self-paced MI-BCI, where the user would initiate the MI events at will, was explored. Since self-initiated MI events were not available, the scenario was partially "simulated" by exploiting the resting-condition recordings and devising a scheme that would mark a time instance as the beginning of an MI event, regardless of the class, only when the temporal patterning in the streaming connectivity-data was deviant from the patterning in the baseline (rest) condition. In this direction, 20 trials were selected from each patient's resting-state recording and "baseline" time-resolved connectivity patterns (extending for 8 s) were formulated, based on an identical signal-analytic pipeline used in the case of MI trials. The algorithmic steps required for the self-paced MI decoder are presented in Figure 9.11.

A training set consisting of 10 trials from each class ("left", "right," and "rest") was formed and used in a two-stage data-learning process. During the first stage, only the MI-related single-trial connectivity patterns ($\{vec(\mathbf{W}[t]^{i}_{left})\}_{i=1:10}$ alongside with $\{vec(\mathbf{W}[t]^{i}_{right})\}_{i=1:10}$) were employed for (i) the feature-selection process, (ii) the training of all "instantaneous" SVMs, and (iii) the "instantaneous" SVM selection. The feature selection step is exemplified in Figure 9.12, for subject P2's connectivity patterns from α_1 rhythm, while SVMs' selection is exemplified in Figure 9.10(a). Finally, the application of the SVM ensemble in a number of trials (from all recording conditions) is depicted in Figure 9.10(b), where the vectors of successive predictions appear as columns in the shown heat-maps. The rightmost panels in Figure 9.10(b) include the corresponding trial-averaged heat-maps, where a "diagonal" pattern can be identified in both cases of "triggered" MI events but not in the case of resting-state. It was exactly this discrepancy that the stratified combination of the outputs of the SVMs participating in the ensemble tried to reveal in a computationally tractable way by means of (9.9).

In the second stage, the temporal traces corresponding to the single-trial "instantaneous" readouts from the SVM ensemble were derived for the abovementioned MI-related connectivity patterns and, in addition, for the baseline-related ones $\{vec(\mathbf{W}[t]^{i}_{rest})\}_{i=1:10}$. Figure 9.13 shows the estimated traces of classification index, $z_{ensemble}(t)$, in continuation of the example shown previously in Figures 9.10 and 9.12. An evident peak can be identified, just after the third second (onset), for both the "left" and "right" conditions. On the contrary, the traces derived from the rest condition trials do not depict any comparable peak. In an attempt to quantify these observations, and simultaneously complete the design of a totally self-paced MI-decoding scheme, we employed these 30 profiles for training purposes, crafting a decision rule, that could provide evidence and therefore decide if the observed temporal patterning in classification index can be correlated with the baseline condition

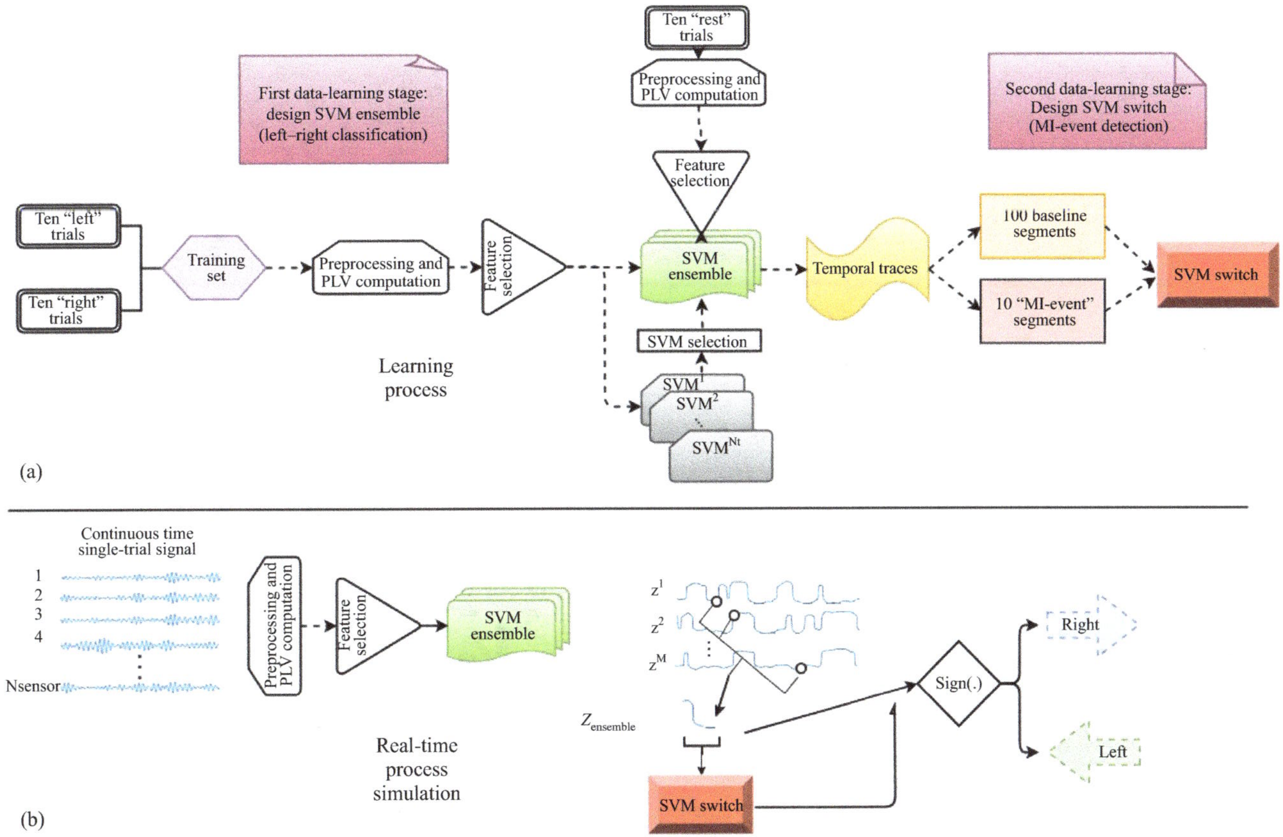

Figure 9.11 *(a) Flowchart of the data-leaning process for the self-paced MI decoding. (b) Graphical depiction of the decoding procedure from the streaming connectivity patterns during one single trial*

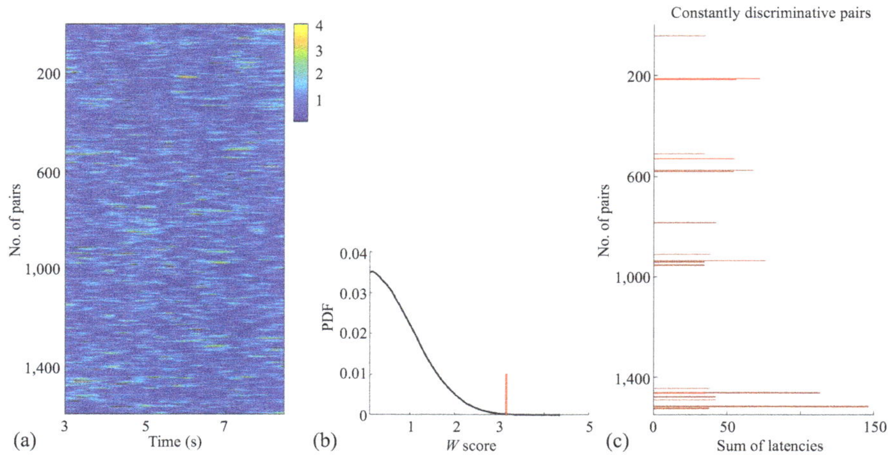

Figure 9.12 Feature-selection procedure: (a) the latency-dependent Wilcoxon score for all sensor pairs. (b) The definition of a "global" threshold based on the distribution of Wilcoxon scores in randomized data. (c) The selected subset of couplings that continuously exceed this threshold for intervals longer than 100 ms (i.e., temporally consistent discriminative couplings)

or with an MI event and, hence, should trigger the command associated with the sign of the trace from the SVM ensemble. To accomplish the data-learning task, multiple 0.5-s segments were extracted from the singe-trial traces shown in Figure 9.13 and confined within the intervals indicated via vertical dotted lines. These 100 segments were corresponding to the "MI-event" class (regardless direction). Moreover, an equal number of segments, with no time restriction, were extracted from the baseline condition, constituting the "baseline" class patterns. Both type of segments were employed for training a binary-SVM (with a radial basis function kernel) to discriminate an MI event from the baseline state. The trained "SVM switch" was then fed with the streaming SVM-ensemble readouts, $z_{ensemble}(t)$, resulted from the testing set of trials. Figure 9.14 exemplifies this step by first illustrating the "instantaneous" single-trial readouts form the SVM ensemble (formed in Figure 9.10(a)) for all recording conditions (Figure 9.14(a)) and, then, the corresponding single-trial traces of the instantaneous confidence of the SVM switch (Figure 9.14(b)). Using as threshold, the confidence level of 0.5, only two FP detections were identified in all three recording conditions (please notice that this would have also been the case if a high confidence level had appeared within the first 3 s interval of a MI trial), and no false negative ones. After referencing these counts to the number of trials, two probabilistic indices regarding the observed probabilities of FN and FP (here 0/30 and 2/30, respectively) were estimated. The overall procedure was repeated after different randomized partitions of available data (i.e., Monte Carlo cross validation scheme), and

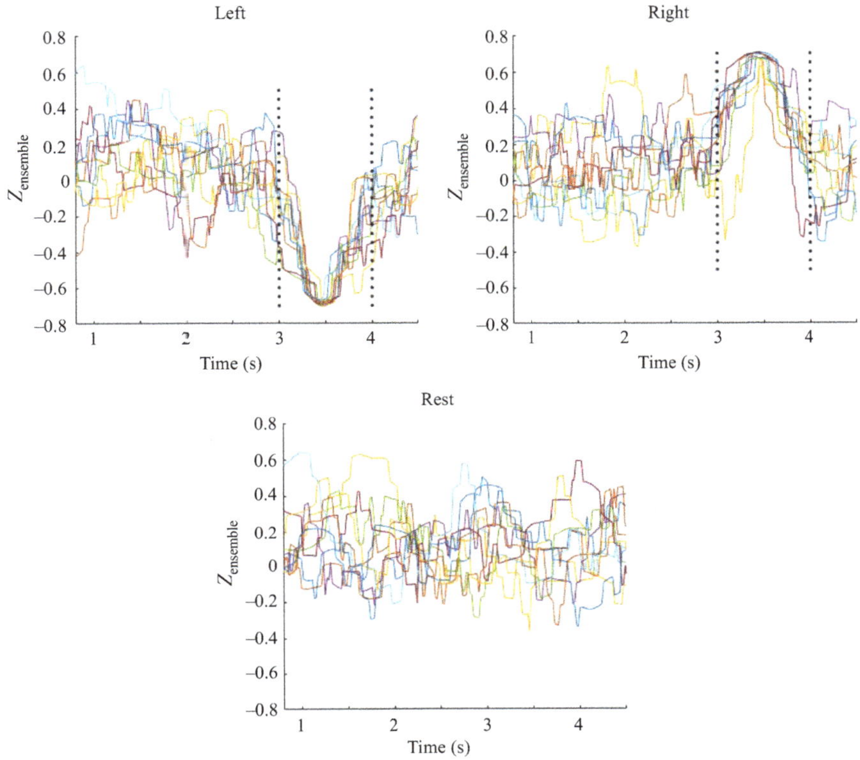

Figure 9.13 *The latency-dependent classification index $Z_{ensemble}$ as derived by means of the time-lagged combination of the SVM-ensemble readouts for a training set of trials*

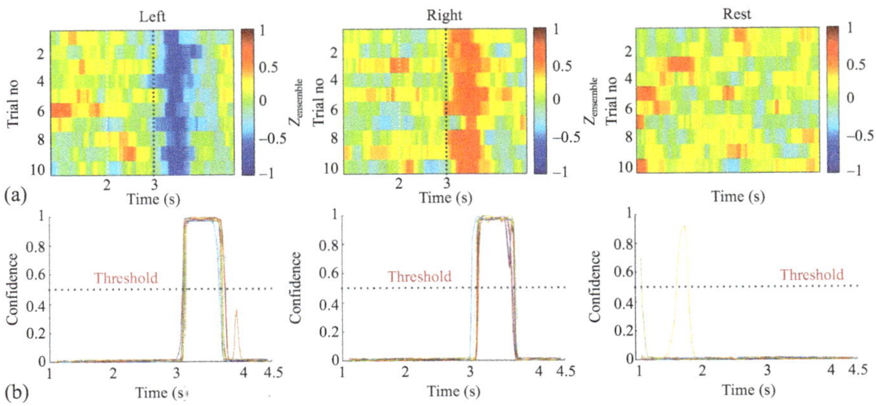

Figure 9.14 *(a) The latency-resolved multitude of instantaneous SVM predictions is shown for a test-set trials, (b) The latency-dependent confidence of the SVM decider (trained based on temporal patterns extracted from Figure 9.13)*

Table 9.2 Subject demographics

Participant ID	P1	P2	P3	P4	P5	P6
Brain rhythm	α_2	α_1	β_2	α_2	α_2	α_2
FP	2.2%	3.5%	7.2%	3.9%	7.2%	2.7%
FN	1.4%	2.0%	5.3%	2.1%	3.3%	1.4%

the averaged results for each subject were tabulated in Table 9.2. The brain rhythms had been selected according to the performance levels shown in Figure 9.9.

The very low probabilities of misdetection and false alarm, in conjunction with the impressive performance of the individual MI decoders participating in the ensemble, constitute the combined scheme (SVM ensemble and SVM switch) potentially suitable for self-paced MI-decoding (see Figure 9.11).

9.6 Summary

NMD is a condition that gradually affects the musculature and eventually leads to the loss of any voluntary muscle control. The reflections of NMD on the electroencephalographic brain activity, under the perspective of establishing efficient BCIs, have rarely been studied. It was the scope of this chapter to examine the differences in the functional brain organization between NMD patients and healthy individuals in an MI paradigm that, traditionally, is considered fruitful for endogenous BCIs. Special attention was paid to dynamic patterns of functional connectivity aiming to determine faster ways to perform MI decoding and provide solutions independent from external triggering.

References

[1] Hinterberger T, Schmidt S, Neumann N, *et al.* Brain–computer communication and slow cortical potentials. IEEE Transactions on Biomedical Engineering. 2004;51(6):1011–1018.

[2] Nakayashiki K, Saeki M, Takata Y, *et al.* Modulation of event-related desynchronization during kinematic and kinetic hand movements. Journal of NeuroEngineering and Rehabilitation. 2014;11(1):90.

[3] Andrade J, Cecílio J, Simões M, *et al.* Separability of motor imagery of the self from interpretation of motor intentions of others at the single trial level: an EEG study. Journal of NeuroEngineering and Rehabilitation. 2017;14(1):63.

[4] Nam CS, Jeon Y, Kim YJ, *et al.* Movement imagery-related lateralization of event-related (de) synchronization (ERD/ERS): motor-imagery duration effects. Clinical Neurophysiology. 2011;122(3):567–577.

[5] Solis-Escalante T, Müller-Putz G, Pfurtscheller G. Overt foot movement detection in one single Laplacian EEG derivation. Journal of Neuroscience Methods. 2008;175(1):148–153.

[6] Ge S, Wang R, Yu D. Classification of four-class motor imagery employing single-channel electroencephalography. PLoS One. 2014;9(6):e98019.

[7] Deng S, Srinivasan R, Lappas T, *et al.* EEG classification of imagined syllable rhythm using Hilbert spectrum methods. Journal of Neural Engineering. 2010;7(4):046006.

[8] Dimitriadis S, Sun Y, Laskaris N, *et al.* Revealing cross-frequency causal interactions during a mental arithmetic task through symbolic transfer entropy: a novel vector-quantization approach. IEEE Transactions on Neural Systems and Rehabilitation Engineering. 2016;24(10):1017–1028.

[9] Yuan H, He B. Brain–computer interfaces using sensorimotor rhythms: current state and future perspectives. IEEE Transactions on Bio-medical Engineering. 2014;61(5):1425–1435.

[10] Formaggio E, Storti SF, Boscolo Galazzo I, *et al.* Modulation of event-related desynchronization in robot-assisted hand performance: brain oscillatory changes in active, passive and imagined movements. Journal of NeuroEngineering and Rehabilitation. 2013;10:24.

[11] Oikonomou VP, Nikolopoulos S, Petrantonakis P, *et al.* Sparse kernel machines for motor imagery EEG classification. In: 2018 40th Annual International Conference of the IEEE Engineering in Medicine and Biology Society (EMBC). IEEE; 2018. p. 207–210.

[12] Ramoser H, Muller-Gerking J, Pfurtscheller G. Optimal spatial filtering of single trial EEG during imagined hand movement. IEEE Transactions on Rehabilitation Engineering. 2000;8(4):441–446.

[13] Ang KK, Chin ZY, Zhang H and Guan C. Filter bank common spatial pattern (FBCSP) in brain-computer interface. In 2008 IEEE International Joint Conference on Neural Networks (IEEE World Congress on Computational Intelligence); 2008. p. 2390–2397.

[14] Robinson N, Guan C, Vinod AP, *et al.* Multi-class EEG classification of voluntary hand movement directions. Journal of Neural Engineering. 2013;10(5):056018.

[15] Oikonomou VP, Georgiadis K, Liaros G, lNikopoulos S and Kompatsiaris I. A comparison study on EEG signal processing techniques using motor imagery EEG data. In 2017 IEEE 30th international symposium on computer-based medical systems (CBMS); 2017. p. 781–786.

[16] Stavrinou ML, Moraru L, Cimponeriu L, *et al.* Evaluation of cortical connectivity during real and imagined rhythmic finger tapping. Brain Topography. 2007;19(3):137–145. Available from: https://doi.org/10.1007/s10548-007-0020-7.

[17] Caramia N, Lotte F and Ramat S. Optimizing spatial filter pairs for EEG classification based on phase-synchronization. In 2014 IEEE International Conference on Acoustics, Speech and Signal Processing (ICASSP); 2014. p. 2049–2053.

[18] Fallani FDV and Bassett DS. Network neuroscience for optimizing brain-computer interfaces. Physics of life reviews; 2019.

[19] Uehara T, Tanaka T, Fiori S. Robust averaging of covariance matrices by Riemannian geometry for motor-imagery brain–computer interfacing. In: Wang R,

Pan X, editors. Advances in Cognitive Neurodynamics (V). Advances in Cognitive Neurodynamics. Singapore: Springer Singapore; 2016. p. 347–353.

[20] Brunner C, Scherer R, Graimann B, *et al.* Online control of a brain–computer interface using phase synchronization. IEEE Transactions on Biomedical Engineering. 2006;53(12):2501–2506.

[21] Song L, Gordon E, Gysels E. Phase synchrony rate for the recognition of motor imagery in brain–computer interface. In: Weiss Y, Schölkopf B, Platt JC, editors. Advances in Neural Information Processing Systems 18. MIT Press; 2006. p. 1265–1272. Available from: http://papers.nips.cc/paper/2849-phase-synchrony-rate-for-the-recognition-of-motor-imagery-in-brain-computer-interface.pdf.

[22] Georgiadis KI, Laskaris N, Nikolopoulos S, *et al.* Connectivity steered graph Fourier transform for motor imagery BCI decoding. Journal of Neural Engineering. 2019;16(5), p.056021.

[23] Tanaka T, Uehara T and Tanaka Y. Dimensionality reduction of sample covariance matrices by graph Fourier transform for motor imagery brain-machine interface. In 2016 IEEE Statistical Signal Processing Workshop (SSP); 2016. p. 1–5.

[24] Kalantar G, Sadreazami H, Mohammadi A, and Asif A. Adaptive dimensionality reduction method using graph-based spectral decomposition for motor imagery-based brain–computer interfaces. In: 2017 IEEE Global Conference on Signal and Information Processing (GlobalSIP); 2017. p. 990–994.

[25] Higashi H, Tanaka T, Tanaka Y. Smoothing of spatial filter by graph Fourier transform for EEG signals. In: Signal and Information Processing Association Annual Summit and Conference (APSIPA), 2014 Asia-Pacific; 2014. p. 1–8.

[26] Georgiadis K, Laskaris N, Nikolopoulos S, *et al.* (2019, July). Using Discriminative Lasso to Detect a Graph Fourier Transform (GFT) Subspace for robust decoding in Motor Imagery BCI. In 2019 41st Annual International Conference of the IEEE Engineering in Medicine and Biology Society (EMBC) (pp. 6167–6171). IEEE.

[27] Moghimi S, Kushki A, Guerguerian AM, *et al.* A review of EEG-based brain–computer interfaces as access pathways for individuals with severe disabilities. Assistive Technology: The Official Journal of RESNA. 2013;25(2): 99–110.

[28] Gurkok H, Nijholt A, Poel M. Brain–computer interface games: towards a framework. In: Nakatsu R, Rauterberg M, Ciancarini P, editors. Handbook of Digital Games and Entertainment Technologies. Singapore: Springer Singapore; 2017. p. 133–150. Available from: https://doi.org/10.1007/978-981-4560-50-4_5.

[29] Kosti MV, Georgiadis K, Adamos DA, *et al.* Towards an affordable brain–computer interface for the assessment of programmers' mental workload. International Journal of Human–Computer Studies. 2018;115:52–66. Available from: https://linkinghub.elsevier.com/retrieve/pii/S1071581918300934.

[30] Leamy DJ, Kocijan J, Domijan K, *et al.* An exploration of EEG features during recovery following stroke – implications for BCI-mediated neurorehabilitation therapy. Journal of NeuroEngineering and Rehabilitation. 2014;11(1):9. Available from: https://doi.org/10.1186/1743-0003-11-9.

[31] King CE, Wang PT, Chui LA, *et al.* Operation of a brain–computer interface walking simulator for individuals with spinal cord injury. Journal of NeuroEngineering and Rehabilitation. 2013;10(1):77. Available from: https://doi.org/10.1186/1743-0003-10-77.

[32] Shindo K, Kawashima K, Ushiba J, *et al.* Effects of neurofeedback training with an electroencephalogram-based brain–computer interface for hand paralysis in patients with chronic stroke: a preliminary case series study. Journal of Rehabilitation Medicine. 2011;43(10):951–957.

[33] Conradi J, Blankertz B, Tangermann M, *et al.* Brain–computer interfacing in tetraplegic patients with high spinal cord injury. Int. J. Bioelectromagn. 2009;11(2):65–8.

[34] Kübler A, Nijboer F, Mellinger J, *et al.* Patients with ALS can use sensorimotor rhythms to operate a brain–computer interface. Neurology. 2005;64(10): 1775–1777.

[35] Bai O, Lin P, Huang D, *et al.* Towards a user-friendly brain–computer interface: initial tests in ALS and PLS patients. Clinical Neurophysiology: Official Journal of the International Federation of Clinical Neurophysiology. 2010;121(8):1293–1303. Available from: https://www.ncbi.nlm.nih.gov/pmc/articles/PMC2895010/.

[36] Heremans E, Nieuwboer A, Spildooren J, *et al.* Cued motor imagery in patients with multiple sclerosis. Neuroscience. 2012;206:115–121.

[37] Cincotti F, Mattia D, Aloise F, *et al.* Non-invasive brain–computer interface system: towards its application as assistive technology. Brain Research Bulletin. 2008;75(6):796–803.

[38] Georgiadis K, Laskaris N, Nikolopoulos S, *et al.* Exploiting the heightened phase synchrony in patients with neuromuscular disease for the establishment of efficient motor imagery BCIs. Journal of NeuroEngineering and Rehabilitation. 2018;15(1):90. Available from: https://doi.org/10.1186/s12984-018-0431-6.

[39] Nikolopoulos S, Petrantonakis PC, Georgiadis K, *et al.* A multimodal dataset for authoring and editing multimedia content: the MAMEM project. Data in Brief. 2017;15:1048–1056.

[40] Lachaux JP, Rodriguez E, Martinerie J, *et al.* Measuring phase synchrony in brain signals. Human Brain Mapping. 1999;8(4):194–208. Available from: https://onlinelibrary.wiley.com/doi/abs/10.1002/%28SICI%291097-0193%281999%298%3A4%3C194%3A%3AAID-HBM4%3E3.0.CO%3B2-C.

[41] De Vico Fallani F, Richiardi J, Chavez M, *et al.* Graph analysis of functional brain networks: practical issues in translational neuroscience. Philosophical Transactions of the Royal Society of London Series B, Biological Sciences. 2014;369(1653).

[42] Latora V, Marchiori M. Efficient behavior of small-world networks. Physical Review Letters. 2001;87(19):198701.

[43] Dimitriadis SI, Laskaris NA, Tsirka V, *et al.* Tracking brain dynamics via time-dependent network analysis. Journal of Neuroscience Methods. 2010;193(1):145–155. Available from: http://www.sciencedirect.com/science/article/pii/S0165027010004875.

[44] Dimitriadis SI, Laskaris NA, Tzelepi A. On the quantization of time-varying phase synchrony patterns into distinct functional connectivity microstates (FCstates) in a multi-trial visual ERP paradigm. Brain Topography. 2013;26(3):397–409.

[45] Benjamini Y, Hochberg Y. Controlling the false discovery rate: a practical and powerful approach to multiple testing. Journal of the Royal Statistical Society Series B (Methodological). 1995;57(1):289–300. Available from: https://www.jstor.org/stable/2346101.

[46] Scherer R, Schloegl A, Lee F, *et al.* The self-paced Graz brain–computer interface: methods and applications. Computational Intelligence and Neuroscience. 2007:1–9.

[47] Chae Y, Jeong J, Jo S. Toward brain-actuated humanoid robots: asynchronous direct control using an EEG-based BCI. IEEE Transactions on Robotics. 2012;28(5):1131–1144.

[48] Leeb R, Friedman D, Müller-Putz GR, *et al.* Self-paced (asynchronous) BCI control of a wheelchair in virtual environments: a case study with a tetraplegic. Computational Intelligence and Neuroscience. 2007:1–8.

[49] Müller-Putz GR, Kaiser V, Solis-Escalante T, *et al.* Fast set-up asynchronous brain-switch based on detection of foot motor imagery in 1-channel EEG. Medical & Biological Engineering & Computing. 2010;48(3):229–233.

Chapter 10
Graph signal processing analysis of NIRS signals for brain–computer interfaces

Panagiotis C. Petrantonakis[1] *and Ioannis Kompatsiaris*[1]

Graph signal processing (GSP) is an emerging field in signal processing that aims at analyzing high-dimensional signals using graphs. The GSP analysis is intended to take into account the signals' inner graphical structure and expand traditional signal processing techniques to the graph–network domain. In this chapter, we present a GSP analysis framework for the implementation of brain–computer interfaces (BCI) based on function near-infrared spectroscopy (NIRS) signals. Firstly, a GSP approach for feature extraction is presented based on the Graph Fourier Transform (GFT). The aforementioned approach captures the spatial information of the NIRS signals. The feature extraction method is applied on a publicly available dataset of NIRS recordings during mental arithmetic task and shows higher classification rates, up to 92.52%, as compared to the classification rates of two state-of-the-art feature extraction methodologies. Moreover, in order to better demonstrate the spatial distribution of the NIRS information and to quantify the smoothness or not of the NIRS signals across the channel montage we present a GSP, Dirichlet energy-based analysis approach of NIRS signals over a graph. The application of the proposed measure on the same NIRS dataset further shows the spatial characteristics of the NIRS data and the efficiency of this GSP approach to capture it. Moreover, Dirichlet energy-based approach shows high classification rates, >97%, when used to extract features from NIRS signals. In sum, the presented methods show the efficacy of the GSP-based analysis of NIRS signals for BCI applications and pave the way for more robust and efficient implementations.

10.1 Introduction

Brain–computer interfaces (BCIs) have received great attention in the last two decades as they play crucial role for communication and rehabilitation of people with motor impairments [1–4]. Despite the fact that several signal processing techniques have

[1]The Multimedia Knowledge and Social Media Analytics Laboratory, Information Technologies Institute, Centre for Research and Technology-Hellas (CERTH), Thessaloniki, Greece

been developed for BCI algorithms [5–10], one of the most fundamental aspects of such systems is the imaging modality used to capture the brain activity. Several such brain–recordings techniques have been used to acquire brain activity with the most popular, noninvasive modalities being the electroencephalography (EEG) [1], the magnetoencephalography (MEG) [11–13], functional magnetic resonance imaging (fMRI) [14–16], and near-infrared spectroscopy (NIRS) [17–20].

NIRS is a relatively new modality that exhibits promising features for efficient BCI systems, such as portability (compared to fMRI and MEG), low noise (compared to EEG and MEG that are severely affected by various electrical artifacts, i.e., eye movements, face muscles etc.), user friendliness, i.e., no conductive gel is needed during NIRS sensor placement in contrast with the EEG case and, finally, due to recent technological advancements on novel wireless and wearable recording devices [21, 22] it allows for noninvasive, portable and efficient BCI systems.

NIRS technique exploits the permeability of the tissue from near-infrared lights. In essence, light source–detector pairs in the near-infrared range (650–1000 nm wavelength) are used for the estimation, via the light absorption, of the concentration changes (CCs) of the chromophore hemoglobin, specifically of its two main variants, oxyhemoglobin (oxy-Hb) and deoxyhemoglobin (deoxy-Hb) [23–25]. Since hemoglobin is an oxygen carrier, changes in concentration levels of oxy-Hb and deoxy-Hb during changes in activation levels of neuronal populations can be related to mental task-specific responses and, thus, used for the realization of BCI systems.

The three main stages of a BCI system comprises (i) the evocation of the task-related brain activity, (ii) the preprocessing and feature extraction stage, and (iii) the classification with respect to the task performed by the subject during the brain activity recording experiment. Mental tasks that are exploited to trigger neural activation for NIRS-based BCI systems refer mainly to two brain regions, i.e., motor cortex (MC) and prefrontal cortex (PFC) [17]. MC-related tasks correspond to different kinds of motor imagery [26–28], whereas most prominent PFC-related tasks refer to mental arithmetic, i.e., mental calculation performing (e.g., sequential subtraction, 97–4, 93–4, etc.) [29–35] and music imagery, i.e., mental music analysis without any auditory stimulus [20, 36].

In regard with the three main steps of BCI systems, in this work we use a public NIRS dataset that was recorded during mental arithmetic task as far as the first step is concerned. Regarding the second step, we present a Graph Fourier Transform (GFT)-based and a Dirichlet energy-based [37, 38] analysis for the extraction of the feature vector whereas for the third step a support vector machine (SVM) classifier [39], one of the most popular classification techniques for NIRS data classification in BCI systems [40], is utilized for classification. The GSP-based feature vectors are compared with two other state-of-the-art approaches that depend on either the estimation of the mean CC values of the oxy-Hb and deoxy-Hb signals during certain periods of the task execution or the respective slope (S) estimation of the signals during the whole period of the task execution in order to capture the aforementioned hemodynamic response [17, 41–44].

The proposed GSP approaches seem to exploit the potential, spatial patterns of hemoglobin activity throughout the graph motif constructed by the different NIRS measurement channels and pave the way of channel-dependent analysis for NIRS signals.

The rest of the chapter is structured as follows. Section 10.2 presents the NIRS dataset used to evaluate the GFT and Dirichlet energy approaches. Section 10.3 elaborates on the methodologies presented in this chapter and the implementation issues concerning their application on the NIRS dataset. Section 10.4 presents the results of the application of the aforementioned methodologies over the NIRS dataset. Section 10.5 discusses various aspects of the presented algorithms. Finally, Section 10.6 concludes the chapter.

10.2 NIRS dataset

The dataset used in this study is available online* and was firstly reported in [31]. The dataset comprises eight subjects (three male and five female) of age 26 ± 2.8 (mean $\pm$ SD) that showed spatial, antagonistic hemodynamic, task-related patterns [31]. Subjects were instructed to perform a cue-guided mental arithmetic task (MAT), i.e., subtract sequentially a one-digit number from a two-digit one for 12 consecutive seconds after a cue. The task-related period was followed by a 28-s rest period (for details, see online dataset documentation). Thus, each trial lasted for 40 s. Subjects 1–3 performed 18 trials, whereas subjects 4–8 performed 24 trials of MAT. The recordings were performed by an NIRS recording set (ETG-400, Hitachi Medical Co., Japan) comprising 16 light detectors and 17 light sources resulting in a 52-channel grid; see Figure 10.1 for channel montage (this montage will be used

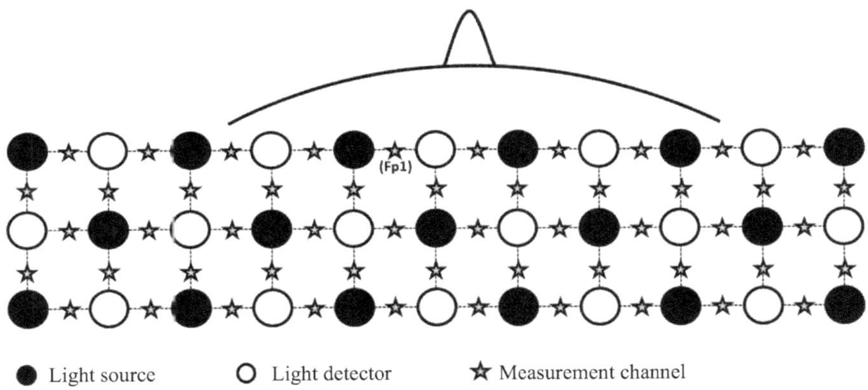

Figure 10.1 *Source–detector configuration. Black circles: light sources, white circles: light detectors, and star symbol: measurement channels (Fp1 association with the measurement channels is illustrated)*

*http://bnci-horizon-2020.eu/database/data-sets.

as a graph in our analysis, channels correspond to vertices of the graph). The distance between light source and light detector was 3 cm, whereas the lowest line of measurement channels was aligned with the Fp1–Fp2 lines (Fp1 channel alignment is shown in Figure 10.1). The sampling rate was 10 Hz. In all signals in the dataset, a fourth-order, 0.01 Hz, high-pass Butterworth filter was applied to remove the baseline drift [29, 30]. Finally, the trial-related signal of all channels was referred to the 10-s baseline interval prior to the task before any analysis of the signals was performed.

10.3 Materials and methods

10.3.1 Graph signal processing basics

GSP theory refers to analysis of signals recorded from sensors comprising a network that resembles a graph. A graph is defined as the pair $\mathcal{G} = (\mathcal{V}, W)$, where $\mathcal{V} = \{v_1, \ldots, v_n\}$ is the set of n vertices (e.g., the sensors) and $W \in \mathbb{R}^{n\times n}$ is the adjacency matrix of the graph with $w_{ij} \geq 0$ denoting the weight of the edge (i, j) between the vertices $v_i, v_j \in \mathcal{V}$. Graphs $\mathcal{G}$ are also considered to be symmetric, i.e., $\forall(i, j)$, $w_{ij} = w_{ji}$. The degree matrix $D \in \mathbb{R}^{n\times n}$ of a graph is a diagonal matrix with diagonal elements $D_{ii} = \sum_{j=1}^{n} w_{ij}$, $i = 1, \ldots, n$. Moreover, the Laplacian matrix $L \in \mathbb{R}^{n\times n}$ of a graph $\mathcal{G}$ is defined as $L = D - \mathrm{W}$.

The Laplacian of a graph is a real, symmetric, positive semidefinite matrix and can be decomposed as

$$L = U\Lambda U^H \tag{10.1}$$

where Λ *is the diagonal eigenvalue matrix* $\Lambda = diag(\lambda_0, \lambda_1, \ldots, \lambda_{n-1})$, *where* $\{\lambda_k\}, k = 0, 1, \ldots, n-1$ *is the corresponding set of the eigenvalues of the Laplacian matrix and* $U = [\boldsymbol{u}_0, \boldsymbol{u}_1, \ldots, \boldsymbol{u}_{n-1}]$ *is the eigenvector matrix.* U^H *is the Hermitian (conjugate transpose) of matrix* U*. It is assumed that the eigenvalues of the Laplacian matrix are arranged in ascending order so that,* $0 = \lambda_0 \leq \lambda_1 \leq \cdots \leq \lambda_{n-1}$.

The eigenvector matrix is used to define the GFT [45]. A signal over a graph $\mathcal{G}$ is a vector $\boldsymbol{x} \in \mathbb{R}^n$ that is interpreted as scalar values observed in each vertex, i.e., each sensor (in this work an NIRS channel), $v_i \in \mathcal{V}, i = 1, \ldots, n$. Thus, given a signal x and a Laplacian L, the GFT of the signal is defined as

$$\tilde{\boldsymbol{x}} = U^H \boldsymbol{x} \tag{10.2}$$

and the inverse Graph Fourier Transform (iGFT) is defined as

$$\boldsymbol{x} = U\tilde{\boldsymbol{x}}. \tag{10.3}$$

Thus, each pair $(\boldsymbol{x}, \tilde{\boldsymbol{x}})$ forms a GFT pair. For the NIRS data case, vertices correspond to measurement channels (see Figure 10.1) and, thus, $\boldsymbol{x}$, corresponds to all measurements acquired over the graph in a specific timestamp.

In essence, GFT encodes the notion of variability over vertices as the traditional Fourier Transform encodes variability of the temporal properties of the signals.

This variability is encoded on the eigenvectors and their fluctuations over the graph $\mathcal{G}$ vary for different frequency ranges, i.e., rapid fluctuations for high frequencies (last coordinates of $\tilde{x}$) and more smooth fluctuations for lower ones (first coordinates of $\tilde{x}$ [46].

Here, we present the results of three different frequency, f, ranges, i.e., low, medium, and high. As f ranges from 1 to 52, low-, medium-, and high-frequency ranges were chosen to correspond to approximately equal frequency ranges, i.e., 1 to 18, 19 to 35, and 36 to 52, respectively, across the whole frequency range.

10.3.2 Dirichlet energy over a graph

To better quantify the distribution of useful information across the channel montage and provide a measure for smoothness (or nonsmoothness) of the NIRS signals over the channel grid, we also estimated the Dirichlet energy of the graph NIRS signals. In GSP theory, the notion of smoothness of a graph signal $\boldsymbol{x}$ is expressed via the p-Dirichlet form of $\boldsymbol{x}$ which is defined as [37, 38]

$$E_p(\boldsymbol{x}) = \frac{1}{p} \sum\nolimits_{i \in \mathcal{V}} \|\nabla_i \boldsymbol{x}\|_2^p. \tag{10.4}$$

When $p = 2$ then:

$$E_2(\boldsymbol{x}) = \sum\nolimits_{(i,j) \in \mathcal{E}} w_{ij}(x_i - x_j)^2 = \boldsymbol{x}^T L \boldsymbol{x}, \tag{10.5}$$

that is, the graph Laplacian quadratic form, which is denoted here as the Dirichlet Energy (E) of $\boldsymbol{x}$. If $\boldsymbol{x}$ is constant across all vertices of the graph, then $E = 0$. In general, E is small when the signal $\boldsymbol{x}$ has similar values in neighboring vertices, i.e., the signal $\boldsymbol{x}$ is smooth. In essence, E is a measure of how much a graph signal changes with respect to the network of the nodes (here the network of the NIRS channels). In the NIRS case, it is a measure of how much an NIRS signal changes with respect to the channel montage.

Thus, it is expected that E will be approximately zero when no mental task is executed, thus no oxy-Hb or deoxy-Hb CCs are detected across NIRS channels. In the contrary, E is expected to maximize within the task execution period [47].

10.3.3 Graph construction algorithm

To construct the NIRS graph, i.e., to estimate the W matrix of the weights between the nodes, the Semilocal (SL) approach was adopted here. It has been proved that the SL approach is the most efficient among other approaches of W estimation for brain signals [46]. Moreover, the whole approach is computationally efficient as it is mainly based on estimating the covariance between two NIRS channels. In particular, a W matrix for an SL graph connects only close NIRS measurement channels (with the Euclidian notion of proximity). The weights used for the connection of the nodes though, correspond to the absolute value of covariance between the measurements of the two channels, i.e.,

$$w_{ij}^{cov} = \begin{cases} |cov(\boldsymbol{x}_i, \boldsymbol{x}_j)| & if\, d(v_i, v_j) \leq T_w \\ 0 & otherwise \end{cases} \tag{10.6}$$

where $\boldsymbol{x}_i, \boldsymbol{x}_j$ are the signals that correspond to the measurements over time of the NIRS, i, j, channels (vertices of the graph) for a specific trial of the task. $d(v_i, v_j)$ is the Euclidean distance between the NIRS measurement channels vertices (Figure 10.1). Finally, T_w is the distance threshold used to determine the proximity criterion.

10.3.4 Feature extraction

In the following subsections, the GFT-based feature vector [48] along with the baseline feature vector extraction approaches, S and CC, are described.

10.3.4.1 GFT feature vector

The GFT feature vector is based on the decomposition of a trial, second-dependent, measurement $\boldsymbol{x}'$ over the graph $\mathcal{G}$. In particular, for a specified second t (t is actually a set of sample moments corresponding to 1 s period) within a trial, the signal $\boldsymbol{x}'$ is defined as the mean signal over all 10 measurements (sampling frequency of 10 Hz) during the specified second of the trial. The GFT $\tilde{\boldsymbol{x}}'$ is computed using (12.2) and the respective GFT feature vector (FV) is defined as

$$FV_{GFT} = \tilde{\boldsymbol{x}}'_f \tag{10.7}$$

where f denotes the frequency range, i.e., low, medium, high. The number of the frequencies in each range determines the number of the coefficients of $\tilde{\boldsymbol{x}}'$ that are used for the construction of the respective feature vector. In addition, except for the low-, medium-, and high-frequency ranges, the $f =$ *"all"* frequencies option is also presented here. Vectors $\boldsymbol{x}'$ that correspond to seconds $t_a = 1, \ldots, 14$ s after cue onset are considered to belong to MAT class [29], whereas vectors corresponding to seconds $t_b = 20, \ldots, 30$ s after cue onset are considered to belong to the Relax class [29] as far as the GFT vector is concerned.

10.3.4.2 S feature vector

Here we also present the results of the slope-based S feature vector that is used in the majority of the NIRS-based BCI systems to capture the fluctuations of oxy-Hb and deoxy-Hb exhibited during mental tasks. For a specified time interval window (I) the slope of the regression line fit of the respective measurements is estimated. To better capture the increase/decrease effects, a sliding window I with overlap of 33% [32] was used with $I = 10$ (equal to sampling frequency). Again the time period between seconds 1 to 14 was considered as MAT class. Thus, the S_t^a, $t = 1, \ldots, 14$ s is the S feature vector for class MAT estimated for seconds 1 to t. For instance, S_1^a is the S feature vector estimated only for the first second after onset whereas, e.g., S_5^a is estimated for the first 5 s after onset.

Similarly, S_t^b, $t = 17, \ldots, 30$ s, feature vectors were estimated for the class Relax. In essence, $t = 17$ is considered the first second of the Relax state. Thus, S_{17}^b is the

S feature vector estimated for the first second of the Relax state whereas, e.g., S_{21}^{b} is estimated for the first 5 s of the Relax state. Ranges of t for MAT and Relax classes, i.e., 1–14 and 17–30, respectively, were selected in order to result in equal numbers of features for the two classes depending on the seconds used for their estimation [48].

10.3.4.3 CC feature vector

Apart from the slope-based feature vector, the CC feature of mean CCs is extensively used in NIRS-based BCI systems [17]. Here, CC-based feature is also used for baseline comparison. The CC feature adopted in this work was recently proposed [29, 30] based on the antagonistic pattern detected during MATs. In particular, CCs (averaged over 1 s period) at seconds $t = 10, \ldots, 14$ were labeled as MAT class. On the contrary, mean CC values at seconds $t = 26, \ldots, 30$ after onset were used for the class Relax.

10.3.5 Classification

In the majority of the NIRS-based BCI system studies, LDA and SVMs [49] are the most commonly used classifiers [17]. Nevertheless, it was recently shown [30] that linear SVM is the most effective classifier for the dataset used here, compared to a variety of other classification methods. Thus, in this work the linear SVM classifier was used in order to test the reliability of the GSP-based feature vectors compared to the baseline ones, S and CC feature vectors.

10.3.6 Implementation issues

For the estimation of the W^{cov} matrix, we investigated four different threshold T_w values, i.e., $T_w = 3$, 4.5, 6, and 7.5 cm, taking into account that the distance between a source–detector pair is 3 cm. For each subject the value that led to the most efficient recognition of the arithmetic mental task was used. Moreover, the time window used for the W^{cov} corresponded both to the measurements during only the task execution period, i.e., 12 s (seconds 1 to 12 after cue onset), and to measurements during all 40 s, i.e., for seconds -10 to 30 in respect to the cue onset were used. Again the most efficient choice was used for each subject.

In addition, for the W^{cov} estimation two trials $r = 1, 2$ (approximately 10% of the trials) from each subject were used. and the mean between the two trial-referred matrices was used for the final weight matrix of the graph [48]. Finally, weights were normalized in the range [0, 1]. For the analysis of GSP theory the GSPBOX toolbox was used [50].

As far as the classification process is concerned, a leave-p-out cross validation framework was adopted. In each classification iteration, 70% of the trials of each subject were randomly chosen to be used for training and the rest (p observations) were kept for testing. The mean classification rate (CR) across 10 iterations of the leave-p-out process was used as an evaluation criterion throughout the rest of this paper. It must be stressed out that the trials used for the estimation of the matrix W^{cov} were not considered for the classification stage. Thus, 16 trials from subjects 1, 2, and 3 and 22 trials from subjects 4, 5, 6, 7, and 8 were used for classification.

10.4 Results

The evaluation of the GFT-based feature vector went through a thorough experimentation phase were the optimal parameter set, in terms of the corresponding classification accuracy, of the $W^{cov-win}$, T_w, t_a, t_b and f variables was determined for each subject. The same process for parameter optimization took place as far as the S and CC feature vectors are concerned. The results concerning the subject dependent classification task and the respective mean CR values across all 10 iterations are presented in Table 10.1 for the oxy-Hb signals and for all three feature vectors.

The mean CR across all subjects was 88.39% (±9.36), 82.60% (±9.68), and 92.52% (±6.67) for oxy-Hb and S, CC, and GFT feature vectors, respectively. Thus, the GFT base vector outperforms the rest state of the art approaches for all subjects except for S1 and S3 where the S-slope approach exhibits better CR values. Nevertheless, the S and CC approaches result in feature vector lengths of 936 and 260 features, respectively, whereas the GFT-based vector is of mean length of approximately 21 features [48]. Hence, the GFT approach accomplishes better classification performance with much shorter feature vector.

Furthermore, Table 10.2 presents the CR values for all three feature vector extraction approaches for the deoxy-Hb signals. Again the GFT-based approach outperforms the other two state-of-the-art methods. Although GFT and CC approaches show poorer performance as compared to the oxy-Hb signals, the S methodology shows advanced performance. Nevertheless, S feature vector performance is poorer than the GFT ones. In particular, the mean CRs across all subjects were 90.35% (±5.71), 78.94% (±12.01), and 90.98% (±5.76) for deoxy-Hb and S, CC, and GFT feature vectors, respectively. Moreover, it should be stressed out that the S and CC approaches result again in a lengthy feature vector, i.e., the feature vectors are of lengths of 988

Table 10.1 CR values for the S, CC, and GFT-based feature vectors for oxy-Hb

Feature vector	**S1**	**S2**	**S3**	**S4**	**S5**	**S6**	**S7**	**S8**	**Mean**	**SD**
S	**91.67**	90.00	**98.33**	92.92	72.50	75.00	93.33	93.33	88.39	9.36
CC	87.22	78.33	89.44	90.83	66.25	70.83	86.25	91.67	82.60	9.68
GFT	86.88	**95.63**	98.12	**99.55**	**84.55**	**82.73**	**95.91**	**96.82**	**92.52**	6.67

Table 10.2 CR values for the S, CC, and GFT-based feature vectors for deoxy-Hb

Feature vector	**S1**	**S2**	**S3**	**S4**	**S5**	**S6**	**S7**	**S8**	**Mean**	**SD**
S	87.78	88.33	**96.67**	92.92	**85.42**	81.67	91.25	**98.75**	90.35	5.71
CC	83.33	72.78	78.33	88.33	74.58	55.42	83.33	95.42	78.94	12.01
GFT	**90.00**	**96.25**	92.50	**94.55**	78.18	**89.09**	**91.82**	95.45	**90.98**	5.76

and 260 features, respectively, whereas the GFT-based vector is of mean length of approximately 26 features [48]. Hence, the GFT approach accomplishes again better classification performance with much shorter feature vector.

As previously stated, the superiority of the GFT approach over the state-of-the-art ones is assumed to be based on the spatial characteristics of the NIRS signals that the GFT approach takes into consideration. To further validate this assumption, we estimated the Dirichlet energy (E) over each trial of the BCI experiment.

In Figure 10.2, the E values over the oxy-Hb values of the trial 12 of subject 2 are presented. It is obvious that before the cue onset, E values are very small detecting low divergence across the NIRS channel grid (smooth graph signal). On the contrary, after the cue onset, E starts to increase depicting associated increase of the divergence of the NIRS values across the respective measurement channels (nonsmooth graph signal).

Indeed, this can be proved by inspecting the concentrations changes of oxy-Hb for all NIRS channels across time for the respective trial.

Figure 10.3 demonstrates the aforementioned CCs. Thus, despite the fact that the graph NIRS signal x seems to be smooth during the pretask period (10-s period before the dashed line) CCs start to vary after cue onset, leading to a more diverge texture of the graph signal. This spatial, channel-oriented divergence is captured by the Dirichlet energy measure (see Figure 10.2), which starts to increase after the cue onset and decreases later on. It is noteworthy that despite the fact that the task execution period lasts for 12 s after cue onset, the spatial divergence lasts even after that time frame. Nevertheless, it starts to diminish after the 15th s after the cue onset.

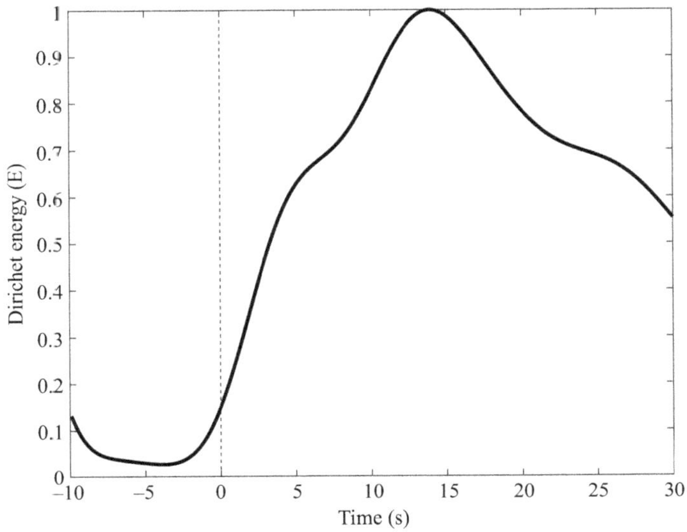

Figure 10.2 *The estimated Dirichlet energy (normalized in the range* [0, 1]) *for trial = 12 of subject 2. The dashed vertical line corresponds to the cue onset*

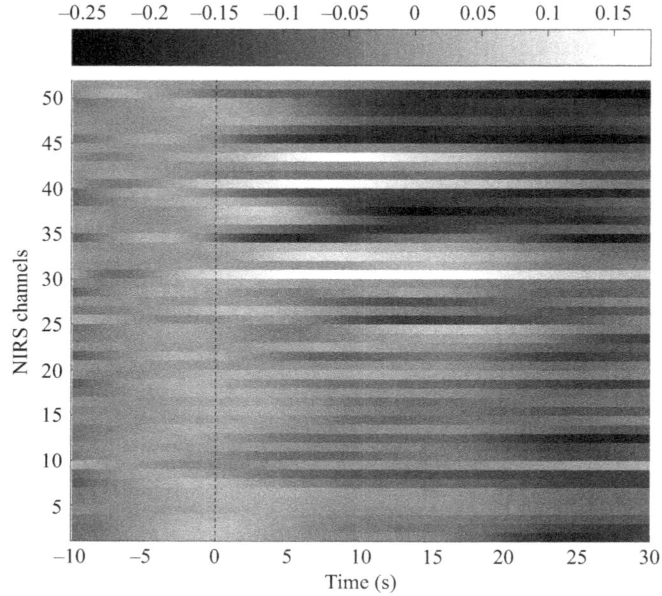

Figure 10.3 Concentration changes of oxy-Hb of trial = 12 of subject 2 of each one of the 52 channels vs. time. The dashed vertical line corresponds to the cue onset

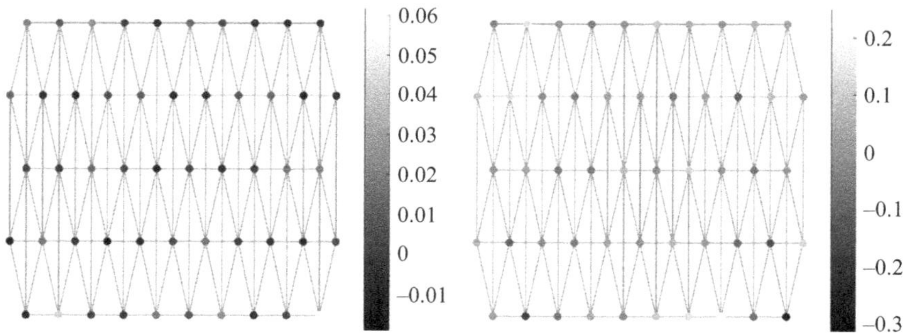

Figure 10.4 The graph signal x across all measurement channels for second −4 (left) and second 12 (right) for subject 2, trial 12 (see Figure 10.3). The vertices of the graphs are associated with the fNIRS channels in Figure 10.1

To better illustrate this texture-like change of the NIRS signals across the channel grid, Figure 10.4 shows the graph signals of seconds −4 (4 s before the cue onset) and 12 (12 s after the cue onset, i.e., the second that the MAT ends). For the former case, the NIRS graph vector over the measurement grid seems to be smooth, i.e., the values of neighboring channels slightly diverge. Moreover, the value range of the

NIRS signals is narrow, i.e., values range from approximately −0.01 to 0.06. On the contrary, the graph signal of the 12th second shows a more intense divergence of NIRS values and the corresponding values range from approximately −0.3 to >0.2.

In an attempt to test the discrimination ability of the Dirichlet energy feature between the Relax phase of the experiment and the phase where the MAT takes place, we also investigate the E feature under a classification task. As a feature vector, we use only two values, i.e., the mean Dirichlet energy (E) values of two consecutive 5-s windows. In particular, for the Relax phase the two mean values of the Dirichlet energy of window −10 to −5 and of the window −5 to 0 are used.

On the contrary, for the MAT phase three different cases were adopted. Case 1: the first feature is the mean E value of the time window 0 to 5 and the second is the mean E value of the time window from 5 to 10 s. Case 2: the first feature is the mean E value of the time window 10 to 15 and the second is the mean E value of the time window from 15 to 20 s. Finally, Case 3: the first feature is the mean E value of the time window 20 to 25 and the second is the mean E value of the time window from 25 to 30 s. Figure 10.5 shows the abovementioned feature sets for all three cases for the oxy-Hb signal.

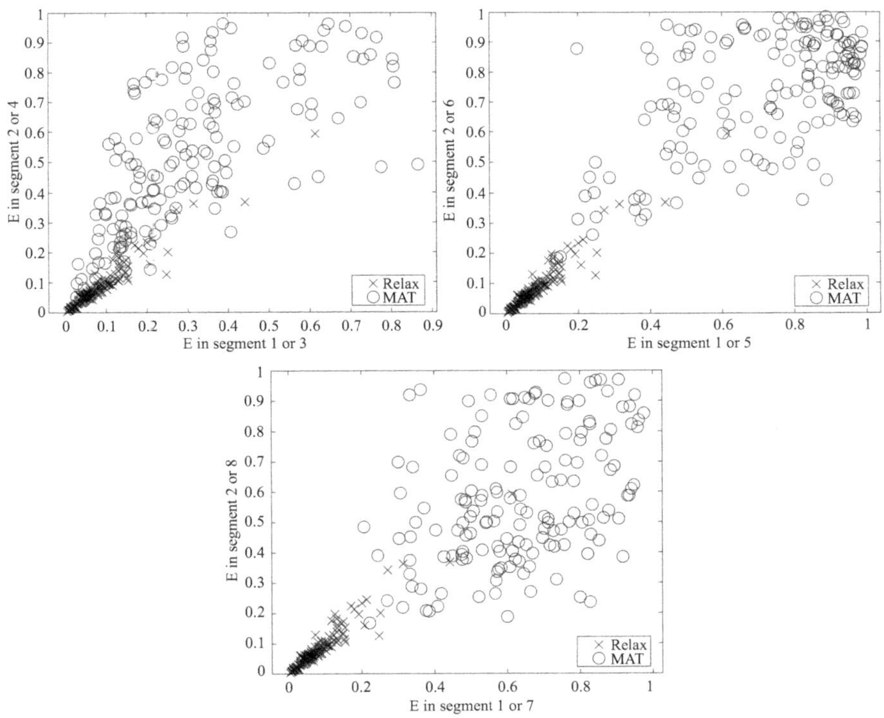

Figure 10.5 The Dirichlet energy-based features of Case 1 (upper left), Case 2 (upper right), and Case 3 (bottom) for the oxy-Hb signals

For the sake of completeness, we also investigated the abovementioned Dirichlet energy features for the deoxy-Hb signals. The corresponding features are depicted in Figure 10.6.

For the classification of the proposed E-based features, the same classifier that was used with the GFT-based feature vector was used and the same classification framework was followed as described in Section 10.3.6. The CR values of the classification outcome are tabulated in Table 10.3 for both oxy- and deoxy-Hb signals.

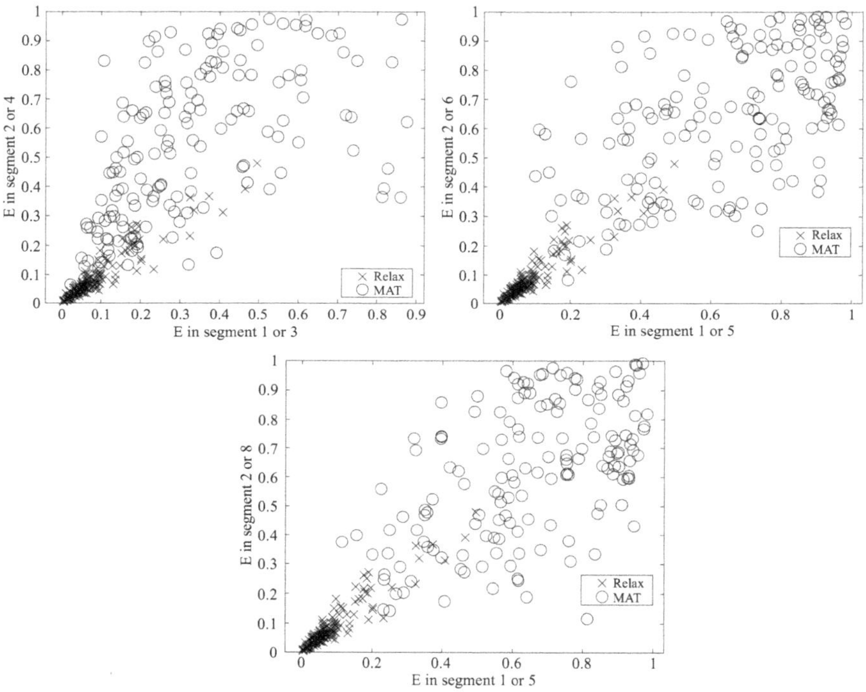

Figure 10.6 The Dirichlet energy-based features of Case 1 (upper left), Case 2 (upper right), and Case 3 (bottom) for the deoxy-Hb signals

Table 10.3 CR values for the subject independent Dirichlet energy-based feature vectors for oxy and deoxy-Hb

Signals	Case 1	Case 2	Case 3
Oxy-Hb	92.97	97.49	97.97
Deoxy-Hb	90.50	95.74	94.84

Despite the fact that the proposed feature vector is comprised of only two values, i.e., the E values of two consecutive, 5-s, time windows, the CR rates are high. In particular, the best CR is accomplished for the oxy-Hb, Case 3 framework with 97.97% of success. Taking into account that the reported CR values are for subject independent classification, it is evident that the Dirichlet energy measure is very efficient on capturing the transition from the Relax phase to the MAT one.

10.5 Discussion

The results presented in the previous section show the efficiency of the GSP theory to extract features from NIRS signals for BCI applications. The efficiency of the GSP approach is based on the incorporation into the feature vector of the spatial characteristics across the NIRS channel montage, an aspect that has been extensively studied in EEG signals [51] but not for NIRS ones. The different texture of the graph NIRS signals as depicted in Figure 10.3 is detected between the different phases of the BCI experiment, i.e., Relax vs. MAT, and provide the NIRS BCI system with spatial information.

Until now, the NIRS-based BCI systems relied mostly on slope oriented features, and thus the information transfer rates remained low as slope-related features require several seconds to be expressed. GFT-based and Dirichlet energy-based feature vectors exploit the spatial information of the hemodynamic response during a mental task, e.g., the spatial patterns during an MAT [31] which are shown to be expressed during even the first 5 s of the MAT phase leading this way to faster BCI applications.

Furthermore, it is noteworthy that the spatial differentiation over the measurement channels between the pre- and post-cue onset threshold may indicate a task-related graph pattern (TRGP) that is expressed within the first seconds after the task onset and probably vanishes thereafter. Further experimentation of the proposed feature vectors should include validation of the proposed approach to more NIRS datasets and investigation of the TRGP behavior as expressed in the GFT space of the NIRS signals of the first seconds after onset.

10.6 Summary

In this work, GSP-based feature extraction methods for NIRS signals were presented. The proposed approaches outperform widely used feature extraction approaches for NIRS-based BCI systems. Moreover, the efficiency of the proposed features is illustrated in real NIRS signals where it is shown to discriminate relax phase from task-related periods even within a time frame of the first 5 s after onset. Nevertheless, the efficacy of the proposed approaches should be further tested on more datasets for NIRS-BCI systems.

References

[1] J. R. Wolpaw, N. Birbaumer, D. J. McFarland, G. Pfurtscheller, and T. M. Vaughan, "Brain–computer interfaces for communication and control.," *Clin. Neurophysiol.*, vol. 113, no. 6, pp. 767–91, 2002.

[2] U. Chaudhary, N. Birbaumer, and A. Ramos-murguialday, "Brain–computer interfaces for communication and rehabilitation," *Nat. Rev. Neurol.*, vol. 12, no. 9, 2016.

[3] M. A. Lebedev and M. A. L. Nicolelis, "Brain-machine interfaces: from basic science to neuroprostheses and neurorehabilitation," *Physiol. Rev.*, vol. 97, no. 2, pp. 767–837, 2017.

[4] I. Lazarou, S. Nikolopoulos, P. C. Petrantonakis, I. Kompatsiaris, and M. Tsolaki, "EEG-based brain–computer interfaces for communication and rehabilitation of people with motor impairment: A novel approach of the 21st century," *Front. Hum. Neurosci.*, vol. 12, 2018.

[5] F. Cincotti, D. Mattia, F, Aloise F, *et al.*, "Non-invasive brain–computer interface system: Towards its application as assistive technology," *Brain Res. Bull.*, vol. 75, no. 6, pp. 796–803, 2008.

[6] G. Schalk and E. C. Leuthardt, "Brain–computer interfaces using electrocorticographic signals," *IEEE Rev. Biomed. Eng.*, vol. 4, pp. 140–154, 2011.

[7] L. F. Nicolas-Alonso and J. Gomez-Gil, "Brain computer interfaces, a review," *Sensors*, vol. 12, no. 2, pp. 1211–1279, 2012.

[8] A. Ramos-Murguialday, D. Broetz, M. Rea, *et al.*, "Brain-machine-interface in chronic stroke rehabilitation: A controlled study.," *Ann. Neurol.*, vol. 74, no. 1, pp. 100–108, 2013.

[9] S. D. Stavisky, J. C. Kao, P. Nuyujukian, S. I. Ryu, and K. V. Shenoy, "A high performing brain-machine interface driven by low-frequency local field potentials alone and together with spikes," *J. Neural Eng.*, vol. 12, no. 3, 2015.

[10] P. C. Petrantonakis and P. Poirazi, "A Novel and Simple Spike Sorting Implementation," *IEEE Trans. Neural Syst. Rehabil. Eng.*, vol. 25, no. 4, 2017.

[11] J. Mellinger, G. Schalk, C. Braun, *et al.*, "An MEG-based brain–computer interface (BCI)," *Neuroimage*, vol. 36, no. 3, pp. 581–593, 2007.

[12] S. T. Foldes, D. J. Weber, and J. L. Collinger, "MEG-based neurofeedback for hand rehabilitation," *J. Neuroeng. Rehabil.*, vol. 12, no. 1, p. 85, 2015.

[13] S. H. Sardouie and M. B. Shamsollahi, "Selection of efficient features for discrimination of hand movements from MEG using a BCI competition IV data set," *Front. Neurosci.*, vol. 6, no. APR, pp. 1–7, 2012.

[14] N. Weiskopf, K. Mathiak, S. W. Bock, *et al.*, "Principles of a brain–computer interface (BCI) based on real-time functional magnetic resonance imaging (fMRI)," *IEEE Trans. Biomed. Eng.*, vol. 51, no. 6, pp. 966–970, 2004.

[15] L. van der Heiden, G. Liberati, R. Sitaram, *et al.*, "Insula and inferior frontal triangularis activations distinguish between conditioned brain responses using emotional sounds for basic BCI communication," *Front. Behav. Neurosci.*, vol. 8, no. July, pp. 1–7, 2014.

[16] C. Zich, S. Debener, C. Kranczioch, M. G. Bleichner, I. Gutberlet, and M. De Vos, "Real-time EEG feedback during simultaneous EEG-fMRI identifies the cortical signature of motor imagery.," *Neuroimage*, vol. 114, pp. 438–447, Jul. 2015.

[17] N. Naseer and K.-S. Hong, "fNIRS-based brain–computer interfaces: a review," *Front. Hum. Neurosci.*, vol. 9, no. January, 2015.

[18] Y. Blokland *et al.*, "Combined EEG-fNIRS decoding of motor attempt and imagery for brain switch control: An offline study in patients with tetraplegia," *IEEE Trans. Neural Syst. Rehabil. Eng.*, vol. 22, no. 2, pp. 222–229, 2014.

[19] P. Y. Lin, S. I. Lin, and J. J. J. Chen, "Functional near infrared spectroscopy study of age-related difference in cortical activation patterns during cycling with speed feedback," *IEEE Trans. Neural Syst. Rehabil. Eng.*, vol. 20, no. 1, pp. 78–84, 2012.

[20] T. H. Falk, M. Guirgis, S. Power, and T. T. Chau, "Taking NIRS-BCIs outside the lab: Towards achieving robustness against environment noise," *IEEE Trans. Neural Syst. Rehabil. Eng.*, vol. 19, no. 2, pp. 136–146, 2011.

[21] D. Wyser, O. Lambercy, F. Scholkmann, M. Wolf, and R. Gassert, "Wearable and modular functional near-infrared spectroscopy instrument with multidistance measurements at four wavelengths," *Neurophotonics*, vol. 4, no. 04, p. 1, 2017.

[22] A. von Lühmann, C. Herff, D. Heger, and T. Schultz, "Toward a Wireless Open Source Instrument: Functional Near-infrared Spectroscopy in Mobile Neuroergonomics and BCI Applications," *Front. Hum. Neurosci.*, vol. 9, no. November, pp. 1–14, 2015.

[23] F. Jobsis, "Noninvasive, infrared monitoring of cerebral and myocardial oxygen sufficiency and circulatory parameters," *Science (80-).*, vol. 198, no. 4323, pp. 1264–1267, 1977.

[24] M. Ferrari and V. Quaresima, "A brief review on the history of human functional near-infrared spectroscopy (fNIRS) development and fields of application," *Neuroimage*, vol. 63, no. 2, pp. 921–935, 2012.

[25] F. Scholkmann, S. Kleiser, A. J. Metz, *et al.*, "A review on continuous wave functional near-infrared spectroscopy and imaging instrumentation and methodology," *Neuroimage*, vol. 85, pp. 6–27, 2014.

[26] S. Coyle, T. Ward, C. Markham, and G. McDarby, "On the suitability of near-infrared (NIR) systems for next-generation brain–computer interfaces," *Physiol. Meas.*, vol. 25, no. 4, pp. 815–822, 2004.

[27] S. M. Coyle, T. E. Ward, and C. M. Markham, "Brain–computer interface using a simplified functional near-infrared spectroscopy system," *J. Neural Eng.*, vol. 4, no. 3, pp. 219–226, 2007.

[28] V. Kaiser, G. Bauernfeind, A. Kreilinger, *et al.*, "Cortical effects of user training in a motor imagery based brain–computer interface measured by fNIRS and EEG," *Neuroimage*, vol. 85, pp. 432–444, 2014.

[29] G. Bauernfeind, R. Scherer, G. Pfurtscheller, and C. Neuper, "Single-trial classification of antagonistic oxyhemoglobin responses during mental arithmetic," *Med. Biol. Eng. Comput.*, vol. 49, no. 9, pp. 979–984, 2011.

[30] G. Bauernfeind, D. Steyrl, C. Brunner, and G. R. Muller-Putz, "Single trial classification of fNIRS-based brain–computer interface mental arithmetic data: a comparison between different classifiers," *Conf. Proc. Annu. Int. Conf. IEEE Eng. Med. Biol. Soc. IEEE Eng. Med. Biol. Soc. Annu. Conf.*, vol. 2014, no. 288566, pp. 2004–2007, 2014.

[31] G. Pfurtscheller, G. Bauernfeind, S. C. Wriessnegger, and C. Neuper, "Focal frontal (de)oxyhemoglobin responses during simple arithmetic," *Int. J. Psychophysiol.*, vol. 76, no. 3, pp. 186–192, 2010.

[32] J. Shin, K.-R. Müller, and H.-J. Hwang, "Near-infrared spectroscopy (NIRS)-based eyes-closed brain–computer interface (BCI) using prefrontal cortex activation due to mental arithmetic," *Sci. Rep.*, vol. 6, no. 1, p. 36203, 2016.

[33] L. C. Schudlo and T. Chau, "Dynamic topographical pattern classification of multichannel prefrontal NIRS signals: II. Online differentiation of mental arithmetic and rest.," *J. Neural Eng.*, vol. 11, no. 1, p. 016003, 2014.

[34] S. D. Power, A. Kushki, and T. Chau, "Intersession consistency of single-trial classification of the prefrontal response to mental arithmetic and the no-control state by NIRS," *PLoS One*, vol. 7, no. 7, pp. 7–9, 2012.

[35] N. Naseer, F. M. Noori, N. K. Qureshi, and K.-S. Hong, "Determining optimal feature-combination for LDA classification of functional near-infrared spectroscopy signals in brain–computer interface application," *Front. Hum. Neurosci.*, vol. 10, no. May, pp. 1–10, 2016.

[36] S. D. Power, T. H. Falk, and T. Chau, "Classification of prefrontal activity due to mental arithmetic and music imagery using hidden Markov models and frequency domain near-infrared spectroscopy," *J. Neural Eng.*, vol. 7, no. 2, p. 026002, 2010.

[37] K. Smith, B. Ricaud, N. Sha, *et al.*, "Locating Temporal Functional Dynamics of Visual Short-Term Memory Binding using Graph Modular Dirichlet Energy.," *Sci. Rep.*, vol. 7, no. January, p. 42013, 2017.

[38] D. I. Shuman, S. K. Narang, P. Frossard, A. Ortega, and P. Vandergheynst, "The Emerging Field of Signal Processing," no. may, pp. 83–98, 2013.

[39] C. J. C. Burges, "A tutorial on support vector machines for pattern recognition," *Data Min. Knowl. Discov.*, vol. 2, no. 2, pp. 121–167, 1998.

[40] K.-S. Hong, M. R. Bhutta, X. Liu, and Y.-I. Shin, "Classification of somatosensory cortex activities using fNIRS," *Behav. Brain Res.*, vol. 333, no. June, pp. 225–234, Aug. 2017.

[41] K.-S. Hong and M. J. Khan, "Hybrid brain–computer interface techniques for improved classification accuracy and increased number of commands: a review," *Front. Neurorobot.*, vol. 11, no. JUL, Jul. 2017.

[42] M. A. Kamran, M. M. N. Mannan, and M. Y. Jeong, "Cortical signal analysis and advances in functional near-infrared spectroscopy signal: A review," *Front. Hum. Neurosci.*, vol. 10, no. June, pp. 1–12, 2016.

[43] K. Mandrick, G. Derosiere, G. Dray, D. Coulon, J. P. Micallef, and S. Perrey, "Utilizing slope method as an alternative data analysis for functional near-infrared spectroscopy-derived cerebral hemodynamic responses," *Int. J. Ind. Ergon.*, vol. 43, no. 4, pp. 335–341, 2013.

[44] A. Zafar and K.-S. Hong, "Detection and classification of three-class initial dips from prefrontal cortex," *Biomed. Opt. Express*, vol. 8, no. 1, p. 367, 2017.

[45] A. Sandryhaila and J. M. F. Moura, "Discrete signal processing on graphs," *IEEE Trans. Signal Process.*, vol. 61, no. 7, pp. 1644–1656, 2013.

[46] M. Ménoret, N. Farrugia, B. Pasdeloup, and V. Gripon, "Evaluating graph signal processing for neuroimaging through classification and dimensionality reduction," *arXiv Prepr. arXiv1703.01842*, Aug. 2017.

[47] P. C. Petrantonakis and I. Kompatsiaris, "Detection of mental task related activity in NIRS - BCI systems using Dirichlet energy over graphs," in *40th Annual International Conference of the IEEE Engineering in Medicine and Biology Society (EMBC)*, 2018, no. 644780, pp. 85–88.

[48] P. C. Petrantonakis and I. Kompatsiaris, "Single-trial NIRS data classification for brain – computer interfaces using graph signal processing," *IEEE Trans. Neural Syst. Rehabil. Eng.*, vol. 26, no. 9, pp. 1700–1709, 2018.

[49] R. O. Duda, P. E. Hart, and D. G. Stork, "Pattern classification," *New York: John Wiley, Section*. p. 680, 2001.

[50] N. Perraudin, J. Paratte, D. Shuman, V. Kalofolias, P. Vandergheynst, and D. K. Hammond, "GSPBOX: A toolbox for signal processing on graphs," pp. 1–8, 2014.

[51] B. Blankertz, R. Tomioka, S. Lemm, M. Kawanabe, and K. Muller, "Optimizing spatial filters for robust EEG single-trial analysis," *IEEE Signal Process. Mag.*, vol. 25, no. 1, pp. 41–56, 2008.

Part III

Multimodal prototype interfaces that can be operated through eyes and mind

Chapter 11

Error-aware BCIs

Fotis P. Kalaganis[1,2], *Elisavet Chatzilari*[1], *Nikos A. Laskaris*[2,3], *Spiros Nikolopoulos*[1], *and Ioannis Kompatsiaris*[1]

The ability of recognizing and correcting the erroneous actions is an integral part of human nature. Plenty of neuroscientific studies have been investigating the ability of human brain to recognize errors. The distinct neuronal responses that are produced by the human brain during the perception of an erroneous action are referred to as error-related potentials (ErrPs). Although research in brain–computer interfaces (BCIs) has managed to achieve significant improvement in terms of detecting the users' intentions over the last years, in a real-world setting, the interpretation of brain commands still remains an error-prone procedure leading to inaccurate interactions. Even for multi-modal interaction schemes, the attained performance is far from optimal. As a means to overcome these debilities, and apart from developing more sophisticated machine-learning techniques or adding further modalities, scientists have also exploited the users' ability to perceive errors. During the rapid growth of the BCI/Human-Machine Interaction (HMI) technology over the last years, ErrPs have been used widely in order to enhance several existing BCI applications serving as a passive correction mechanism towards a more user-friendly environment. The principal idea is that a BCI system may incorporate, as feedback, the user's judgement about its function and use this feedback to correct its current output. In this chapter, we discuss the potentials and applications of ErrPs into developing hybrid BCI systems that emphasize in reliability and user experience by introducing the so-called error awareness.

11.1 Introduction to error-related potentials

ErrPs are characteristic electroencephalogram (EEG) signals observed after subjects, in various tasks, committing errors. Several types of errors can evoke ErrPs-type signals, e.g., depending on who is responsible for the error (the user or the interface) or how the user perceives the error (e.g. as an erroneous feedback to an action or

[1] Information Technologies Institute, CERTH, Thermi-Thessaloniki, Greece
[2] Department of Informatics, Aristotle University of Thessaloniki, Thessaloniki, Greece
[3] Neuroinformatics Group, Aristotle University of Thessaloniki, Thessaloniki, Greece

as an observation of an error made by an independent operator). Through ErrPs, the EEG sensor offers a natural way to correcting the errors taking place in a natural user interface by the responsibility of either the user or the interface itself. Therefore, utilities like 'undo' or 'backspace' can be performed in an automated way, stemming from user's spontaneous physiological responses, leading to seamless user experience and faster interaction.

ErrPs were first presented in [1] and refer to the human brain's electrophysiological response if an erroneous action is monitored and/or performed. A typical ERP signal consists of two components, a negative deflection followed by a positive one [2]. The first (usually called Ne) has a front-central scalp distribution and peaks of about 50–100 ms following incorrect responses. The second component, the error-related positivity (Pe), is associated with awareness of erroneous actions and consists of a high peak of magnitude. This component, subsequent to the Ne component, is mostly noticeable in the centro-parietal brain area. The most prominent findings concerning the localization of this event demonstrate that the anterior cingulate cortex is mainly activated during erroneous processes [3,4].

ErrPs have been widely employed to enhance the user experience provided in several BCI applications. Particularly, spellers, a widely known category of BCI applications, take advantage of the ErrPs. Authors in [5] exploited the neural correlations associated with error awareness so as to significantly improve Information Transfer Rate (ITR) in a P-300-based BCI. An online adaptation of ErrPs was used in [6] along with code-modulated VEPs. Generally, the results of integrating ERP-based correction in an HMI system are promising. A variety of studies have managed to successfully combine motor-imagery-based interfaces with ERP mechanisms that provided an indication to whether defy certain actions (those recognized as erroneous) or not [7,8]. More recently, researchers presented that the amplitude of ERP components is modulated by the severity of the error and that ErrPs can be detected while continuous feedback was presented, which means that discrete feedback presentation is not mandatory [9]. In this chapter, we present how and when ErrPs can be used so as to offer a significant improvement in both an Steady-State Visual Evoked Potentials (SSVEP)-based BCI for a web-site navigation by cancelling out erroneous actions automatically as well as in the setting of gaze-based keyboard that the 'backspace' functionality is automated.

11.2 Spatial filtering

A common preprocessing step, in EEG signal analysis, concerns spatial filtering, where a transformation is applied to the spatial (channels) domain in order to increase the signal-to-noise ratio (SNR). Spatial filters can be separated into two distinct categories. On one hand, are those that are data-independent (i.e. the weights for each sensor are predetermined), such as the Laplacian filter and the common average re-reference technique [10]. These filters are mostly used in electroencephalography analysis to diminish volume conduction effects. On the other hand, are the filters which are data-dependent (i.e. data are necessary for filter calculation). In particular for use within ERP-BCI paradigms, the xDAWN [11] and canonical-correlation-analysis-related filters [12] appear as the most promising ones [13]. The former aims to maximize the signal-to-signal-plus-noise ratio, while the latter the correlation between the single-trial signals and the average evoked response.

In this section, we present two methods for spatial filtering that resorts to the well-established temporal patterning of ERP responses and maximizes the separability among single-trial responses belonging to distinct brain states. The use of this concept had been initiated successfully in a BCI paradigm that relied on transient visual evoked responses [14]. Therein, a differential sensor montage was introduced based on the bipolar pattern emerged in the topographical distribution of the evoked response. Here, we elaborate initially on a method stemming from the domain of subspace learning and then on a generalized methodology for designing spatial filters based on Fisher's discriminant analysis of single-trial temporal patterning. The main goal of this section is to provide methods that increase the SNR of the recovered response (either directly or indirectly) while enhancing the differences between responses of different types or brain state. We show here that Fisher's separability criterion constitutes the natural extension of a standard SNR estimator suitable for multi-trial ERP responses and can therefore naturally lead to spatial filters conforming to discriminant analysis. The presented filters, named hereafter Collaborative Representation Projection (CRP) and discriminant spatial filters (also called discriminant spatial patterns in [15]), offer a flexible framework and are characterized by computational efficiency. The approaches are validated using independent ERP datasets, concerning ErrPs in various BCI settings, in direct comparison with other widely employed methodologies.

11.2.1 Subspace learning

Dimensionality reduction is a very wide field of statistical analysis that has drawn a lot of interests. It is a common practice in many machine-learning and statistics tasks to employ dimensionality reduction methods as a feature extraction process especially when they succeed in preserving the underlying manifold structure of the data. The graph-embedding methodology, where the data are represented by a pairwise distance matrix according to some distance metrics that captures their relationship, offered a unified framework for multiple dimensionality reduction methods (Principal Component Analysis (PCA), Linear Discriminant Analysis, etc.) [16]. The graph construction is probably the most crucial step in these methods, since it should be able to capture and reveal the underlying data association. Collaborative representation has been recently proposed as an automated way for constructing a graph where data similarity stems from the ability of data to reconstruct each other [17]. Each data sample is potentially represented as a weighted sum of the remaining data points. The weights employed during this reconstruction process represent the similarity of data.

11.2.1.1 Unsupervised collaborative representation projection

The collaborative representation projection method [17] aims into offering an automated way for graph construction that represents the underlying data relationship. During this graph construction process, each data sample is represented as a linear combination of the rest of the samples. Let us denote by $X = [\mathbf{x}_1, \mathbf{x}_2, \ldots, \mathbf{x}_n]$

the data matrix that contains n m-dimensional samples. The problem corresponds to calculating the optimal weights $\mathbf{w_i}$ that offer the best reconstruction for each $\mathbf{x_i}$

$$\mathbf{w_i^*} = \underset{\mathbf{w_i}}{\operatorname{argmin}}\{\|\mathbf{x_i} - X\mathbf{w_i}\|_2^2 + \lambda\|\mathbf{w_i}\|^q\} \tag{11.1}$$

In the previous equation, $\mathbf{w_i} = [w_{i,1}, \ldots, w_{i,j-1}, 0, w_{i,j+1}, \ldots, w_{i,n}]^T$ where each element w_{ij} of $\mathbf{w_i}$ represents the contribution of $\mathbf{x_j}$ in the reconstruction of $\mathbf{x_i}$. The L-2 graph $G(X, W)$ is constructed considering that data samples are the vertices, the collaborative weights as the graph weights and q is set to 2. For L-1 graph, q should be set to 1.

The constructed graph can now be employed so as to calculate a new projection space through a projection matrix P. The L-2 graph represents the reconstruction relationship of the data using weak sparse constraints (a more strict sparsity constraint can be achieved using the L-1 graph). We now aim to find a projection space where local compactness of the graph is minimized. This means that the samples with the ability to accurately represent another sample will preserve that property in the projection space. This criterion is formulated mathematically as

$$C_L = \sum_{i=1}^{n} \left\| P^T\mathbf{x_i} - \sum_{i=1}^{n} w_{ij}P^T\mathbf{x_j} \right\|^2 \tag{11.2}$$

and by setting $S_L = X(I - W - W^T + WW^T)X^T$, where I denotes the identity matrix, (11.2) can be rewritten as

$$C_L = P^T S_L P \tag{11.3}$$

Apart from minimizing the local compactness of the graph (which can also be understood as sparsity preservation), we aim at a projection capable offering maximum separability of the data. This leads to the maximization of the total covariance, in the projection space, that can be achieved through the total scatter and is expressed by

$$C_T = \sum_{i=1}^{n} \|P^T\mathbf{x_i} - P^T\bar{\mathbf{x}}\|^2 \tag{11.4}$$

which can be rewritten in matrix form by setting $S_T = \sum_{i=1}^{n}(\mathbf{x_i} - \bar{\mathbf{x}})(\mathbf{x_i} - \bar{\mathbf{x}})^T$ as

$$C_T = P^T S_T P \tag{11.5}$$

It is consistent that for machine-learning purposes both of the aforementioned criteria need to be held simultaneously. We should minimize the local compactness and in the meantime we need to maximize the total separability. The final optimization problem is formulated as

$$P^* = \underset{P}{\operatorname{argmin}} \frac{P^T S_L P}{P^T S_T P} = \underset{P}{\operatorname{argmax}} \frac{P^T S_T P}{P^T S_L P} \tag{11.6}$$

The solution to (11.6) can be obtained by the generalized eigenvalue decomposition of $S_T P = \lambda S_L P$. Finally, P^* corresponds to the eigenvectors of the largest eigenvalues of the previous problem.

11.2.1.2 Supervised collaborative representation

The classic CRP algorithm operates in an unsupervised manner assuming that the weights will be able to uncover the underlying relationship of the data. In EEG, the desired brain activity is only captured by certain channels, while the activity of the rest channels is being inconsistent (regarding a desired stimulus) due to the nature of human brain signals where each brain region is devoted to a different task (e.g. it is not expected to find the reflection of an auditory stimulus in signals recorded over the motor cortex). Consequently, by using the classic CRP approach we would end up with weights that are not meaningful, although the collaborative representation error will be sufficiently small.

The supervised CRP (sCRP) [18] differentiates from the original version only during the graph construction process. In the presented study, we enforce the data to be represented as a linear combination of the same class samples, that is, $w_{ij} = 0$ if i and j data samples belong to different classes. Equation (11.1) is used to calculate the weights for samples of the same class. Since the CRP objective, described by (11.6), ensures that locality is preserved in the low dimensional space, we end up with a more discriminant representation. Intuitively, the samples of the same class tend to create a local neighbourhood, while the overall scatter, across all classes, is maximized.

As already mentioned, during the graph construction we take into account only the samples of the same class. This reduces the computational cost needed during weight calculation (Equation (11.1)), which is the most computationally intensive task, since it involves the calculation of an inverse matrix as many times as the number of training samples.

11.2.2 Increasing signal-to-noise ratio

An ERP signal is a recorded brain response that is the direct aftereffect of a specific event (e.g. the perception of an erroneous action). Due to low SNR of the recorded signals, several single-trial responses (i.e. repetitions) can be aggregated in the so-called ensemble average waveform. Averaging, typically, ensures that the background noise (i.e. brain's electrical activity that is not related to the stimulus) is cancelled out and only a small fraction survives the averaging. Two assumptions must hold so that averaging will increase the SNR of the ErrPs. First, the signal of interest should consist of phase-locked responses with invariable latency and shape. Second, the background noise should follow a random Gaussian process of zero mean, uncorrelated between different recordings and not time-locked to the stimulus [19].

Typically, SNR is used in order to validate the credibility of one signal. However, the SNR hardly offers any information regarding the separability among groups in a classification problem, where distinct types of responses need to be considered. We discuss next, how Fisher's separability criterion can blend SNR measurements from distinct groups into a single class discrimination score. Finally, we extend this idea to the case of multichannel recordings so as to formulate a suitable cost function for spatial filter design.

11.2.2.1 Preliminaries

Considering that the average waveform is 'practically' free of noise (due to a sufficiently large number of recorded responses), the calculation of SNR may proceed as follows.

Let us denote by $X^i = [s_1^i, s_2^i, \ldots, s_m^i]^T \in R^{m \times p}$, the ith multichannel single-trial response (i.e. a matrix that contains a single brain response recorded from multiple sites) with DC offset removed, where m denotes the number of sensors and p the number of samples. Hence, $\bar{s}_k = (1/N_x)\sum_{i=1}^{N_x} s_k^i$ represents the average response, which is a p-dimensional vector, of N_x single-trial responses for the kth sensor. Similarly, the average multichannel response is expressed as $\overline{X} = (1/N_x)\sum_{i=1}^{N_x} X^i = [\bar{s}_1, \bar{s}_2, \ldots, \bar{s}_m]^T$.

Consequently, the signal power (SP) and noise power (NP) of kth sensor are defined as

$$SP_k = \frac{1}{p}\|\bar{s}_k\|_2^2 \tag{11.7}$$

$$NP_k = \frac{1}{p(N_x - 1)}\sum_{i=1}^{N_x}\|\bar{s}_k - s_k^i\|_2^2 = \frac{1}{p}c_{s_k} \tag{11.8}$$

where c_{s_k} can be interpreted as a measure of dispersion of the single-trials from the average waveform. Formally, this quantity corresponds to the energy of noise. Equation (11.8) implies that every trial is an addition of the ERP signal and noise. Then the corresponding SNR of the kth sensor is the ratio of SP_k over NP_k.

$$SNR_k = \frac{\|\bar{s}_k\|_2^2}{c_{s_k}} \tag{11.9}$$

Observing the previous equation, we could easily conclude that in an ideal scenario we would expect identical ErrPs with zero scatter among them.

In a typical ERP-based BCI, we aim at distinguishing between two types of brain response, ERP vs nonERP, or equivalently to discriminate between two groups of single-trial responses. One could easily separate the two groups if the temporal patterning is consistent within the groups and deviates across groups. In such a case, the corresponding averaged waveforms would differ significantly, while the dispersion for both groups will be low. The first condition can be formulated as high power for the differential averaged signal (i.e. the difference of the average waveforms). The second condition can be expressed as the sum of two terms, expressing the NP of either group. These two conditions ensure that indeed the two groups are distinguishable, while the single-trial responses are identical within each group. By forming the corresponding ratio SP/NP and simplifying the expression, we end up with Fisher's separability criterion used in standard discriminant analysis [20].

Considering two groups of single-trial responses, X^i and Y^j with N_x and N_y trials each, the separability index of the kth sensor becomes

$$J_k = \frac{\|\bar{s}_k(X) - \bar{s}_k(Y)\|_2^2}{c_{s_k}(X) + c_{s_k}(Y)} \tag{11.10}$$

11.2.2.2 Discriminant spatial filters

Having defined all the necessary notations as previously, we are now in a position to formulate the spatial filter design as an optimization problem. Although (11.10) could be used for sensor selection, we are looking for a spatial filter that captures this idea and extends it to multiple sensors. Since surface EEG measurements may reflect the neural response of interest in several electrodes placed over the scalp, and with a varying grade of credibility, we seek the linear combination (i.e. a weighted sum of sensor signals to create a 'virtual' signal) that will maximize Fisher's separability criterion.

Let $w = [w_1, w_2, \ldots, w_m] \in R^m$ be a spatial filter (weight vector). Denoting $C_X = (1/N_x - 1)\sum_{i=1}^{N_x}(X^i - \bar{X})(X^i - \bar{X})^T$, the average response and noise energy for the 'virtual sensor' (VS) for the X^i group of trials can be written as

$$\bar{s}_{vs}(X) = w_1\bar{s}_1 + \cdots w_m\bar{s}_m = w\begin{bmatrix}\bar{s}_1\\ \vdots \\ \bar{s}_m\end{bmatrix} = w\bar{X} \tag{11.11}$$

$$c_{vs}(X) = \frac{1}{N_x - 1}\sum_{i=1}^{N_x}\|w\bar{X} - wX^i\|_2^2 = wC_X w^T \tag{11.12}$$

Similarly, for the Y group, the corresponding average response and noise energy are calculated as

$$\bar{s}_{vs}(Y) = w\bar{Y} \tag{11.13}$$

$$c_{vs}(Y) = wC_Y w^T \tag{11.14}$$

following the similar denotation, as previous, for the Y group where $C_Y = (1/N_x - 1)\sum_{i=1}^{N_x}(Y^i - \bar{Y})(Y^i - \bar{Y})^T$.

Fisher's separability index of the 'VS' can now be expressed as

$$J_{vs} = \frac{\|\bar{s}_{vs}(X) - \bar{s}_{vs}(Y)\|_2^2}{c_{vs}(X) + c_{vs}(Y)} \tag{11.15}$$

Substituting (11.11)–(11.14) to (11.15), J_{vs} can be easily reformulated and hence maximized by

$$w = \arg\max_{w} \frac{w(\bar{X} - \bar{Y})(\bar{X} - \bar{Y})^T w^T}{w(C_x + C_Y)w^T} \tag{11.16}$$

It is easy to notice that the ratio in the right-hand side is a generalized Rayleigh quotient and hence the solution can be obtained by solving the generalized eigenvalue decomposition problem. The desired spatial filter corresponds to the eigenvector associated with the largest eigenvalue. We should note here that the previous formula could be easily adapted to enhance the discriminability in problems that only one group of responses is available (i.e. ERP needs to be recovered from background activity) or problems that concern the simultaneous separation of more than two groups, following an approach similar to [21].

11.3 Measuring the efficiency – ICRT

Our objective in this section is to present the advantage of incorporating an error-detection system (EDS) in a BCI, with respect to the ITR (i.e. the time that is required to perform an action correctly).

In order to calculate the effectiveness of an error-agnostic BCI system that incorporates error detection capabilities, we calculate the inverse of the average time needed such that an individual will complete an action (e.g. to correctly select a specific navigation option) correctly while considering the initial system's (error-agnostic) detection rate as well as the precision and recall values that correspond to the EDS [22]. This quantity, that will be referred to as *inverse correct response time*, is related in a monotonous fashion to the ITR of the system (i.e. information transfer per unit of time). Denoting the number of actions to be completed as s, accuracy of the initial system as Acc (without the error detection feature), the duration required for a user to complete an action as t (i.e. for how many seconds will the boxes flicker in SSVEP setup – refer to Section 11.4.1), the recall of correctly interpreted actions (actions that were interpreted by the system as the user intended) as $Re(c)$, the recall of erroneous actions (actions that where miss-interpreted by the initial system) as $Re(e)$ and the time needed for the user to transition from the erroneous state to the initial state (navigation panel) as d, we calculate the time needed to complete s correct actions in an error-agnostic system as

$$T = s \cdot t + (d + t) \cdot s \sum_{i=1}^{\infty} (1 - Acc)^i \tag{11.17}$$

Equation (11.17) sums the time for s actions plus the extra time needed to repeat the erroneous ones till none erroneous is left. Although it is straight forward to calculate the time needed by an error-agnostic system, the calculation of time required in an error-aware system derives from the addition of four subcomponents. Considering the first stage of a simple system, where the initial system classifies the user's intentions with accuracy Acc. Then the EDS detects the errors with a true positive (TP) rate, a false positive (FP) rate, a false negative (FN) rate and a true negative (TN) rate. In the first case (TP), the user's intention was correctly interpreted by the initial system and the EDS did not detect any erroneous action (e.g. an ERP when the selection was presented to the user). These trials do not need to be repeated. In the case of FP, where the initial system falsely interpreted the user's intention and the EDS did not manage to detect this miss-interpretation, the user needs d time to undo the previous action and t time to repeat the action. In the case of FN, where the initial system correctly classified user's intention but it was considered falsely as a miss-interpretation by the EDS, the user just needs t time to repeat the selection, since the cancelling of the previous action is performed automatically by the EDS. Finally, in the case of TN, where the initial system erroneously interpreted the user's intention but the EDS was able to capture this miss-interpretation, the user needs t time to repeat the selection action. In all the cases, there is an additional time e that is essential for the error detector (in our case, this time corresponds to the time needed for ErrPs to be elicited), which is added to the time t of each trial.

$$T_{TP} = Acc \cdot s \cdot Re(c) \cdot (t + e) \tag{11.18}$$

$$T_{FP} = (1 - Acc) \cdot s \cdot (1 - Re(e)) \cdot (t + e + d) \tag{11.19}$$

$$T_{FN} = Acc \cdot s \cdot (1 - Re(c)) \cdot (t + e) \tag{11.20}$$

$$T_{TN} = (1 - Acc) \cdot s \cdot Re(e) \cdot (t + e) \tag{11.21}$$

Equations (11.18)–(11.21) describe the amount of time needed during the first pass of action (after the operation of the simple system's classifier) modulated by the EDS. In order to calculate the total time required by the error-aware system to successfully interpret s actions, we recursively compute the previous equations substituting s with the number of actions that need to be repeated. This recursive computation leads to the following formula:

$$\begin{aligned} T = {} & s \cdot (t + e) + [Acc \cdot s \cdot (1 - Re(c)) \cdot (t + e) \\ & + (1 - Acc) \cdot s \cdot (1 - Re(e)) \cdot d] \cdot \sum_{i=1}^{\infty} (1 - Acc \cdot Re(c))^{i-1} \end{aligned} \tag{11.22}$$

Finally, ICRT is defined to be the number of actions times the inverse of the already calculated total time

$$ICRT = \frac{s}{T} \tag{11.23}$$

This approach presents a generic framework that quantifies the efficiency of a classifier-based system augmented by an error-detection procedure. It can be easily shown that the sums of (11.17) and (11.23) are geometric series and converge when Acc or $Re(c)$ does not equal to 0 [23].

11.4 An error-aware SSVEP-based BCI

11.4.1 Experimental protocol

The experimental protocol is based on an SSVEP-based paradigm concerning the selection of five boxes in the context of a website with a flickering frequency of 60/9, 60/8, 60/7, 60/6 and 60/5 Hz, respectively (these flickering frequencies are a result of the monitor's display rate of 60 Hz) [24]. Each participant was asked to select one among the five magenta boxes flickering at different frequencies via an auditory indication. After a flickering duration of 5 s, each box would stop flickering and a preview of the box that was selected by the system was presented for 2 s to the participant by turning the colour of the selected box from magenta to green (Figure 11.1). In the case that the previewed box was different than the one the subject was asked to observe, it becomes obvious that we would expect an ERP complex in the recorded EEG signal a few ms (200–800) after the preview. In order

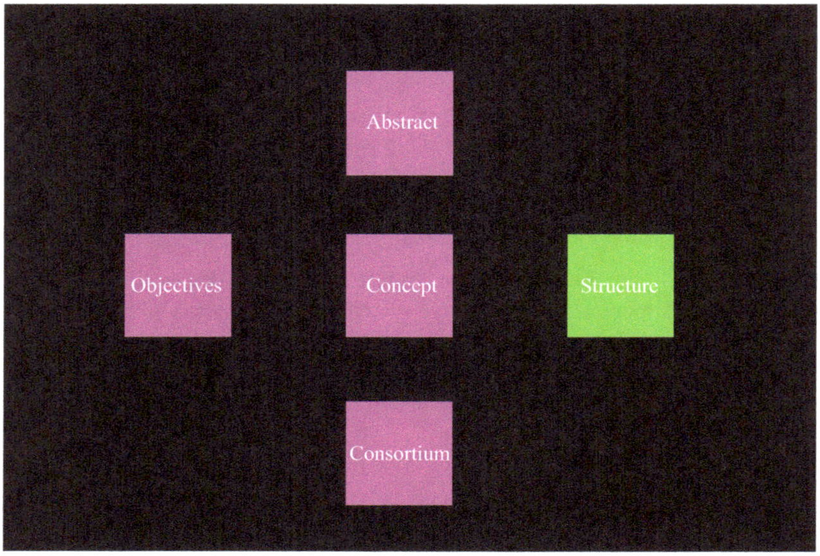

Figure 11.1 Selection preview for the SSVEP-based interface

to maintain a similar ratio of correct and error trials for each participant, we opted for a predetermined classifier to select the boxes (with a ratio of 70% correct and 30% erroneous). The participants were not aware of the pseudorandom classifier so as to get a natural response to the unexpected erroneous trials.

Compared to the standard experimental protocols for ErrPs detection and despite the fact that the participants were instructed to confine any eye movement during the SSVEP detection and the feedback stimulations, their attention unavoidably is drawn towards the selected box that turns green. Thus, during erroneous actions only, there is an unintentional gaze shift towards the selected box (i.e. above, bellow, on the left or on the right). Nevertheless, this issue has been already investigated in [25] leading to the conclusion that the generated ErrPs are not modulated by this occasional gaze shift.

11.4.2 Dataset

The signal recordings were performed with the EBN cap (64 electrodes – Figure 11.2) at a sampling rate of 128 Hz. Five healthy subjects, already experienced in SSVEP-based BCIs, participated voluntarily in the study, all male, right-handed with their age varying from 26 to 37 years. Each subject performed a total of 100 trials (20 trials per flickering frequency), out of which 70 were correct and 30 erroneous. After the starting of the selection preview, which lasted for 2 s, the EEG signal was recorded in order to acquire the corresponding brain responses and then the system would redirect the participant to the selected option so that the user could move on to the next trial.

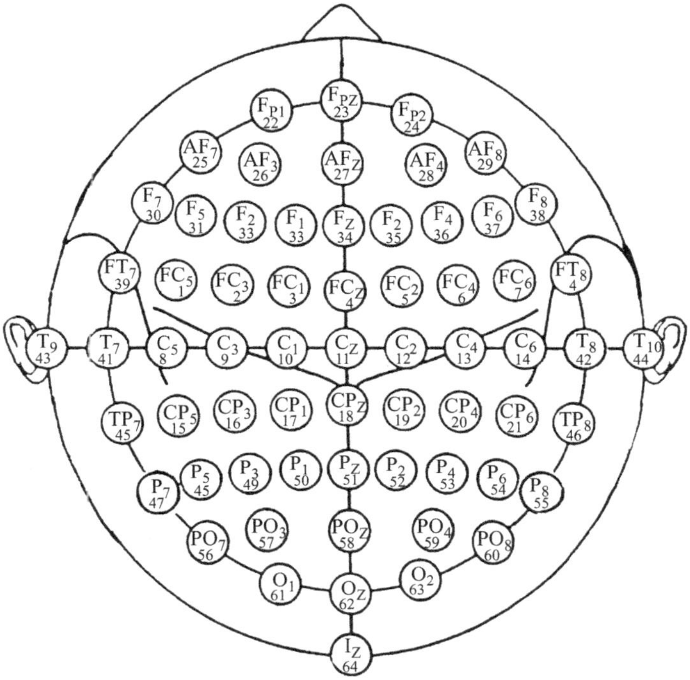

Figure 11.2 Spatial location of electrodes for the EBN headcap

11.4.3 Implementation details – preprocessing

Here, we present information regarding the data analysis. Initially, zero-phase filtering (1–13 Hz so as to include δ, θ and α brain oscillatory activity) was applied on the signals. Since the reference electrode of the EEG device we used is close to the anticipated area of interest, the EEG signals have been re-referenced by removing the average overall electrodes from each electrodes for each time instant, a procedure referred to as common average re-reference. For the ErrPs, the EEG signal between 0.2 and 0.8 s after the preview of the selection was isolated and employed, since this is the most prominent time frame for the ERP detection. Then, we performed an outlier detection procedure by employing the function *robustcov* in MATLAB® that was applied with default parameters. In order to evaluate the spatial filtering methods, six electrodes (F8, FCz, CPz, AF3, AF4, F7) were employed (Figure 11.2), located mainly over the frontal brain areas in an effort to cover the anterior cingulate cortex. As the default classification scheme, we used the support vector machines (SVMs) using polynomial kernel of third degree. The detection rates of SVMs (accuracy, precision and recall) were used in order to evaluate the contribution of each setting. All of the processing steps (namely outlier detection, spatial filtering and classification) were applied and evaluated by means of 10-fold cross validation, in a personalized manner, tailored for each participant individually.

In order to calculate the spatial filters, we treated the EEG as a collection of multiple 6-dimensional vectors (where the dimensionality from the number of employed sensors) defying the information underlying in the temporal domain. Then, these 6-dimensional vectors were considered independent observations for input in the sCRP algorithm. Since sCRP is a computational intensive task during the training process, the signals were subsampled, in order to reject redundant information as well as to reduce the training time.

11.4.4 Results

11.4.4.1 Visual representation

In Figure 11.3, we depict the average response across all trials of one participant and for the six utilized electrodes. The red line corresponds to the average signal of the erroneous trials (i.e. recordings that are expected to contain an ERP signal), the blue line corresponds to the correct trials, while the green line is their relative difference (erroneous minus correct). By visually inspecting the aforementioned signals, we can

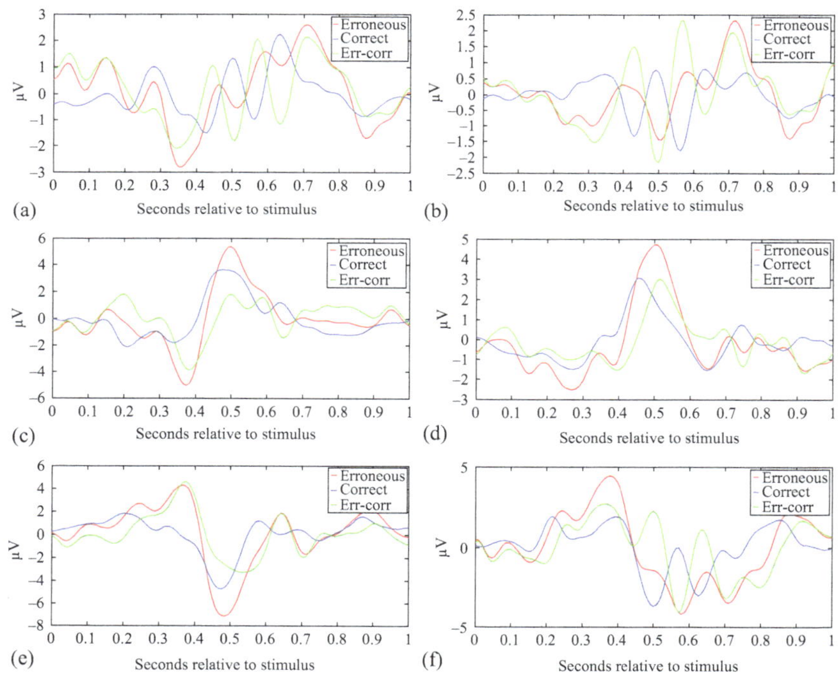

Figure 11.3 Average responses for correct, error trials and their difference. Time (x-axis) is relative to preview onset. (a) AF3, (b) AF4, (c) F7, (d) F8, (e) FCz and (f) CPz

easily find out two peaks at, approximately, 350 and 450 ms, only during erroneous responses, over the frontal area (F7, F8 and FCz). The observed contradiction in polarity (peaks at FCz are characterized by an antithetical sign) is justified by the common average re-reference procedure. By performing a direct comparison of the signals with the literature reports, it becomes evident that the two components are time delayed about 100 ms, which could be possibly explained by the previous condition during which the brain was performing a completely different task and habituated [26]. Although it is clear that the two states (erroneous and correct) have evident differences in their average forms, when SNR is significantly improved, single-trial analysis, which is the case we are mostly concerned of, is much more complicated.

11.4.4.2 Spatial filtering evaluation

Baseline comparison

In an effort to assess the benefit of sCRP as a spatial filtering approach, we initially compare its performance against a baseline configuration, where no spatial filtering is applied. Table 11.1 shows the average performance, across five participants, of the ERP-detection system. Results are presented for three distinct spatial configurations: FCz (which is the most prominent electrode for ERP detection), all used electrodes by means of concatenating the signals and finally when sCRP is applied as a spatial filter. As we can see, in the first two rows, when the six channels are employed, the corresponding results are significantly better in terms of ICRT, compared to the results of the FCz which is the most prominent electrode for the ERP detection. This fact serves as an indication that information, necessary for ERP detection, stems from various channels (at least those surrounding the brain area of interest), although it may not always be evident by visually inspecting their average forms. The third row corresponds to the results when sCRP is employed as a spatial filter. The trade-off between the $Re(c)$ and $Re(e)$ is captured by the ICRT value which is increased in the sCRP case. We should note that the used Acc value for ICRT calculation is selected so as to match the accuracy of the SSVEP system during the data-collection procedure.

Assessing the contribution of error-related potentials

Our next objective in this section is to showcase the benefit of having an EDS in a web-site BCI, with respect to the ICRT. Since the SSVEP system's detection rates are inextricably connected to the duration of the trial, we present results (by means of figures) that show the benefit of the sCRP approach in three different scenarios that diverge in time needed for the SSVEP detection to operate (e was set to 0.25 s

Table 11.1 Baseline comparison

	Pr(c)	Re(c)	Pr(e)	Re(e)	ICRT
FCz	0.772	0.8743	0.5572	0.3933	0.5790
Six ch.	0.7403	0.9678	0.7562	0.1912	0.6062
sCRP	0.7346	0.9844	0.8703	0.1555	0.6107

ICRT settings: for the ICRT calculation we assumed $t = d = 1$, $Acc = 0.7$ and $e = 0.25$

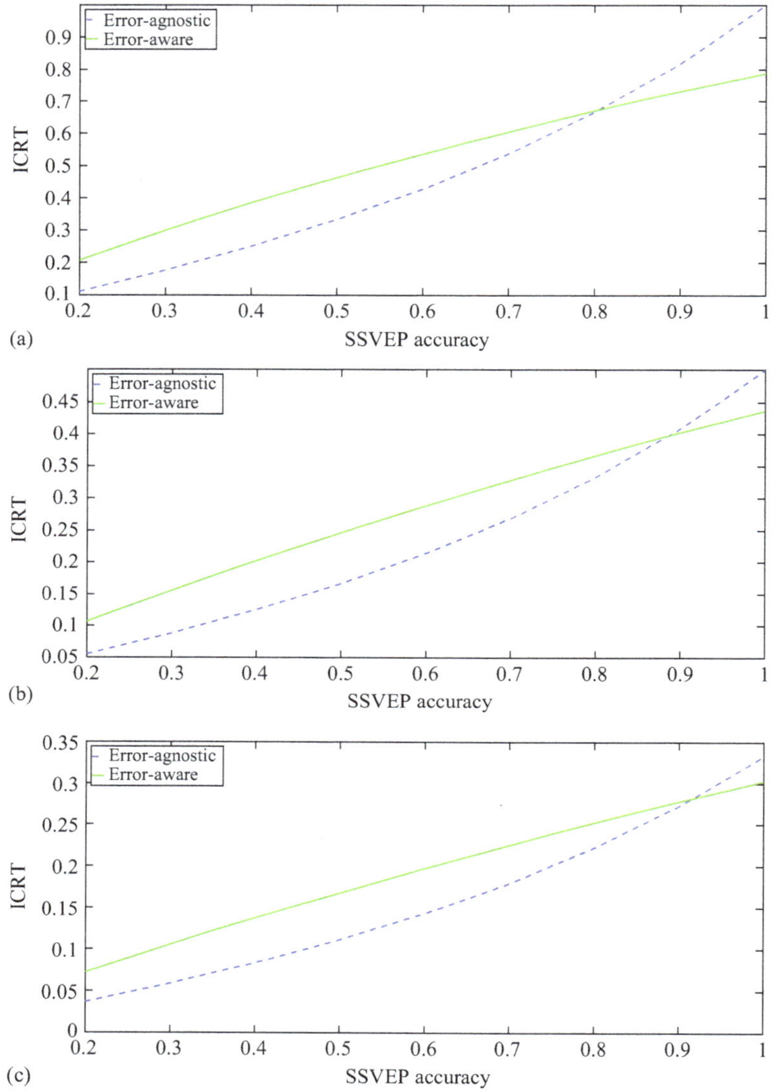

Figure 11.4 The average ICRT across all participants for the error-aware system (green solid line) compared to the error-agnostic (blue dashed line) as function of SSVEP's accuracy: (a) t=1 s, (b) t=2 s and (c) t=3 s

in this case, which is approximately the expected time for ErrPs to be elicited). In order to depict the obtained results, we plot the ICRT of the systems with respect to the accuracy of the SSVEP detection, which serves as an independent variable. The results are shown in Figure 11.4(a)–(c) for the three durations (t) of SSVEP.

As depicted in the figures, the ERP system is significantly improved by means of ICRT in the case where the SSVEP-system accuracy is below a certain threshold, which depends on the SSVEP trial length (t). This threshold corresponds to the intersection of the simple SSVEP system (blue dashed line) and the SSVEP-ERP system (green solid line). The presented results refer to the sCRP spatial filtering method applied to enhance the ERP detection.

Apart from the reliability of both the error-agnostic and EDS, there is one more parameter that plays an important role in the efficiency of an error-aware system. The error awareness introduces a time delay during each action that is unavoided (in the presented case, the system pauses till the elicitation and detection of ErrPs). The shortest the duration an SSVEP system needs to operate (i.e. short-time segments for flickering box detection), the more is affected by this delay. Even in the case of an infallible ERP-detection system, it should not be expected to improve any SSVEP system. Intuitively, the use of an error detection (which in our case is realized by detecting ERP signals) is justified when three conditions are present: (a) the operation of the error-agnostic system (which in our case is a SSVEP interface) is error-prone, (b) the accuracy of EDS is high and (c) the introduced time-delay is insignificant compared to the time needed for the error-agnostic system to operate.

11.5 An error-aware gaze-based keyboard

In this section, we investigate whether EEG can be used combined with an eye-tracker towards the realization of a high-speed gaze-based keyboard. This section proceeds in two distinct directions. Initially, we demonstrate that a specific neurophysiological event associated with error perception is elicited just after the realization of a type-setting error by the user. A related activation pattern, free of eye-related artefacts, can be reliably detected and employed to indicate the errors. Second, we provide evidence that this neurophysiological event, which is a special case of an ERP, constitutes the basis for an automated EDS that stems directly from the user's brain responses and can be complemented by information extracted from the eye-movement patterns to further improve its effectiveness. Apart from verifying the theoretical improvement by means of typing speed in a gaze-based keyboard, we also present experimental results from an online simulation, where the presented approach is compared against a regular gaze-based typesetting scenario, without any error-detection assistance. To our best knowledge, [27] was the first study that attempted to combine gaze-based typesetting with ErrPs' detection and demonstrated that such an integration holds promise for an enhanced user experience.

11.5.1 Methodology

In our experiments, the electrical brain activity and gazing location were recorded in a continuous aspect, while the subject was performing a gaze-based typing task (Figure 11.5). The data acquisition protocol was specifically designed so as to uncover the physiological patterns related to the perception of an erroneous visual key press

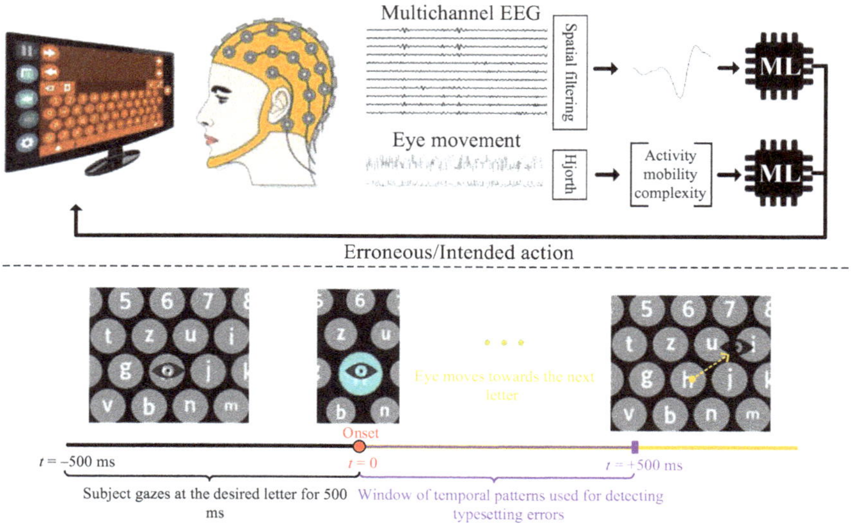

Figure 11.5 (Top) Schematic outline of the error-aware keyboard. A hybrid BCI system, relying on brain activity patterns and eye-movement features, detects and deletes characters that are mistyped. The depicted machine-learning (ML) modules correspond to linear SVMs. (Bottom) Timeline describing the sequence of events during the typesetting experiment. Initially, the participant starts gazing at the desired letter. When he completes a 500 ms time-interval of continuous gazing, the key is registered and simultaneously the associated visual indication is presented. The physiological responses following this indication are used to detect typesetting errors. We note that the 'eye' icon was not presented in the experiments and it is only shown here for presentation clarity purposes

(i.e. unintentional key registration). The discovered patterns that stem from the participant's brain and eye spontaneous activity are then used in a machine-learning scheme so as to implement a gaze-based keyboard with an automatic error-detection, and potentially error-correction, capability.

11.5.2 Typing task and physiological recordings

The data acquisition protocol was based on a standard gaze-based keyboard paradigm that was implemented by an eye-tracker attached to a pc monitor [28,29]. The gazing information, in the form of a densely sampled sequence of *x–y* coordinates corresponding to the screen's eye-trace, was registered simultaneously with the participant's EEG. The motivation behind this experimental setup was to obtain data where patterns in the physiological activity, of either brain or/and eyes, could be associated with the case of an unintentional key registration (due to either the inaccuracy of the eye-tracker or a human mistake). In the study of event-related neurophysiological responses, the precise timing is of paramount importance. For this reason, the operational functionality of the gaze-based typesetting system had to be modified. Typical gaze-based keyboards employ a visual indication in order to continuously inform the user about the corresponding on-screen gaze location. A visual key is registered, only, after the user has continuously gazed at it for a specific amount of time, usually referred to as dwell time. However, this visual feedback informs the user on the typing result at arbitrary times and as such the ErrPs are not time-locked to the registration of the visual key. This feature of continuous visual feedback was deactivated during our experimental protocol so as to ensure that evoked brain responses, time-locked to the perception of an erroneous typesetting, would be elicited. It was only after a stared key had been typed (or, equivalently, gazed at for more than 0.5 s) that appeared as selected. In this way, the perception of a typo could be related to a specific timestamp (i.e. the onset of a wrong selection was the trigger mechanism for an ERP response).

Twenty sentences were provided, in a sequential fashion, to the participants with the instruction to type them with the adjusted gaze-based keyboard. The sentence that had to be typed was not accessible to the subjects during the typesetting; hence, they had to memorize it at the beginning of each attempt. The purpose behind this experimental setup was to bring the subjects closer to the natural way of spontaneous typing. The only difference compared to a regular typesetting mode was that the participants were instructed to refrain from using the backspace button and ignore typos, since we were interested in physiological events associated with error perception and not in those related to reaction. All sentences were written using lower-case letters with a full stop at the end of each sentence. Each session, which consisted of typing one sentence, was followed by a short time-break.

11.5.3 Pragmatic typing protocol

Since it was essential to compare the individuals' typing performance with and without the EDS and concerning the time required to correctly type a given sentence, we performed an additional round of control experiments. During this round, each participant had to visually type the same sentences, but this time using the gaze-based keyboard in its regular mode (without error-detection assistance) and with the instruction that each sentence should be regarded as complete only when it had been correctly typed. In this regular typing mode, participants were allowed to use the backspace button during the typesetting; all other parameters remained the same.

11.5.4 Data analysis

The concurrent data streams (after the essential pre-filtering for the EEG signals) were segmented into epochs. Each epoch corresponded to the physiological responses starting 200 ms before the onset of visual key pressed and lasting for 700 ms. Since the epochs corresponding to the erroneous responses were much less than the correct ones, the two groups (erroneous and correct) were not equally represented. To alleviate this shortcoming, we employed the SMOTE [30] method that creates synthetic samples so as to populate the minority group. In order to analyse the brain signals, the approach of discriminant spatial patterns [15] was employed so as to improve the detectability of ErrPs. For analysing the gaze-related signals, the Hjorth descriptors [31] were employed for characterizing the derived time series that stems from the successive displacements of the gazing position. Two independent models that relied on SVMs with linear kernel [32] were trained in order to discriminate between correct and erroneous typesettings. The one used the feature vectors extracted from the single epochs of brain activity and the other on the feature vectors extracted from the associated epochs of gaze-related activity. The individual outputs were fused to realize the final error detection.

11.5.5 System adjustment and evaluation

Accuracy is typically employed in classification tasks to assess the performance of the system. However, in imbalance class situations (which is the most probable scenario for our EDS), sensitivity/specificity pairs offer a more robust assessment of a given classification model. The Utility metric [33] is a suitable composite measure, which complementary to specificity and sensitivity considers the dwell typing time modulated by the mistyping probability as well. It is widely employed for assessing the performance of BCI-spellers, and here it is used for tuning the EDS along with providing the final unbiased justification of it is gain via a Monte Carlo cross-validation scheme.

After the classification scheme was decided, on data obtained from the first typing task (i.e. using the error-aware keyboard), an error-aware simulation took place. During this simulation, the time interval for typing a given sentence in error-aware mode was calculated and then compared against the time interval using the regular gaze-based keyboard. In order to avoid any biases, in the former case, we initially trained the incorporated EDS in a leave-one-sentence-out manner (i.e. using the epochs corresponding to the rest 19 sentences) and, then, simulated 'off-line' the operation of the error-aware keyboard on the sentence. Then, taking into account both the FPs and negatives of the EDS system (and the associated gain/loss in time), the necessary time interval for our typesetting approach was predicted.

11.5.6 Results

11.5.6.1 Physiological findings

Among the main objectives of this chapter was to understand and identify the physiological responses related to the perception of an error during the gaze-based typing procedure. Figure 11.6 presents our main empirical findings about the neural correlates and the eye-motion patterns that constituted the basis for deriving error-detection patterns from the recording data streams. Using single-subject data, the averaged responses for both brain activity and an eye-movement descriptor are shown for the case of correct and erroneous typing. Concerning the brain activity patterns, it becomes evident that the main components following an erroneous response are a negative deflection approximately 300 ms (with a frontocentral scalp distribution) followed by a positive peak 100 ms later (with a centro-parietal scalp distribution). These latencies have been defined with respect to stimulus onset, which is the time instant that the gazed letter is registered and its preview is shown to the user. The depicted activation pattern topographically complies with the relevant ERP literature [2] but varies slightly in the timing. In the correct typesetting case, the brain activation presents a pattern that differentiates from the anticipated null response. Specifically, a moderate bipolar response can be observed, consisting of a positive deflection approximately at 200 ms and a negative one at about 300 ms. Its topographical representation indicates a cortical source located centrally. The explanation for the observed neural patterns naturally stems by the temporal patterning of the associated eye activity. It can be seen (Figure 11.6; bottom-most traces) that the averaged profile of eye-movement speed is suggestive of evident eye-motion only in the case of a correct typesetting, and particularly well after the gazed letter is registered. It becomes apparent that after a typing error, subjects slightly adjust their gaze, since their intention has been marginally misinterpreted, while in the opposite case, they have to type the next letter which is probably located, on the screen, far from the previously gazed position. Nevertheless, the successive typesetting of locally nearby letters is also a possible scenario (e.g. in the case of the word 'were'). Finally, it should be noted that the presented scalp topographies (presented in rightmost panels) after the correct typesetting are in accordance with cortical generators lying in a brain region that is known to be causally associated with the eye movements [34]. More importantly, they cannot be identified as ocular artefacts generated by the eye-movement.

In an effort to further examine the association between the EEG activity and gaze shifts, we employed a data analytic procedure, in which the variability in eye-movement directionality was initially coarse-grained and then used to condition the brain activity signals' grouping. Figure 11.7 includes the results from analysing the single epoch data of the same participant as in Figure 11.6. Using the aggregated gaze displacement, from −200 to 500 ms around each correct letter registration, we applied the k-means method to cluster the eye-movements into k=4 distinct directions (Figure 11.7; most left). Then, the derived grouping was applied to the eye movement speed profiles additionally to the corresponding EEG traces. Finally, the four uncovered prototypical traces, for both types of physiological data, were presented in a contrasting manner (Figure 11.7; right bottom and top). It becomes apparent that EEG-traces do not show any polarity inversion with the change of eye-movement direction (blue/red waveform corresponds to right/left) as it would have been expected in the case that the EEG signals were contaminated by ocular activity artefacts. On the contrary, the temporal pattern as recorded at the Cz sensor follows the eye movement speed profile.

11.5.6.2 SVM classification for predicting typesetting errors from physiological activity

Having investigated the ability of brain activity and gaze-movement patterns for distinguishing between correct and mistyped (unintentionally) letters, we moved towards implementing the idea of error detection by means of a machine-learning

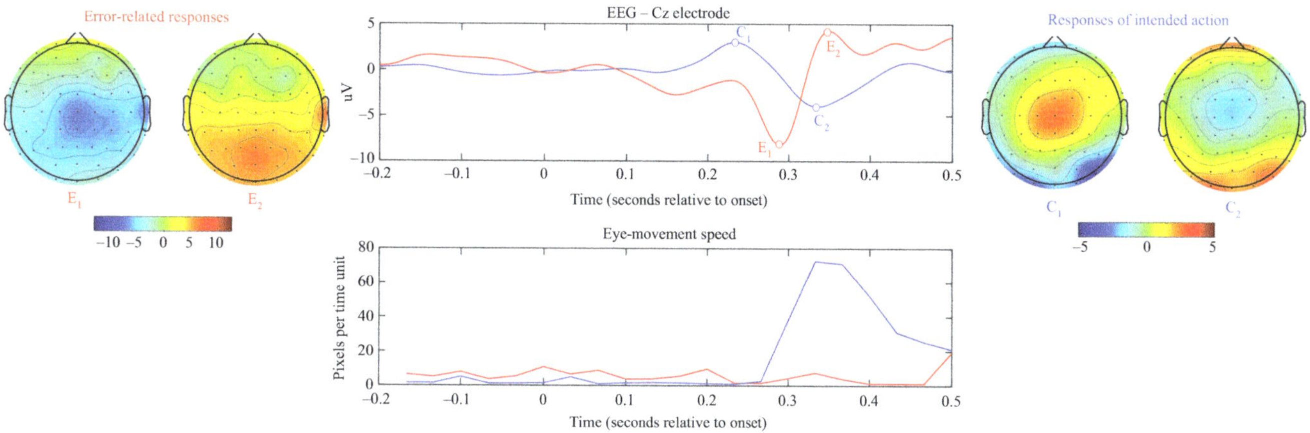

Figure 11.6 *Single-subject averaged brain activation traces for the correct (blue) and wrong (red) selections of buttons are shown in the top middle panel. Particular latencies are indicated on these traces (E: error; C: correct) and the corresponding topographies have been included in the top left/right panels. The traces shown in the bottom middle panel reflect eye-movement activity (derived by averaging correspondingly across the epochs of gaze-related signal). Zero time indicates the instant that the typing of the current letter has been completed and the eyes are free to move towards the next letter*

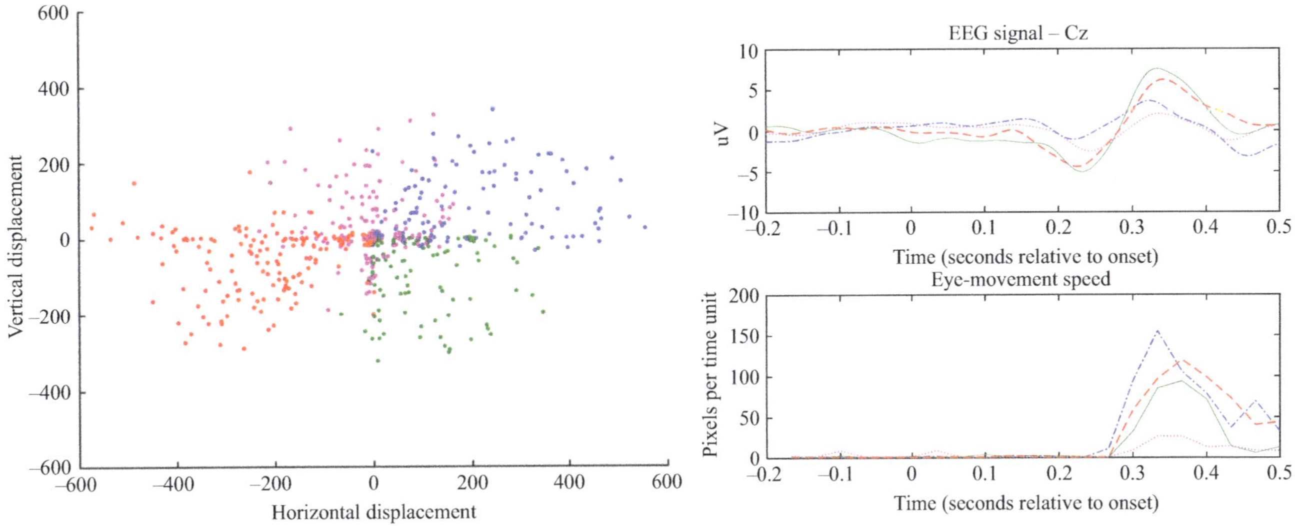

Figure 11.7 *(Left) Scatter-plot of gaze centre displacements (derived by integrating the derivatives of eye position coordinates within a time interval that includes the key registration at 0 latency). Each dot indicates the main direction of the eye after a correct typesetting of a single letter. The point swarm has been partitioned into four groups, and the membership of each dot is indicated by colour. The associated brain-signal and eye-movement activity traces have been grouped accordingly and their (sub)averages are indicated in top right and bottom right, respectively, using the colour code defined in scatter-plot*

algorithm. Features extracted from both the EEG-related epochs or/and the concurrent gaze-based traces were employed in the context of an SVM-classification setting. Based on the type of the features and the way they were combined, four different classification schemes were implemented and assessed using the widely known performance indicators of sensitivity and specificity. In the first setting, only EEG features were taken into account, while in the second only eye movement descriptors were considered. In the other two settings, namely 'early' and 'late' fusion, the error-detection was based on both data streams. The corresponding features were concatenated and fed to a single SVM classifier in the former case, while in the latter case they were fed separately into two separate SVMs where the most dominant confidence score was selected to indicate the corresponding classification result. Table 11.2 shows that the classification using the brain activity patterns tends to emphasize specificity, while eye-movement descriptors emphasize sensitivity.

11.5.6.3 Incorporating the SVM classifier(s) in the gaze-based keyboard

The accurate detection of error responses plays a critical role in the realization of an error-aware gaze-based keyboard. Hence, the classification accuracy of the employed SVM-algorithm, alone, was not enough to fully justify the developed EDS. This becomes more apparent when considering that the typesetting errors occur at low probability (approximately one out of ten characters) and hence the SVM model has to deal with an imbalanced classification task. The Utility metric offers an intuitive way to weight meaningfully both the sensitivity and specificity while, simultaneously, to appropriately consider the cost of typing time and the error chance. This metric served as an (inverse) loss function to optimize the functionality of the linear SVM model, by adjusting the offset of the separating hyperplane. As a result, the classifier that was incorporated in the error-aware keyboard was not designed to perform optimally regarding the accuracy of error detection task but was, instead, designed so as to lead to higher Utility gain (the ratio $\text{Utility}_{\text{EDS}}/\ \text{Utility}_{\text{regular}}$). Figure 11.8(a) depicts the standard ROC curves for the four classification schemes, using as threshold a value within $[-1,\ 1]$ for classifying a sample either as erroneous or correct. It becomes apparent that the late fusion setting shows always superior performance with respect to specificity and sensitivity. Figure 11.8(b) shows the corresponding Utility gain with respect to the threshold. A constantly increasing gain is related to lowering the threshold in all four settings. This trend in gain is followed by an increased specificity, which is apparently more significant than the sensitivity in the case of error detection. As expected, lowering the threshold beyond the value of -1 results in a gain loss. Summarizing, when the two SVMs in the late fusion settings use the most negative threshold value, which is -1, the best performance of the error-aware gaze-based keyboard will be obtained.

Table 11.3 presents a more detailed picture of the results obtained at the optimal threshold by means of Utility gain for all subjects. The following empirical facts need to be underlined. EDS based on features extracted from the EEG responses, but not from the corresponding eye-related activity, may lead to improved user's experience

Table 11.2 Performance metrics for the classification task of discriminating between correct and erroneous typesetting based on EEG traces and gaze-movement patterns (used both separately and jointly). Tabulated are the results averaged from 100 repetitions of Monte Carlo cross validation. Four implementation scenarios have been validated, for each subject independently

Subject ID	**EEG**		**Eye motion**		**Early fusion**		**Late fusion**	
	Sensitivity	**Specificity**	**Sensitivity**	**Specificity**	**Sensitivity**	**Specificity**	**Sensitivity**	**Specificity**
S01	73.41	83.19	97.82	51.39	78.95	85.96	91.51	78.00
S02	75.54	87.97	90.14	71.94	80.48	89.34	85.55	87.00
S03	85.52	93.20	89.36	82.83	87.40	94.00	89.50	93.13
S04	74.76	85.18	87.45	70.26	80.13	85.18	83.84	81.92
S05	76.38	85.90	88.34	75.71	82.58	89.83	85.25	89.96
S06	76.69	80.09	95.68	63.95	88.92	75.82	92.65	73.69
S07	61.61	78.79	83.98	67.89	71.71	77.79	79.37	75.57
S08	69.31	83.76	94.21	71.35	89.87	85.30	90.26	83.76
S09	67.39	82.36	87.73	77.18	78.64	83.94	82.71	83.70
S10	70.47	81.41	80.77	80.56	79.35	80.80	79.76	82.69
Average	73.10	84.18	89.54	71.30	81.80	84.79	86.04	82.94

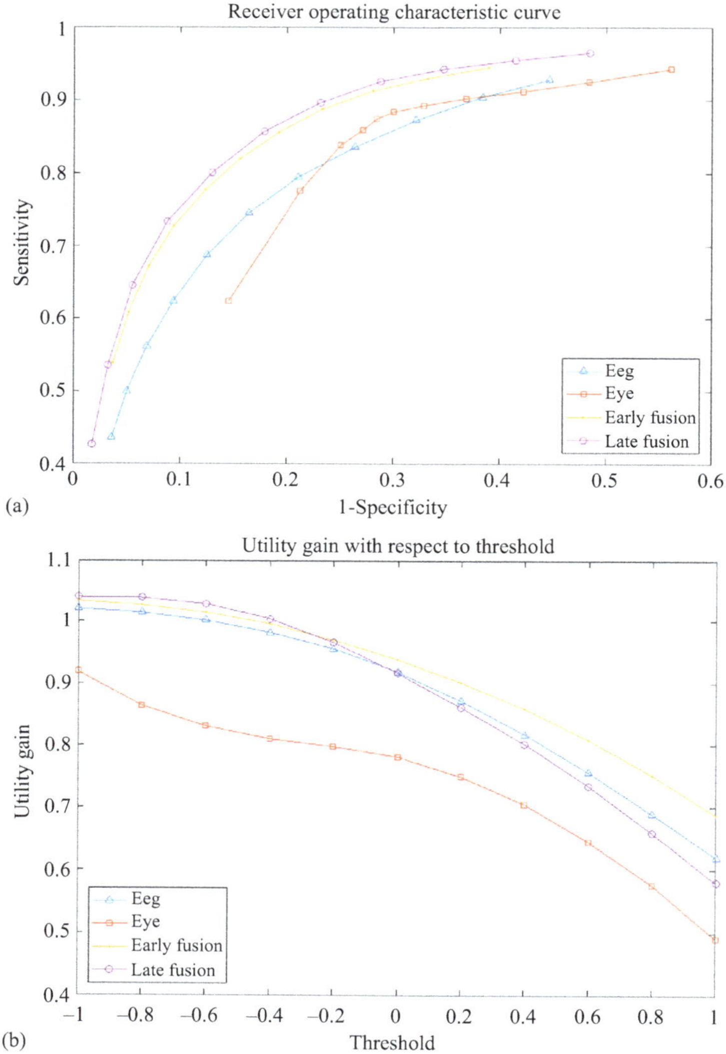

Figure 11.8 The grand average sensitivity and specificity values (a) along with the Utility gain (b), after 100 Monte Carlo cross-validation repetitions, with respect to threshold moving within the normalized SVM margins

by means of time efficiency (Utility gain higher than one). The late fusion EDS scheme systematically provides better results, which on average correspond to a 4% increase in Utility gain. Finally, all accuracies are well above the random chance (which stems from the typing error chance).

Table 11.3 Columns 2–6: chance that the eye-tracker will interpret user's intention falsely and Utility gain for the four classification schemes. Columns 7–9: classification performance metrics at the optimal threshold for the late fusion classification scheme. Tabulated are the results averaged from 100 repetitions of Monte Carlo cross validation

Subject ID	Typing error chance (%)	Utility gain				Specificity	Sensitivity	Accuracy
		EEG	Eye motion	Early fusion	Late fusion			
S01	11.60	1.02	0.69	1.05	1.05	98.64	45.60	92.48
S02	7.92	1.01	0.93	1.02	1.03	98.55	45.40	94.33
S03	6.36	1.04	0.99	1.04	1.04	99.28	71.00	97.48
S04	11.60	1.04	0.93	1.04	1.05	98.53	46.50	92.48
S05	8.87	1.02	0.91	1.04	1.04	98.94	50.60	94.66
S06	13.67	1.03	0.92	1.03	1.03	96.68	39.20	88.81
S07	13.50	0.99	0.92	0.99	1.02	97.97	24.10	87.99
S08	9.24	1.00	0.94	1.03	1.03	98.89	28.70	92.39
S09	6.16	0.98	0.91	0.99	1.01	97.66	35.40	93.82
S10	13.84	1.03	1.01	1.05	1.05	98.84	31.20	89.44
Average	10.27	1.02	0.92	1.03	1.04	98.40	41.77	92.38

11.5.6.4 Simulated implementation of the EDS

In an effort to compare the time efficiency of the error-aware gaze-based keyboard against the error-agnostic one, we contrasted the time needed for users to type a given sentence with our EDS system erasing automatically the erroneously typed letters and then retyping the letters immediately (T1 task), with the time needed for them to use the backspace button in order to correct the errors (T2 task). Table 11.4 shows the average (AVG) essential times regarding the regular gaze-based keyboard, for each subject individually, and the corresponding difference with the error-aware typing scheme. It becomes evident that, on average, users require 2.7 s less to type a sentence which would require 29.36 s approximately. This leads to an improvement of 9.3% in typing speed (which is statistically significant; p-value equals to 0.032 based on the Wilcoxon signed-rank test) that remarkably differs from the theoretical improvement of 4% that was calculated by the Utility metric. The reason for this difference is based on the fact that the Utility metric assumes identical key press times for conventional letter and backspace buttons, which is not the case according to our empirical findings. Moreover, the correlation between the obtained time gain and the user's probability of making a typo is of great interest. Actually, the gain from the EDS is upper bounded by the misinterpretation probability of the eye-tracker [18]. In other words, users who are more prone to errors during the gaze-based typing scheme tend to benefit more from the presented hybrid BCI system.

Table 11.4 Chance that the eye-tracker will interpret user's intention falsely in the two typesetting procedures (T1 and T2), the average gain in time (accompanied by the respective percentage) that is obtained by the EDS system taking advantage of the late fusion classifier and the average time required to type one sentence in both tasks. The results are obtained according to a Leave-one-Sentence-out cross-validation manner

Subject ID	Typing error chance (%)		Gain (T2-T1)		Average sentence time	
	T1	T2	Time (seconds)	Percentage	T1 (seconds)	T2 (seconds)
S01	11.60	6.48	1.23	4.51	26.07	27.30
S02	7.92	4.14	4.17	13.99	25.64	29.81
S03	6.36	8.74	7.33	22.96	24.60	31.93
S04	11.60	8.44	7.53	21.68	27.21	34.74
S05	8.87	9.05	4.92	16.12	25.61	30.53
S06	13.67	3.56	−0.78	−2.90	27.72	26.94
S07	13.50	5.83	0.44	1.52	28.51	28.95
S08	9.24	3.36	−0.11	−0.42	26.20	26.09
S09	6.16	2.55	−1.38	−5.41	26.88	25.50
S10	13.84	9.50	3.74	11.74	28.11	31.85
Average	10.27	6.17	2.70	9.23	26.67	29.36

11.6 Summary

Brain–computer interfaces have been widely employed for supporting alternative communication and control options in a wide variety of applications, including information recommender systems [35], spellers [36,37], robotic devices [38] and wheelchair controllers [39]. Guided by advances in machine learning, and in conjunction with the ever-increasing availability of consumer EEG scanners, BCIs are currently incorporated into multimodal systems as well, leading to improved, adaptable, versatile and natural interfaces.

The main aim of this chapter was to introduce novel brain-based error-detection paradigms (by means of multimodal BCIs) that have the potential to improve the overall user experience. We have shown that certain physiological events, associated with both brain and eye activity, can mark an unintended action and therefore serve as the basis for an error correction system. It is important to note at this point that the essence of the presented work lies in detecting users' intentions. From this perspective, there is a major difference between the presented EDSs and an automatic error-correction system (such as a proofing system) that may offer even better user experience.

References

[1] Falkenstein M, Hohnsbein J, Hoormann J, *et al.* Effects of crossmodal divided attention on late ERP components. II. Error processing in choice reaction tasks. Electroencephalography and Clinical Neurophysiology. 1991;78(6): 447–455.

[2] Chavarriaga R, Sobolewski A, Millán JdR. Errare machinale est: the use of error-related potentials in brain–machine interfaces. 2014. Frontiers in neuroscience 8 (2014): 208.

[3] Holroyd CB, Dien J, Coles MG. Error-related scalp potentials elicited by hand and foot movements: evidence for an output-independent error-processing system in humans. Neuroscience Letters. 1998;242(2):65–68.

[4] Gehring WJ, Willoughby AR. The medial frontal cortex and the rapid processing of monetary gains and losses. Science. 2002;295(5563):2279–2282.

[5] Spüler M, Bensch M, Kleih S, *et al.* Online use of error-related potentials in healthy users and people with severe motor impairment increases performance of a P300-BCI. Clinical Neurophysiology. 2012;123(7):1328–1337.

[6] Spüler M, Rosenstiel W, Bogdan M. Online adaptation of a c-VEP brain–computer interface (BCI) based on error-related potentials and unsupervised learning. PLoS One. 2012;7(12):e51077.

[7] Kreilinger A, Neuper C, Pfurtscheller G, *et al.* Implementation of error detection into the graz-brain–computer interface, the interaction error potential. In: AAATE'09: 9th European Conf. for the Advancement of Assistive Technology (Florence, Italy); 2009. p. 1–5.

[8] Ferrez PW, Millán JdR. Simultaneous real-time detection of motor imagery and error-related potentials for improved BCI accuracy. In: Proceedings of the 4th International Brain–Computer Interface Workshop and Training Course. CNBI-CONF-2008-004; 2008. p. 197–202.

[9] Spüler M, Niethammer C. Error-related potentials during continuous feedback: using EEG to detect errors of different type and severity. Frontiers in Human Neuroscience. Lausanne, Switzerland; 2015;9:155.

[10] McFarland DJ, McCane LM, David SV, *et al.* Spatial filter selection for EEG-based communication. Electroencephalography and Clinical Neurophysiology. 1997;103(3):386–394. Available from: http://www.sciencedirect.com/science/article/pii/S0013469497000222.

[11] Rivet B, Souloumiac A, Attina V, *et al.* xDAWN algorithm to enhance evoked potentials: application to brain–computer interface. IEEE Transactions on Biomedical Engineering. 2009;56(8):2035–2043.

[12] Spüler M, Walter A, Rosenstiel W, *et al.* Spatial filtering based on canonical correlation analysis for classification of evoked or event-related potentials in EEG data. IEEE Transactions on Neural Systems and Rehabilitation Engineering. 2014;22(6):1097–1103.

[13] Roy RN, Bonnet S, Charbonnier S, *et al.* A comparison of ERP spatial filtering methods for optimal mental workload estimation. In: Engineering in Medicine and Biology Society (EMBC), 2015 37th Annual International Conference of the IEEE. Piscataway, NJ: IEEE; 2015. p. 7254–7257.

[14] Liparas D, Dimitriadis SI, Laskaris NA, *et al.* Exploiting the temporal patterning of transient VEP signals: a statistical single-trial methodology with implications to brain–computer interfaces (BCIs). Journal of Neuroscience Methods. 2014;232(Supplement C):189–198. Available from: http://www.sciencedirect.com/science/article/pii/S0165027014001484.
[15] Liao X, Yao D, Wu D, *et al.* Combining spatial filters for the classification of single-trial EEG in a finger movement task. IEEE Transactions on Biomedical Engineering. 2007;54(5):821–831.
[16] Yan S, Xu D, Zhang B, *et al.* Graph embedding and extensions: a general framework for dimensionality reduction. IEEE Transactions on Pattern Analysis and Machine Intelligence. 2007;29(1):40–51.
[17] Yang W, Wang Z, Sun C. A collaborative representation based projections method for feature extraction. Pattern Recognition. 2015;48(1):20–27.
[18] Kalaganis FP, Chatzilari E, Nikolopoulos S, *et al.* A collaborative representation approach to detecting error-related potentials in SSVEP-BCIs. In: Proceedings of the on Thematic Workshops of ACM Multimedia 2017. New York, NY: ACM; 2017. p. 262–270.
[19] Luck SJ. An Introduction to the Event-Related Potential Technique. MIT Press; 2014.
[20] Duda RO, Hart PE, Stork DG. Pattern Classification (2nd Edition). Hoboken, NJ: Wiley-Interscience; 2000.
[21] Rao CR. The utilization of multiple measurements in problems of biological classification. Journal of the Royal Statistical Society Series B (Methodological). 1948;10(2):159–203. Available from: http://www.jstor.org/stable/2983775.
[22] Kalaganis F, Chatzilari E, Georgiadis K, *et al.* An error aware SSVEP-based BCI. In: 2017 IEEE 30th International Symposium on Computer-Based Medical Systems (CBMS). Piscataway, NJ: IEEE; 2017. p. 775–780.
[23] Spivak M. Calculus. Houston, TX: Publish or Perish Inc.; 1994.
[24] Volosyak I, Cecotti H, Gräser A. Impact of frequency selection on LCD screens for SSVEP based brain–computer interfaces. Bio-Inspired Systems: Computational and Ambient Intelligence. In: Proceedings of the 5th International Work-Conference on Artificial Neural Networks. Springer, Berlin, Heidelberg. 2009;706–713.
[25] Ferrez PW, Millán JdR. Error-related EEG potentials generated during simulated brain–computer interaction. IEEE Transactions on Biomedical Engineering. 2008;55(3):923–929.
[26] Schmid S, Wilson DA, Rankin CH. Habituation mechanisms and their importance for cognitive function. Frontiers in Integrative Neuroscience. 2014;8.
[27] Kalaganis FP, Chatzilari E, Nikolopoulos S, *et al.* An error-aware gaze-based keyboard by means of a hybrid BCI system. Scientific Reports. 2018; 8(1):13176.

[28] Menges R, Kumar C, Sengupta K, *et al.* eyeGUI: a novel framework for eye-controlled user interfaces. In: Proceedings of the 9th Nordic Conference on Human–Computer Interaction. New York, NY: ACM; 2016. p. 121.

[29] Menges R, Kumar C, Müller D, *et al.* GazeTheWeb: a gaze-controlled Web browser. In: Proceedings of the 14th Web for All Conference. W4A'17. New York, NY: ACM; 2017. Available from: http://dx.doi.org/10.1145/3058555.3058582.

[30] Chawla NV, Bowyer KW, Hall LO, *et al.* SMOTE: synthetic minority over-sampling technique. Journal of Artificial Intelligence Research. 2002;16: 321–357.

[31] Hjorth B. EEG analysis based on time domain properties. Electroencephalography and Clinical Neurophysiology. 1970;29(3):306–310.

[32] Parra LC, Spence CD, Gerson AD, *et al.* Recipes for the linear analysis of EEG. NeuroImage. 2005;28(2):326–341.

[33] Dal Seno B, Matteucci M, Mainardi LT. The utility metric: a novel method to assess the overall performance of discrete brain–computer interfaces. IEEE Transactions on Neural Systems and Rehabilitation Engineering. 2010;18(1):20–28.

[34] Stuphorn V, Brown JW, Schall JD. Role of supplementary eye field in saccade initiation: executive, not direct, control. Journal of Neurophysiology. 2010;103(2):801–816.

[35] Eugster MJ, Ruotsalo T, Spapé MM, *et al.* Natural brain-information interfaces: recommending information by relevance inferred from human brain signals. Scientific Reports. 2016;6:38580.

[36] Dal Seno B, Matteucci M, Mainardi L. Online detection of P300 and error potentials in a BCI speller. Computational Intelligence and Neuroscience. 2010;2010:11.

[37] Bin G, Gao X, Wang Y, *et al.* A high-speed BCI based on code modulation VEP. Journal of Neural Engineering. 2011;8(2):025015.

[38] Meng J, Zhang S, Bekyo A, *et al.* Noninvasive electroencephalogram based control of a robotic arm for reach and grasp tasks. Scientific Reports. 2016;6.

[39] Matsuzawa K, Ishii C. Control of an electric wheelchair with a brain–computer interface headset. In: Advanced Mechatronic Systems (ICAMechS), 2016 International Conference on. Piscataway, NJ: IEEE; 2016. p. 504–509.

Chapter 12

Multimodal BCIs – the hands-free Tetris paradigm

Elisavet Chatzilari[1], Georgios Liaros[1], Spiros Nikolopoulos[1], and Ioannis Kompatsiaris[1]

In this chapter, we will explore ways to integrate three natural sensory modalities, i.e. vision, brain commands and stress levels into a single visceral experience that allows simultaneous control of various interface options. In this direction, we present MM-Tetris, the multimodal reinvention of the popular Tetris game, modified to be controlled with the user's eye-movements, mental commands and bio-measurements. MM-Tetris is intended for use by motor-impaired people who are not able to operate computing devices through the regular controllers (i.e. mouse and keyboard). In the proposed version of the game, the use of eye-movements and mental commands works in a complementary fashion, by facilitating two different controls, the horizontal movement of the tiles (i.e. tetriminos) through the coordinates of the gaze and the tile rotation through sensorimotor rhythm (SMR) signals detection, respectively. Additionally, bio-measurements provide the stress levels of the player, which in turn determines the speed of the tiles' drop. In this way, the three modalities smoothly collaborate to facilitate playing a game like Tetris. Eventually, the design of the game provides a natural gamified interface for user training in generating more discriminative SMR signals for better detection of imaginary movements.

12.1 Introduction

Loss of the voluntary muscular control while preserving cognitive functions is a common symptom of neuromuscular disorders leading to a variety of functional deficits, including the ability to operate software tools that require the use of conventional interfaces like mouse, keyboard or touchscreens. As a result, the affected individuals are marginalised and unable to keep up with the rest of the society in a digitised world. MAMEM,* an EU H2020 funded project, aims to integrate these people back into society by increasing their potential for communication and exchange in

[1]Information Technologies Institute, Centre for Research and Technology Hellas, 57001 Thermi, Thessaloniki, Greece

*www.mamem.eu

leisure (e.g. social networks, gaming) and non-leisure context (e.g. workplace). In this direction, MAMEM delivers the technology to enable interface channels that can be controlled through bio-signals (i.e. eye-movements, brain activity and skin conductance). This is accomplished by replacing the mouse and keyboard with novel control interfaces based on three sensors; eye-tracker, electroencephalogram (EEG) recorder and galvanic skin response (GSR) sensor. Then, pattern recognition and tracking algorithms are employed to jointly translate these signals into meaningful control and enable a set of novel paradigms for multimodal interaction. In the MAMEM context, our approach in this chapter is to modify the popular Tetris game in its hands-free version by employing signals from the three sensors. In this new version of Tetris, gaze input is able to control the horizontal movement of the Tetris tiles [1,2], EEG signals are used for tile rotation through the detection of imaginary movement in the SMR brain waves [3], while GSR readings are used to adapt the speed of the game based on the stress levels of the user [4].

Our motivation for reinventing Tetris is that in the context of brain–computer interface (BCI) applications (and especially the ones relying on SMR), it is imperative to not only train the system to recognise the events related to the signals but also the user to produce distinguishable signals. In this respect, the use of an application that will allow users to train themselves needs to precede the actual use of a BCI. Typical SMR training approaches consist of presenting the user with arrows as cues to indicate what type of movement they are asked to perform mentally while providing feedback as to what movement the system has detected. This feedback allows the users to adapt their mental strategy by optimising their mental movement command in order to maximise the detection performance of the feedback mechanism. However, this is a tedious process and the users get bored and frustrated before mastering their SMR-signal generation to a sufficient extent. Alleviating this problem, we reinvent the popular Tetris game to serve as an SMR training application that will engage the user to keep training themselves. In the following, we present the gameplay of MM-Tetris and how we replace the keyboard-based controls with bio-measurement-based controls (Section 12.2), the used algorithms and the challenges (Section 12.3) and our experiments towards generating a satisfying experience for the user (Section 12.5).

12.2 Gameplay design

Tetris is a popular puzzle video game that relies on the placement of tiles on a board. These tiles have different geometric shapes composed of four square blocks each and are also known as tetriminos. A random sequence of tetriminos fall down the board. The objective of the game is to manipulate these tetriminos, by moving each one sideways and/or rotating by quarter-turns, so that they form a solid horizontal line without gaps. When such a line is formed, it disappears and any blocks above it fall down to fill the space. Usually, as the game progresses and the user gets to a higher score, the falling speed of the tetriminos increases. In decomposing the gameplay, we can see that there are three controls of the game; the horizontal movement of the tetriminos typically controlled though the left–right arrows of the keyboard, the rotation of the tetriminos through the use of the up arrow and the drop speed of the tetriminos that can be controlled through the down arrow for each tetrimino.

MM-Tetris is the multimodal reinvention of the Tetris game, modified to be controlled without the conventional means (i.e. keyboard and/or mouse) but instead with the users' eye-movements, mental commands and bio-measurements. In maintaining the same gameplay, our objective is to replace the arrow clicks in the aforementioned controls with eye, mental and bio-based commands. In the proposed version of the game (Figure 12.1), the use of these new modalities works in a complementary fashion, allowing for the replacement of each control independently. In more detail, the horizontal movement of the tiles is enabled through the coordinates of the user's gaze, the tile rotation through the user's mental state (in our case SMR signals detection) and the drop speed through the user's stress levels. In this way, the three modalities smoothly collaborate to facilitate playing a game like Tetris. The design of MM-Tetris incorporates three sensors in the following way so as to implement the required controls of the game (Table 12.1).

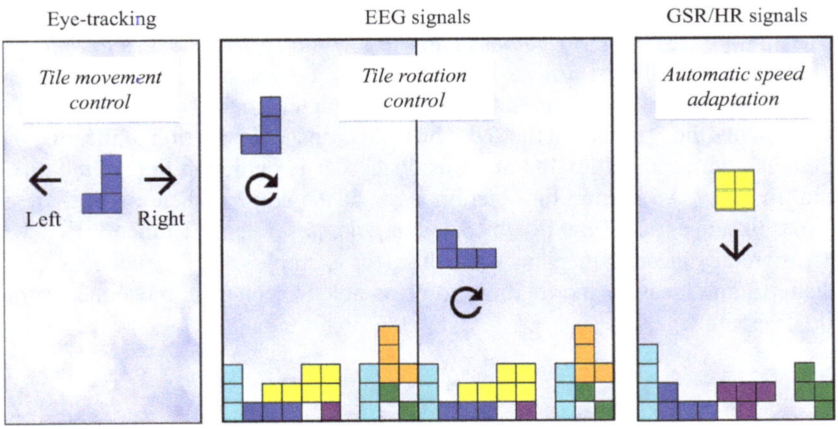

Figure 12.1 The MM-Tetris interface and commands

Table 12.1 Multimodal controls for MM-Tetris

Device type	**Signal type**	**Trigger**	**Control**
Eye-tracker	Gaze coordinates	Eye movement	Horizontal tetrimino movement
EEG recorder	SMR signals	Imaginary hand movement	Tetrimino rotation
GSR recorder	Skin conductance	Stress level	Tetrimino drop speed

In order to demonstrate an MM-Tetris playing session, we describe the following scenario that showcases the *beginners* gaming mode of MM-Tetris. The eye-tracker is first connected and a calibration procedure is performed, during which the participants are asked to fixate their gaze on nine dots appearing on multiple locations on the screen. Next, the game starts with the first tetrimino appearing at a locked position (tetrimino is not moving) on the top of the screen. A few seconds period is dedicated in order for the user to rotate the tetrimino at the desired angle by brain activity during imaginary movement. The exact duration of the locked position is determined on the basis of imaginary task classification output. Classification runs simultaneously on the background. Specifically, the user-specific classifier, as will be explained in detail next (Section 12.3.2.3), is used to classify the EEG signals online and the corresponding scores are communicated each second to the Tetris game platform. If the score exceeds a certain threshold, the tetrimino is rotated. The tetrimino is unlocked when a decrease of the classification score is observed for three consecutive classification iterations, since the last time the tetrimino was rotated (i.e. three consecutive classification scores are below the threshold). Afterwards, the rotation angle of the tetrimino is locked and it starts to drop at a dropping speed calculated on the basis of the user's stress level. When the tetrimino starts dropping, the user can move it across the horizontal axis using their eye gaze (captured by the eye-tracker system). The tetriminos are being moved constantly following the participant's gaze location on the screen one step at a time. If the participants look at the same location for 5 s continuously, then the tetrimino becomes completely locked and drops down fast until it reaches the 'ground' in the Tetris board. For visual inspection of the imaginary movements during the Tetris game, a feedback bar is depicting the strength of the mental imagery task during the rotation stage of the game. When the users master the controls and are confident with controlling the new interface functions, they can select more advanced gameplays that allow, for example, the simultaneous control of the tetrimino location and rotation, so being able to rotate and move the tetrimino while dropping.

12.3 Algorithms and associated challenges

12.3.1 Navigating with the eyes

Eye-trackers are easy-to-use input devices for users who retain control of their eye movements. Eye gaze has several desirable characteristics, such as being natural and fast pointing. Eye-tracking techniques measure the person's eye movements so that the gaze point at any time and the eyes' shifting are established accurately. Commercially available eye-trackers having high resolution give liberty to the user to move their head freely. Saccade measurements and the shortest latencies with fully remote, fiducial-free and contact-free setup are possible, even with less compliant subject groups. The fully automatic calibration that is included in the eye-trackers' software takes only a few seconds and maintains drift-free accuracy throughout the use. It is an easy procedure, where the user is asked to look at or follow predefined number of specific points on the screen, also known as calibration dots. Eventually, this software provides an indication of the quality of the calibration to the user so that they can recalibrate with more dots if needed.

Eye-trackers provide access to real-time eye signals, of which the main signal is the gaze coordinates of the users' eyes on the screen. In our case, the gaze or eye-tracking modality provides us with information on where our current visual attention is directed at. Our approach is to acquire a user's eye gaze information in real-time by an eye-tracking device that can be used to generate gaze events, and to analyse the data to deduce more high-level events. We focus on this explicit eye input in MM-Tetris to implement the gaze-based command and control of horizontal tile movement. More specifically, the MM-Tetris board is divided into invisible columns, each of which has the size of the tetriminos' square block. Then, if the x-axis coordinates of the user's gaze are in the right/left of the current tetrimino, then a right/left-arrow event is actuated, respectively. On the other hand, if the user's gaze x-axis coordinates are exactly on the current tetrimino for a number of consequent measurements, the horizontal position placement of the current tetrimino is locked (i.e. it cannot change) so that the user can rest their eyes for the time remaining until the next tetrimino appears.

12.3.2 Rotating with the mind

12.3.2.1 Sensorimotor rhythm

The second control we need to replace is that of the tetriminos' rotation. In this direction, we use mental commands extracted from brain signals through an EEG recorder. The EEG signal can be roughly defined as the signal which corresponds to the mean electrical activity of the brain cells in different locations of the head. It can be acquired using either intracranial electrodes inside the brain or scalp electrodes on the surface of the head. In our case, we use non-invasive EEG recorders that place electrodes on the surface of the head. To ensure reproducibility among studies, an international system for electrode placement, the 10–20 international system, has been defined [5]. In this system, the electrodes' locations are related to specific brain areas. For example, electrodes O1, O2 and O*z* are above the visual cortex. Each EEG signal can therefore be correlated to an underlying brain area.

A BCI system translates the recorded electric brain activity to output commands. Different electrophysiological sources for BCI control include event-related synchronisation/desynchronisation, Visually Evoked Potential (VEP), Steady state visually evoked potentials (SSVEP), slow cortical potentials, P300-evoked potentials, error-related potentials (ErrPs) and μ and β rhythms. In MM-Tetris, we selected the option of SMRs (specifically mu and beta) which provide a more natural way to complement the missing functionalities from a gaze-based interaction system and particularly, the rotation command. SMRs are brain waves that appear on EEG recordings from areas of the brain, which are associated with planning, control and execution of voluntary movements. When a user is in a resting state (i.e. no movement is occurring), the neurons in these motor areas generate synchronised electrical activity resulting in high amplitudes of EEG recordings. Movement or even preparation and imagination of a movement triggers an event that de-synchronises the electrical activity of the neurons in the motor areas resulting in reduced amplitude or power of a specific frequency band in the EEG. When the state of the user is reversed to the idle state, the electrical activity of the neurons is again synchronised. The detection of the de-synchronisation period is associated with a tile rotation command for our MM-Tetris application. According to the 10–20 system, brain activity produced by left- and right-hand movement is most prominent over locations C4 (right motor region) and C3 (left motor region), respectively, while foot movement invokes activity on the central electrode (C*z*).

The main advantage of SMR-based BCIs is that they do not depend on external stimuli for invoking brain patterns (e.g. as in the case of SSVEPs). This provides more freedom to the user, since the system is continuously available and they can decide freely when they wish to generate the control signal. As a result, SMRs offer an asynchronous interaction capability, which is particularly fit for our gaming scenario, i.e. the players can generate a rotation event with their own volition. In MM-Tetris, the upper arrow command corresponding to the tetrimino rotation is replaced with imaginary hand movements. When the player imagines a hand movement, the game detects their SMR signal de-synchronisation event and actuates an upper arrow event that rotates the current tetrimino clock-wise. Similarly to the gaze-based interaction, if the game does not detect any SMR-signal de-synchronisation event for a number of consequent measurements, the rotation ability of the current tetrimino is locked (i.e. it cannot change) so that the users can rest their thinking for the time remaining until the next tetrimino appears.

12.3.2.2 Challenges

A key limitation in SMR-based systems is the amount of training required by the user to be finally able to use the system. While stimulus-based BCIs (e.g. SSVEPs) require little to no amount of training, since most of the users can learn the simple task of focusing on a target letter or symbol within a few minutes, SMR-based BCIs highly depend on how well users can produce signals that are easily detected. Thus, such systems typically include training sessions for the user, so that they can produce appropriate signals. Eventually, asynchronous BCI systems require both training of the system, since each user may provide different signals on the performance of the imaginary movement, and user training, since the user needs to be trained so as to generate signals that can be robustly identified by the SMR classifier. The user training is typically done using a feedback mechanism that provides real-time information to the user with respect to what was detected based on their EEG signal, thus allowing them to continuously modify their thoughts so as to optimise the detection accuracy as shown in the feedback mechanism. In MM-Tetris, this is facilitated by the proposed gamified interface that includes a feedback bar on the output of the SMR classifier. In this gamified environment, the users are more eager to train themselves for optimising the SMR detection by playing a game compared to the typically tedious interfaces with simple arrows or smileys that are found in the literature.

Furthermore, while SMR offers this asynchronous interaction capability, it is much more complicated to implement such a system than a stimulus-based BCI, since asynchronous systems must continuously be able to detect whether the user is on an idle state or wishes to execute a command. This results in many false positives, meaning that a lot of control commands are passed on the BCI system unintentionally. In designing a functional MM-Tetris game, the first challenge we face is to provide an algorithm for asynchronous, i.e. self-paced, SMR-based movement detection. Most existing works in the literature focus on maximising the accuracy of detecting SMR signals by choosing the 'time window' of the trials that maximises the accuracy during testing, knowing when the user started to think of imaginary hand movements. However, this does not simulate a real application scenario where the user may do an imaginary movement at any arbitrary time and the system should detect this intention for movement in an asynchronous way. Motivated by this, the objective of this chapter is to investigate what is the optimal methodology for self-paced imaginary movement detection. Towards this goal, first we present common spatial pattern (CSP) algorithm based on filter bands (CSPFB), the EEG processing algorithm for training a classifier that detects imaginary movements in SMR signals in Section 12.3.2.3. Finally, in Section 12.5, we present the methodology for testing this classifier in the Tetris scenario, by presenting extensive experiments in the direction of self-paced movement detection (i.e. using the CSPFB-based classifier for detecting imaginary movements in signals with arbitrary length).

12.3.2.3 Algorithm for SMR detection

The CSP algorithm is effective in constructing optimal spatial filters that discriminate two classes of EEG measurements in SMR-based BCI. However, the performance of this spatial filter is dependent on its operational frequency band. To avoid this effect in [6], a variation of CSPFB was proposed. This approach is adopted in our study. More specifically, the CSPFB consists of four basic steps: frequency filtering, spatial filtering, feature extraction and classification. The general architecture of the adopted approach is described in Figure 12.2.

The first stage employs a filter bank that bandpass filters the EEG measurements into multiple bands. The second stage performs spatial filtering on each of these bands using the CSP algorithm [7]. As we see, each pair of bandpass and spatial filter yields CSP features that are specific to the frequency range of the bandpass filter. In the third stage, we extract the CSP features from the filter bank. The fourth stage uses a classification algorithm Support Vector Machine (SVMs) to model and classify the selected CSP features.

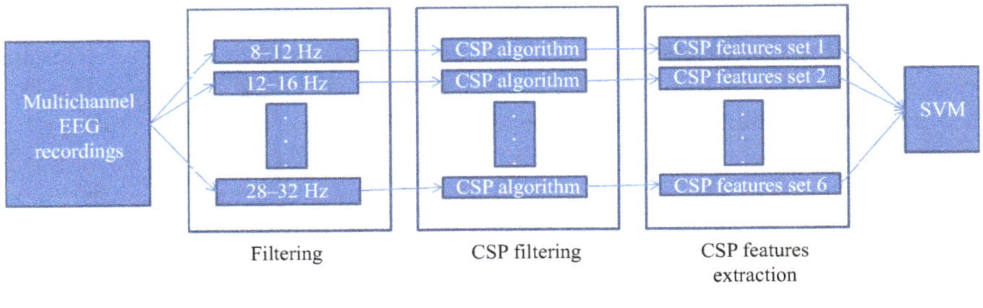

Figure 12.2 Architecture of CSPFB

12.3.3 Regulating drop speed with stress

12.3.3.1 Galvanic skin response

The third and final control we need to replace is that of the tetriminos' drop speed. In this direction, we use the user's stress levels to replicate this control, by decreasing/increasing the tetriminos' dropping speed as the user gets more stressed/relaxed. The stress levels are extracted from their GSR measurements. GSR, also known as electrodermal activity, is the property of the human body that causes continuous variation in the electrical characteristics of the skin. GSR is also a good measure to indicate stress. In general, sympathetic activation increases when someone experiences excitement, or something important is happening or about to happen. It also increases with stressors – whether physical, emotional or cognitive. However, apart from such stimuli and emotion states that have an impact on skin conductance values, there are also external factors which influence GSR values, such as ambient temperature.

There is a wide availability of wearable devices that measure skin conductivity and such devices are usually easy to find and affordable to buy (e.g. Shimmer 3 GSR+[†]). The devices are lightweight and allow real-time collection of high quality and scientifically validated data (i.e. measurements of the electrical characteristics or skin conductance) that are transmitted wirelessly. The way that they gather such measurements is by monitoring skin conductivity between two reusable electrodes attached to two fingers of one hand. The finger probe used to capture skin conductance is an unobtrusive sensor that can be easily attached to patients.

Skin conductance is characterised by great variability among different individuals, as it depends on various factors, including age, gender, ethnicity and hormonal cycles. Thus, comparing absolute skin conductance levels really makes sense only when the measurements originate from the same individual. In order to overcome the individual variability problem, we developed an algorithm that establishes a personal baseline by observing the skin conductance levels for each individual for a specific amount of time. This baseline stores the typical skin conductance values for an individual, and how often they occur. Afterwards, it is used to create stress level estimates from low to high, using a classifier that is tuned for each individual. In the following, we present the details of the developed algorithms.

12.3.3.2 Signal preprocessing

The raw signal of skin conductance is often noisy, as it contains many artefacts. Therefore, in our proposed methodology, a critical step is to remove such noisy observations. First, artefacts that result from poor contact between the electrodes and the skin are eliminated. Values that correspond to moments that the sensor lost skin contact, either intentionally or unintentionally due to movement, are zero or very close to zero. Thus, if the 90% of values within a 5 s window do not exceed the lower bound threshold (experimentally found to be 0.001 nS), they are removed. Furthermore, skin conductance levels do not tend to change very abruptly. Namely, each second, the values of skin conductance were experimentally found do not change more than

[†]https://www.shimmersensing.com/products/shimmer3-wireless-gsr-sensor

20% when increasing or 10% when decreasing. Thus, as the sensor used is capable of acquiring the GSR values more than once per second, a moving 1-s median filter is used to even out the signal. Windows where the signal changes more rapidly than these values, in either direction, are also marked as noise.

Subsequently, in order to obtain the slowly changing component of Skin Conductance Level (SCL), a low-pass filter is applied in the form of a sliding window filter, with a window size of 1 min, only on time intervals which do not include noisy observations more than 60% of the window. The purpose of this filter is to remove the fast-changing component of the skin conductance signal, and to derive the more slow-changing skin conductance level. The final step of the proposed methodology concerns the reconstruction of the signal in order to fill the noisy gaps and to smoothen out the skin conductance level in these areas. For this purpose, linear interpolation is applied between the 'accepted' observations, in order to generate the new values of the skin conductance for the eliminated measurements.

12.3.3.3 Personalised stress level detection

The proposed stress level detection method entails the computation of skin conductance thresholds that indicate five different levels. Considering the monotonous relationship between skin conductance values and stress level, the thresholds are calculated on the basis of minimum (l_0) and maximum (l_5) skin conductance levels corresponding to the non-stressful and the highest stressful status, respectively. These thresholds are expected to be different for each individual, as it was explained earlier in Section 12.3.3.1. Thus, we need to define a personalised methodology that will compute them for each individual separately. Towards this goal, our algorithm proposes a monitoring period for each participant, during which the different thresholds are identified.

According to the bibliography [8], a 5-min rest period ensures sufficient recordings in order to successfully detect the lowest level of skin conductance. Generally, we assume that, in the monitoring period, the person had at least one period of being calm for at least 5 min. We use the min–max algorithm for overlapping 5-min windows to find this calmest period and the maximal SCL value in it. The detected value in this 5-min time-window is called zero level (l_0).

Besides the zero-level value, the next step is to compute a threshold that indicates high stress. This point is called the l_5 level (because the algorithm is set to discriminate five different distinct stress levels). In order to compute this value, we capture the signal for a longer time than what is required for the zero-value. Then, based on this recording, we calculate a baseline histogram, which is used to derive which skin conductance levels are common for that individual, and which levels are unusually low or unusually high. Its form is a 600-bin histogram, holding the information concerning the occurrences of skin conductance level observations calculated over a specific amount of time for each individual. Then, l_5 is computed from the baseline histogram, as the first level of skin conductance for which no values in the baseline histogram have been observed.

For the developed methodology, the establishment of a correct baseline histogram is very critical. Many problems could occur from sparse baseline histograms, i.e. if there are many skin conductance levels for which there are not yet any observations. The most important problem is the underestimation of the l_5 threshold, as the sparsity of the baseline histogram may lead to the placement of the l_5 value at a low level. In order to avoid this, stress levels cannot be computed if more than 40% of the observed skin conductance levels in the baseline are larger than l_5. Taking this into account, the proposed algorithm incorporates an additional online adaptation process, which continuously updates the histogram and the l_5 level, while the participant wears the sensor. Note that the windows with a large amount of noisy measurements ($\geq$ 40%) are excluded for this step. The whole signal (including the reconstructed parts) is used only in the classification step described afterwards as well as for the visualisation of the signals.

Afterwards, based on the already calculated l_0 and l_5 values, the remaining stress level thresholds are computed according to the following formulas:

$$\delta = \frac{l_5 - l_0}{4} \tag{12.1}$$

$$l_1 = l_0 + \frac{1}{2} \times \delta \tag{12.2}$$

$$l_i = l_{i-1} + \delta, \quad for \quad i = 2, \ldots, 4 \tag{12.3}$$

Finally, after the thresholds have been set and in order to classify the stress level for each period of time, the algorithm computes the mean skin conductance level of the processed signal within the time-window for which we want to detect the stress level (e.g. at 5 s time intervals). The mean values are afterwards compared to the predefined thresholds and assigned to the proper category. Notably, a given value is classified within each stress level boundary and a single value is not selected to serve as a threshold. As a result, the algorithm does not actually highlight events of stress but rather defines a stress level from one to five for each time segment. Indeed, as thresholds can yield different outcomes, this allows to further tailor the method at later stages, e.g. by setting up different actions at level 5 stress than level 4. In MM-Tetris, the various stress levels vary the dropping speed of the tetriminos, with the stress level 5 setting the drop speed to the slowest and the stress level 0 to the fastest.

12.4 Experimental design and game setup

12.4.1 Apparatus

In our experiments, we used the following three sensors in order to acquire the bio-signals necessary to operate the MM-Tetris game.

- Eye-tracking (MyGaze[‡]) provides the gaze coordinates of the user.
- EEG signal recorder (ENOBIO 8[§]) provides the SMR signal of the user upon imaginary movement.

[‡]http://www.mygaze.com/
[§]http://www.neuroelectrics.com/products/enobio/enobio-8/

- GSR measurements (Shimmer+[||]) provide information for the SGR signal and the stress level of the user.

12.4.2 Events, sampling and synchronisation

The controls of the game are treated as timed events (i.e. move the tile to x coordinate, or rotate the tile, or decrease game speed), based on the signals that are sampled at a frequency of 30, 500 and 32 Hz from the three devices, MyGaze, ENOBIO 8 and Shimmer+, respectively. In order to have a universal timing reference for the identified events, the three signal streams and the events are synchronised through the library SensorLib,[¶] which relies on the software-based synchronisation library LabStreamingLayer.[**]

12.4.3 EEG sensors

The EEG signals were recorded by an ENOBIO headset using eight wet electrodes that were placed on the CP1, CP2, C3, C1, C2, C4, FC1, FC2 sites of the 10–20 international system. The sampling frequency of the system was 500 Hz.

12.4.4 Calibration

The selected eye-tracker myGaze offers an easy interface for calibration to adapt the unique characteristics of the user's eyes so as to achieve the best possible gaze-tracking accuracy. The software of myGaze offers various calibration options and the user can select between 1, 2, 5 and 9-point calibration (i.e. the number of fixation points on the display that the user focuses on in succession to calibrate their eye gaze) depending on the results. Finally, the user also has the option to select the fixation point style between a circle, a crosshair, a star or an image.

For calibrating the EEG sensor, we need to develop a subject-specific classifier for the SMR signals. In getting the necessary training data for this classifier, a specifically designed experiment for prompted imaginary movement was conducted using the OpenVIBE framework. Note that OpenVIBE is only used for the calibration phase in order to enable a user to play MM-Tetris in the first place by generating a classifier for detecting movements, while MM-Tetris is designed to train the user afterwards, while playing, for generating more discriminative signals. For the calibration, an interface of visual cues is presented indicating whether the participants should perform the imagination of movement as well as which of the hands to move. In particular, the experimental session starts with a black screen that lasts for 40 s, during which the participant is asked to remain calm without imagining any movement in order to use the respective recording as a baseline signal. Afterwards, an iteration of 40 trials is initiated consisting of 20 trials for the imagination of left-hand movement and 20 for the imagination of right-hand movement. Each trial is initiated

[||] http://www.shimmersensing.com/
[¶] https://github.com/MAMEM/SensorLib
[**] https://github.com/sccn/labstreaminglayer

by a green cross that lasts for 1 s and allows the participants to get prepared for the execution of imaginary movement. Subsequently, a red arrow pointing towards the left or right direction instructs the participants to imagine the respective movement. After a 5-s period, the arrow disappears and a black-screen period that lasts for a random duration of 5–20 s follows. The whole duration of the calibration session is approximately 20 min. A classifier is then trained based on this calibration session using the algorithm presented in Section 12.3.2.3.

Finally, the calibration of the GSR sensor also entails the development of a subject-specific classifier for the GSR signal based on the algorithm presented in Section 12.3.3.3. In gathering the necessary training data without extending the calibration time, the user wears the GSR sensor during the EEG calibration phase and their signal throughout this session is captured. Based on this signal and using the algorithms presented in Section 12.3.3, the stress levels of the users are identified and are ready to be used for classifying their GSR signals in the game.

12.5 Data processing and experimental results

In this section, our aim is to show how to generate a subject-specific classifier for the self-paced SMR detection problem so as to provide the optimal user experience during playing MM-Tetris. In this direction, first, we discuss the way to segment the signals in an asynchronous scenario (i.e. when not knowing the starting time of a trial). Then, we propose an evaluation metric tailored to reflect the user experience during gameplay and perform experiments on eight subjects using the designed evaluation metric. Finally, we present the methodology for online classification of SMR signals, which is used during an MM-Tetris session.

12.5.1 Data segmentation

The EEG signals recorded during the calibration phase are segmented by overlapping windows of 2 s duration. The step of each window was set to 100 samples or 0.2 s. Each window from now on referred to as trial consisted of 1,000 samples, each of which was annotated as 1 or 0 based on whether the sample was recorded when the participant was performing a mental movement regardless of the direction. The label of each trial was then set based on the majority of the 1,000 samples to either 1 (movement) or 0 (no movement). After the segmentation, we have n_m and n_n number of trials for the movement and no movement labels, respectively. Considering that the users were asked to perform no movement for more time compared to the movement case, the number of n_n trials is bigger than that of movement trials n_m ($n_n > n_m$), and as such, the two classes are imbalanced.

12.5.2 Offline classification

The generated dataset during the calibration session is divided in half so as to produce a training and test set of $(n_m + n_n)/2$ trials. In order to tackle the class imbalance

Table 12.2 Offline experiments for evaluating Tetris performance

Subject	Precision	Recall	Accuracy
S1	39.65	76.66	89.06
S2	41.09	100	93.28
S3	23.80	83.33	83.59
S4	35.21	83.33	88.9
S5	37.83	82.35	88.12
S6	29.50	60	83.9
S7	32.5	86.66	88.4
S8	40.67	80	89.84
Avg. acc.			83.75
Baseline acc. (do nothing)			76.56

problem, which can hinder the learning process of the classification model, we balance the training set by randomly under-sampling the trials that correspond to the majority class (no movement in this case), so as each class will have exactly the same number of trials. The testing set on the other hand is left unbalanced to simulate a real-world application scenario, where the classification model will be applied in a continuous EEG signal stream, which is expected to be imbalanced. Each trial is then filtered with a Butterworth band-pass filter in the ranges between 7 and 36 Hz and the CSPFB features are extracted for each trial. The filter banks that were used for extracting the features were in the frequency ranges of 8–12, 12–16, 16–20, 20–24, 24–28 and 28–32 Hz. As previously mentioned, the features from the first half of each session were used to train a linear kernel SVM classifier and the other half was used for testing.

The testing was performed as follows; the unknown to the classifier portion of the recorded session was scanned with a sliding window of length 2 s and step 0.2 s. Each trial was passed to the previously trained classifier in order to retrieve a score. After 1 s, meaning that five scores are generated by the classifier, the sign of the median of the scores was used to determine whether to issue a rotation command or not. If the sign is positive, a rotation command is simulated and the classification is paused for 5 s. This pause is inserted so as to avoid constant rotations of a tetrimino based on the same imaginary hand movement event.

The results of this method can be found in Table 12.2, which were performed by eight in-house participants. The precision indicator refers to the number of the correct rotation commands that were issued (i.e. detected during the period that the subject was performing the movement) divided by the total rotation commands. The recall measurement refers to the number of correct rotation commands divided by the total number of movements that was performed during the experiment. Finally, the accuracy metric represents the percentage of the seconds that the classifier behaved according to the user's will. To calculate this accuracy metric, we used (12.4) where n_s refers to the number of seconds of the testing set, n_{move} were the number of times that the participant was asked to perform a movement by the experiment interface, *TP* was the number of rotations that were detected by the system and were during the time the participant was performing a mental movement and *FP* was the number of rotations detected by the system during the time the participant was not performing any action.

$$acc = \frac{n_s - 5 \cdot (n_{move} - TP) - FP}{n_s} \tag{12.4}$$

Intuitively, this accuracy metric calculates the number of seconds that the system was performing according to the intention of the user and places it on the numerator. The total number of seconds is placed on the denominator; thus, an accuracy of 80% for a 100-s experiment means that the system was performing correctly for a total of 80 s. A baseline accuracy of a classifier that does not detect any rotations is also reported on this table as 76.56%. An example of the performance of our classifier can be seen in Figure 12.3, where the blue pulses represent the intention of the user, i.e. the times the user intended to perform the mental movement, the green lines are when the system detected a rotation command and the red lines are when the system did not detect any mental movement. In this example, a blue pulse corresponds to a single rotation intention, and the aforementioned 5-s pause in the classifier ensures that multiple rotations per pulse are avoided.

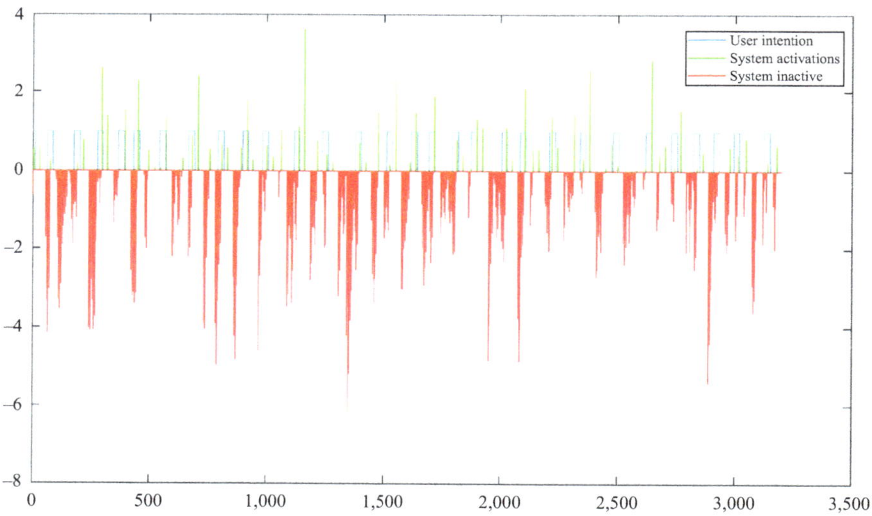

Figure 12.3 Subject 2 performing mental movement (blue lines) and the result of our classifier coloured as green (rotation command) and red (no rotation command)

12.5.3 Online classification framework

For the online classification, we used the same classifier as for the offline case trained with the whole calibration session instead of splitting the dataset in half. The EEG signal is again sampled with a moving sliding window which moves for 0.2 s each time and produces a classification score. After each second, the median classification scores of the past five windows are calculated and communicated to the Tetris game. Since the scores are negative if the window segment corresponds to the no movement class and positive for the movement class, this practically means that in order to have a movement outcome, at least three out of the five past windows must produce a positive classification score.

12.6 Summary

In this chapter, we presented MM-Tetris, the hands-free adaptation of the popular Tetris game, that can be played through the player's bio-signals. In this direction, the controls of the keyboard-based Tetris game were replaced by bio-signals of multiple modalities. In more detail, the horizontal movement of the tetriminos that is controlled through the left–right arrows of the keyboard was replaced by the player's gaze coordinates, the rotation of the tetriminos typically controlled with the up arrow was replaced by the player's brain signals and the dropping speed of the tetriminos controlled by the down arrow in the keyboard version was replaced with the stress levels of the player. Concerning the new controls of MM-Tetris, the most challenging is that of brain-based control, though the SMR signals elicited by imaginary movements, and more specifically detecting such commands in arbitrary time (i.e. self-paced). Towards self-paced SMR detection, we have presented a series of experiments and a new accuracy-based metric that is tailored to measure the percentage of time that the system works as intended. These initial experiments were performed by a limited number of eight participants. In our future plans, our aim is to increase the number of participants so as to be able to reach safer conclusions. Furthermore, we plan to develop an online training system to accompany MM-Tetris targeting performance improvements in the detection of tetrimino rotations. This online training system will rely on the detection of errors during gameplay. These errors may occur both due to SMR detection imperfections and also due to user's wrong decisions for the tetriminos' moves. ErrPs [9] offer a natural way to detect and distinguish between these two cases. The incorporation of the ErrPs due to system errors in the hands-free Tetris is in our future plans for enhancing the game accuracy while playing in two ways. First, ErrPs can be used to annotate the SMR detection events as correct or error and thus be included as additional training samples in the calibration-trained classifier in order to update the classifier through an incremental learning scenario. Second, ErrPs can be also used to cancel the rotations that have been made through erroneous movement detection from the SMR classifier. Eventually, the incorporation of this novel error-based online system in MM-Tetris will offer bidirectional training (i.e. training of the system to detect SMR signals and training of the user to provide distinguishable signals).

References

[1] Kumar C, Menges R, Müller D, and Staab S. 2017. Chromium based Framework to Include Gaze Interaction in Web Browser. In Proceedings of the 26th International Conference on World Wide Web Companion (WWW '17 Companion). International World Wide Web Conferences Steering Committee, Republic and Canton of Geneva, Switzerland, 219–223. DOI: https://doi.org/10.1145/3041021.3054730

[2] Menges R, Kumar C, Müller D, and Sengupta K. 2017. GazeTheWeb: A Gaze-Controlled Web Browser. In Proceedings of the 14th Web for All Conference on The Future of Accessible Work (W4A '17). ACM, New York, NY, USA, Article 25, 2 pages. DOI: https://doi.org/10.1145/3058555.3058582

[3] Oikonomou VP, Georgiadis K, Liaros G, Nikolopoulos S and Kompatsiaris I, "A Comparison Study on EEG Signal Processing Techniques Using Motor Imagery EEG Data," 2017 IEEE 30th International Symposium on Computer-Based Medical Systems (CBMS), Thessaloniki, 2017, pp. 781–786. doi: 10.1109/CBMS.2017.113

[4] Kikhia B, Stavropoulos TG, Meditskos G, *et al.* Utilizing Ambient and Wearable Sensors to Monitor Sleep and Stress for People With BPSD in Nursing Homes. Journal of Ambient Intelligence and Humanized Computing. 2018;9:261. DOI: https://doi.org/10.1007/s12652-015-0331-6

[5] Nikolopoulos S and Kompatsiaris I, "A Comparison Study on EEG Signal Processing Techniques Using Motor Imagery EEG Data," 2017 IEEE 30th International Symposium on Computer-Based Medical Systems (CBMS), Thessaloniki, 2017, pp. 781–786. doi: 10.1109/CBMS.2017.113

[6] Ang KK, Chin ZY, Wang C, *et al.* Filter Bank Common Spatial Pattern Algorithm on BCI Competition IV Datasets 2a and 2b. Frontiers in Neuroscience. 2012;6:39. Available from: http://journal.frontiersin.org/article/10.3389/fnins.2012.00039.

[7] Lotte F, Guan C. Regularizing Common Spatial Patterns to Improve BCI Designs: Unified Theory and New Algorithms. IEEE Transactions on Biomedical Engineering. 2011;58(2):355–362.

[8] Kocielnik R, Sidorova N, Maggi FM, *et al.* Smart Technologies for Long-Term Stress Monitoring at Work. In: Proceedings of the 26th IEEE International Symposium on Computer-Based Medical Systems. IEEE; 2013. p. 53–58.

[9] Kalaganis K, Chatzilari E, Georgiadis K, et al. "An Error Aware SSVEP-based BCI," 2017 IEEE 30th International Symposium on Computer-Based Medical Systems (CBMS), Thessaloniki, 2017, pp. 775–780. doi: 10.1109/CBMS.2017.44

Chapter 13

Conclusions

Chandan Kumar[1], Spiros Nikolopoulos[2], and Ioannis Kompatsiaris[2]

13.1 Wrap-up

Interaction with computer applications is usually performed using conventional input devices such as mouse or keyboard. However, people lacking fine motor skills are often not able to use these devices, which limits their ability to interact with computer applications and thus excludes them from the digital information spaces that help us stay connected with families, friends, and colleagues. The evolution of eye-tracking systems and brain–computer interfaces (BCI) has given a new perspective on the control channels that can be used for interacting with computer applications. This book presented a study on end user characteristics and their needs for such control channels, and the knowledge on how it can be fulfilled with eye tracking and BCI interaction using signal-processing algorithms and interface adaptations. The contributors of various chapters are researchers, engineers, clinical experts, and industry practitioners, who collaborated in the context of the 3-year research and innovation action "MAMEM"—"Multimedia Authoring and Management using your Eyes and Mind" (mamem.eu). Hence, the book covers the underlying challenges of eye and mind interaction, and possible solutions that identify future directions to encourage the researchers around the world. In the following, we summarize the key outcomes and takeaways of this book:

- **End user requirements:** The necessity of using eye and mind as alternative interaction channels has been mainly motivated in the context of assisting people with disabilities. Hence, it is imperative to identify and understand the requirements of target-group users to imply the usefulness and usability of novel interaction channels. In this context, the authors have conducted several studies to characterize these requirements and generate valuable insights and guidelines for future research. More specifically, Chapters 2–4 elaborated on how computer use habits and difficulties of people with motor impairment differs significantly from that of able-bodied, what are the specific requirements and expectations from

[1]Institute for Web Science and Technologies, University of Koblenz-Landau, Koblenz, Germany
[2]Informations Technology Institute, Centre for Research and Technology Hellas, Thermi-Thessaloniki, Greece

assistive interfaces, and how persuasion theories and intervention frameworks can stimulate a positive attitude toward the usage of eye and mind-based interface.

- **Signal processing algorithms:** One of the most challenging aspects of eye and mind-based interaction is to analyze and process the physiological signals recorded from eye-tracking and BCI devices and convert them to meaningful commands. In this regard, the book provides a comprehensive knowledge on various aspects of signal acquisition, preprocessing, enhancement, feature extraction, and classification. More specifically, with respect to eye-tracking-based interaction, Chapter 5 describes how to acquire the user's eye gaze information in real-time by an eye-tracking device, filter and process this information to generate eye gaze as input mechanism. With respect to Electroencephalography (EEG) based interaction, Chapters 7–10 cover state-of-the-art methods and machine-learning algorithms to understand and process EEG, steady-state-visual-evoked potentials (SSVEPs), motor imagery, error-related-potentials (ERRPs), and sensorimotor rhythm (SMR) for interaction process.
- **Application interface developments:** Contrary to the traditional emulation approach of operating standard application interfaces with eye and mind-based commands; this book argues the need of adapted interface developments for improved interaction experience. In this regard, Chapter 5 not only outlines the principles of GUI development for eye gaze interaction but also avails the eyeGUI and GazeTheWeb framework for developers. eyeGUI supports the development aspects, like rendering, layout, dynamic modification of content, support of graphics, and animation. GazeTheWeb as an open-source browser* provides a framework for researchers to investigate methods for improved interaction with eye gaze and other input modalities in the web environment. In this direction, the prototype of GazeTheWeb has already been used to integrate EEG for multimodal error-aware BCI and Tetris game described in Chapters 10 and 11.
- **Evaluation methods:** The book also presents a comprehensive guideline and examples of conducting evaluations to assess usability, performance, and feasibility of eye gaze and EEG-based interaction algorithm and interfaces. This would help the readers to quickly recognize and solicit suitable study design, procedure, baseline, measures, and analysis methods. Chapter 6 reviews the common methodology of human–computer interaction experiments and adapts it for eye-tracking research. It describes how eye gaze pointing, selection, and typing can be evaluated, and how the performance and feasibility of eye-controlled interfaces can be assessed using lab and field studies. Chapters 7–10 elaborates on how EEG, SSVEPs, SMR, and ERRPs-based algorithms can be evaluated and compared for better performance and accuracy.

13.2 Open questions

Based on the studies and methods presented in this book, there is no hesitation in acknowledging the significance, purpose, and applicability of eye and mind-based

*https://github.com/MAMEM/GazeTheWeb

interactions. However, it also uncovers the challenges and open questions that are imperative for the success of these novel interaction methods and applications.

- **Hardware challenges:** Many hardware improvements are required to make the system work outside the laboratories. As we found out in MAMEM trials that eye-tracking accuracy is subjective to various environmental variables such as lighting condition and distance to screen. Eye-tracker device gets overheated, and sometimes it is required to unplug and restart the system. BCI systems are even more sensitive and do not work properly outside the suitable environment like at noisy places. Furthermore, both eye tracking and BCI involve a tedious calibration procedure before starting the interaction.
- **Device usability:** Most often, end user wants the system to be easily approachable. A complex system that is not easy to handle could be disliked. In MAMEM trials, we experienced that some participants found it cumbersome to use the system and hence the usage frequency was very low over the month period. Eye-tracker accuracy is affected if the head position is not stationary, which makes the interaction physically demanding. EEG process includes the application of gel to the electrodes to measure the exact level of volt potential and requires lot of training to be able to interact intuitively. The new EEG technique with dry electrodes does not need the application of gel but still involves a cumbersome training procedure. Moreover, common problem with wearing BCI is sweating, and large energy consumption is also a challenge.
- **Ethical issues:** Although the reported studies in the book follow the ethical and clinical protocol. The long-term usage of these novel systems is subjective to further privacy issues. While using eye tracking for interaction, the recorded data might contain sensitive eye gaze information of user attention and might reveal every piece of information the user has looked into. The ethical issues of BCI also include sensitive physiological signals and cognitive information. Furthermore, BCI headset with very high range of signal detection may read the mind of others, which would also be an ethical issue. Currently, there are no universal ethical guidelines for eye and mind-based interaction procedures.

13.3 Future perspectives

A chief focus of this book was on the applicability of eye tracking and BCI signals as a computer-input mechanism to support people with motor impairment. However, the proposed interaction methods, algorithms, and analytic insights have the potential to expand into a variety of further promising scenarios. For example, supporting healthy older adults, integration into public displays, as well as in medical professions, such as surgeons in the operating theatre. We might use gaze and mind-based applications for a hands-free control of devices or looking up information. Furthermore, in aviation applications that may include gaze and mind-based cockpit control for pilots, attention, and tiredness monitoring for control room-operators.

- **Entertainment applications:** In the entertainment field, eye tracking and BCI have lot of promise in the form of gaming. This kind of gaming uses player's eye movements, brain signals, player's heartbeat, or facial expressions instead (or in addition to) of traditional user controllers to control the game. In this book, we discussed some interesting gaming applications such as eye-controlled 3D game Schau Genau! in Chapter 5, and the multimodal Tetris game in Chapter 11. In these games, a user can get a good score by controlling a virtual object by eye movement or imagining movements. Future efforts in these directions can enhance the scope of eye and mind applications in the mainstream market. Commercial organizations like Tobii[†] have already introduced gaze attention and control in several gaming applications. Furthermore, cognition-based games such as "NeuroRacer" or "NeuroMage" have also received considerable success, indicating the future potential of BCI-based gaming applications.

- **Security applications:** The physiological signals from eyes and brain could help detecting irregular behavior and suspicious objects. In this direction, ERP, EEG, and eye movement could help detect signal distortions in the event of any suspicious event, i.e., to identify the potential vulnerable targets. There are already many defense and pharmaceutical research institutes started working on eye tracking and BCI, as both the signals in unison can help in criminal investigation and a better understanding of the human body. For example, the physiological functions to detect a lie could be very helpful in crime cases.

[†]https://gaming.tobii.com/

Index

www.ingramcontent.com/pod-product-compliance
Lightning Source LLC
LaVergne TN
LVHW011848200926
848484LV00002B/423